Understanding FIDIC

Understanding FIDIC explains in simple and practical terms what is often seen as a very complex range of international engineering and construction contracts.

Covering the FIDIC 2017 Red, Yellow and Silver Books (referred to as "The Rainbow Suite"), the book gives an overview of all three contracts, including coverage of changes between the 1999 contracts and the present 2017 suite. FIDIC contracts are widely used as far afield as Europe, the Middle East, Asia and Australia, and this book provides a practical yet thorough guide to the key elements that practitioners preparing and administering these contracts would need to be aware of.

In his approachable and readable style, Kelvin Hughes covers:

- The obligations and responsibilities of the Employer, the Employer's Representative, the Engineer and the Contractor
- Quality and Defects Liability
- Design Responsibility and Liability
- Variations, Measurement and Payment Procedures
- Progress, Delays, Extensions of Time and Completion
- Suspension and Termination
- Insurances
- Employer's and Contractor's Claims
- The Dispute Avoidance/Adjudication Board and the Resolution of Disputes
- Tendering

Anyone working with FIDIC contracts whether as the Employer, Employer's Representative, Engineer or Contractor will benefit greatly from this easy-to-read guide to the Rainbow Suite. Students on professional courses or researching the contracts for project work will also find this book extremely useful.

Kelvin Hughes is a leading authority on construction contracts and has had substantial theoretical and practical experience of all contracts, including FIDIC, over a 46-year working life, including 4 years as a full-time consultant to a major project management company drafting a suite of contracts for use on public projects in Qatar, and was also head of contracts/claims for another major project management company in Qatar. He has written eight previous practical books on construction contracts.

Understanding Construction

Understanding Quality Assurance in Construction
A practical guide to ISO 9000
H. W. Chung (1999)

Understanding Active Noise Cancellation
Colin H. Hansen (2001)

Understanding the CDM Regulations
Owen Griffiths (2006)

Understanding Building Failures, 4th edition
James Douglas and W. H. Ransom (2013)

Understanding NEC3: Professional Services Contract
Kelvin Hughes (2013)

Understanding NEC3: Engineering and Construction Short Contract
Kelvin Hughes (2014)

Understanding The Building Regulations 6th edition
Simon Polley (2014)

Understanding JCT Standard Building Contracts 10th edition
David Chappell (2017)

Understanding NEC4: Term Service Contract
Kelvin Hughes and Patrick Waterhouse (2018)

Understanding the NEC4 ECC Contract: A Practical Handbook
Kelvin Hughes (2018)

Understanding the NEC4 Professional Service Contract
A Practical Handbook
Kelvin Hughes (2020)

Understanding FIDIC
The Rainbow Suite
Kelvin Hughes (2021)

Understanding FIDIC

The Rainbow Suite

Kelvin Hughes

Routledge
Taylor & Francis Group
LONDON AND NEW YORK

First published 2021
by Routledge
2 Park Square, Milton Park, Abingdon, Oxon OX14 4RN

and by Routledge
52 Vanderbilt Avenue, New York, NY 10017

Routledge is an imprint of the Taylor & Francis Group, an informa business

British Library Cataloguing-in-Publication Data
A catalogue record for this book is available from the British Library

Library of Congress Cataloging-in-Publication Data
Names: Hughes, Kelvin (Engineering consultant) author.
Title: Understanding FIDIC: the rainbow suite / Kelvin Hughes.
Description: Abingdon, Oxon; New York, NY: Routledge, 2021. |
Series: Understanding construction |
Includes bibliographical references and index.
Identifiers: LCCN 2020033749 (print) | LCCN 2020033750 (ebook) |
ISBN 9780367427917 (hardback) | ISBN 9780367427924 (paperback) |
ISBN 9780367855161 (ebook)
Subjects: LCSH: Construction contracts. | Engineering contracts. |
Architectural contracts. | Standardized terms of contract. |
International Federation of Consulting Engineers.
Classification: LCC K891.B8 H84 2021 (print) |
LCC K891.B8 (ebook) | DDC 343.07/8624—dc23
LC record available at https://lccn.loc.gov/2020033749
LC ebook record available at https://lccn.loc.gov/2020033750

ISBN: 978-0-367-42791-7 (hbk)
ISBN: 978-0-367-42792-4 (pbk)
ISBN: 978-0-367-85516-1 (ebk)

Typeset in Times New Roman
by codeMantra

Contents

Preface vi
Acknowledgements ix

1 Overview of FIDIC and its contracts 1

2 "The old and the new" – FIDIC 1999 becomes FIDIC 2017 24

3 Obligations and responsibilities of the Employer, the Employer's Representative, the Engineer and the Contractor 79

4 Quality and Defects Liability 118

5 Design responsibility and liability 142

6 Variations, Measurement and Payment Procedures 161

7 Progress, Delays, Extensions of Time and Completion 181

8 Suspension and Termination 203

9 Care of the Works and Indemnities, Exceptional Events and Insurance 213

10 Employer's and Contractor's Claims 229

11 The Dispute Avoidance/Adjudication Board and the Resolution of Disputes 255

12 Tendering using the FIDIC 2017 Contracts 266

Appendix: Errata to the FIDIC Conditions of Contract 2017 305

Index 311

Preface

I have been involved with the FIDIC contracts for over 25 years, in the capacity of contract adviser, contract consultant and trainer, including a period spent in a senior management role with a major project management consultant in Qatar.

In that time, I have advised on numerous projects using the FIDIC contracts from as far back as FIDIC 1987 and as far afield as Europe, the Middle East, the Far East and the Indian subcontinent.

This book provides an overview of FIDIC and its wide range of contracts; considers the obligations and responsibilities of the Employer, and those that represent the Employer, in the form of the Engineer or the Employer's Representative, and also the Contractor; and then compares the previous FIDIC contracts with the FIDIC 2017 Contracts published in December 2017, followed by a detailed commentary on the new contracts.

Whilst the FIDIC 2017 Contracts are structurally similar to the FIDIC 1999 Contracts they are considerably larger, with far more pages than the previous editions. For example, the General Conditions within the Red Book now comprise 106 pages, whereas the FIDIC 1999 version was 62 pages, and there is a significant increase in administrative requirements, mostly to achieve greater certainty of project outcome over a wider range of projects in a wider range of locations.

The author is of the opinion that the FIDIC 2017 Contracts are significantly better than the FIDIC 1999 Contracts, particularly in respect of clarity of wording, but also in the new provisions in terms of programme, quality management, claims and disputes. The newer contracts are also quite prescriptive, which lends itself to a more clarified and disciplined approach to managing the contracts. There are those that say that the FIDIC 2017 Contracts will be administratively burdensome in use, but these are early days, so there is no clear evidence that this is true.

Certainly, the movement towards the use of the FIDIC 2017 Contracts appears to be quite slow to date, time will tell as to whether FIDIC users will move from the previous editions of the contract, particularly the FIDIC 1987 and 1999 Contracts to the newer Contracts, and also whether users of other contracts will migrate to using FIDIC 2017 Contracts.

Though there are several contracts in the FIDIC family some of which are briefly reviewed within this book, the book focuses primarily on the "Rainbow Suite":

- Conditions of Contract for Construction (The "Red Book") – for building or engineering works where most of the design is provided by the Employer or his Representative.

- Conditions of Contract for Plant and Design-Build (The "Yellow Book") – for the provision of electrical and/or mechanical plant, and for the design and execution by the Contractor of building or engineering works.
- Conditions of Contract for EPC/Turnkey Project (The "Silver Book") – for the provision on a turnkey basis of a process or power plant, of a factory or similar facility or of an infrastructure project or other type of development, where the Contractor takes total responsibility for the design and execution of the project.

Whilst many practitioners still use previous editions of the FIDIC contracts – for example, the FIDIC 1995 Orange Book, with which the author was personally involved on projects in India – this book is primarily aimed at giving guidance to FIDIC 2017 Contract users, and also to a certain extent, FIDIC 1999 Contract users.

Although the structure and content of the FIDIC contracts was changed in 1999, there are some broad similarities with the previous contracts, and to that end much of the advice given in this book is of use to all FIDIC contract users.

The book is intended to be of benefit to professionals who are actually using the FIDIC contracts on live projects, but also to students who need some awareness of FIDIC contracts as part of their studies. I have also, as with my other books, in several of the chapters explained subjects such as quality, programme and claims, which may seem obvious to experienced practitioners, but I have done so in order that students and less experienced practitioners understand the concepts and principles before reviewing those provisions within the FIDIC contracts.

Readers may note the absence of case law within the text of the book. As with my previous books on NEC Contracts, this is a deliberate policy on my part as I am not a lawyer, my background being in senior commercial positions within major building contractors and more recently in senior management with a major project management company in Qatar, so I felt, and readers may probably agree, that I was unqualified to quote, and to attempt, a legal review of the contracts with a detailed commentary on any case law, particularly as the FIDIC contracts are used in such a wide range of countries and legislations.

In terms of the structure of the book, Chapter 1 provides an overview of the FIDIC contracts; Chapter 2 provides a comparison of the FIDIC 2017 Contracts with the previous FIDIC 1999 Contracts; Chapter 3 considers the obligations and responsibilities of the Employer, the Employer's Representative, the Engineer and the Contractor; then Chapters 4–11 review the FIDIC 2017 Rainbow Contracts in some detail, highlighting the changes from the 1999 to 2017 Contracts, with Chapter 12 considering tendering under the FIDIC 2017 Contracts.

Note, it is not my intention to provide a detailed "clause by clause" analysis of each of the FIDIC contracts, but to provide a practical review of the areas covered by the Contracts, with some commentary on some of the changes within the newer FIDIC 2017 Contracts.

I had originally intended to deal with the FIDIC Red, Yellow and Silver Books as separate chapters, but the final structure I adopted lessens the risk of repetition, and I hope lends itself to a more user friendly and readable volume.

Within each chapter I have referred to the Red, Yellow and Silver Books at the same time, for example:

The Engineer (Employer under the Silver Book), the Engineer (under the Red and Yellow Book), etc.

I have tended not to quote specific Clause/Sub-Clause numbers unless it is totally relevant as I am comparing the provisions under the FIDIC Red, Yellow and Silver Books, and it would be cumbersome to read if you refer simultaneously to the different clause numbers under the three contracts that, in many cases, have exactly the same wording.

I have used the Red Book as the baseline, then added Yellow Book and Silver Book references if they are different to the Red Book. If there is no mention of the Yellow Book and Silver Book then the reader can assume that the provisions and the relevant Clauses/Sub-Clauses for these two contracts are the same as for the Red Book.

Kelvin Hughes
July 2020

Acknowledgements

As with all my previous books, I would like to extend my continued and sincere thanks to my wife Lesley, who as always gives me the love, the time, the inspiration and the support to fulfil my life's ambitions.

Also love and thanks to my parents, Maureen and Dennis, who made me who I am, and I hope in return I continue to make them proud.

I would also like to acknowledge the wisdom, support and friendship of Mr. Seah Ban Kiat during my time in Qatar, when we dealt with FIDIC contracts, and contracts derived from FIDIC, and various issues arising from them. We were also part of a team which drafted our own suite of contracts.

My thanks also got to Seah's family for their friendship, while my wife and I were in Qatar.

Finally, I would like to thank Patrick Hetherington of Taylor & Francis Group for his support and unfailing patience during the writing of this book as I juggled writing, running a business and my family, amid the COVID-19 lockdown.

1 Overview of FIDIC and its contracts

This chapter provides an overview of FIDIC, and some of the various contracts that have been published more recently, prior to the FIDIC 2017 Contracts, which are the subject of this book.

What is FIDIC?

"FIDIC" is the acronym for "Federation Internationale des Ingenieurs-Conseils" (in English "The International Federation of Consulting Engineers"), which is the leading body for developing forms of contract for use in the international construction industry between Employers and Contractors, Contractors and Subcontractors, and Employers and Consultants and Consultants and Subconsultants.

Of all the contracts in use today, FIDIC is the nearest to a truly "international" form of contract and is the contract mostly recognised by international funders such as the World Bank.

The International Federation of Consulting Engineers was founded on 22 July 1913, following a search for independent consultants for the World Fair Exhibition.

There were 59 participants at the inaugural meeting, with delegates from Austria, Belgium, Canada, Denmark, France, Germany, Hungary, the Netherlands, Russia, Switzerland, the United Kingdom and the USA.

Three countries such as Belgium, France and Switzerland decided to found the Federation, the other countries maintained provisional links during the initial years. However, due to the World Wars and other major political disturbances, FIDIC development was slow until the late 1940s. The number of member countries changed constantly, and all came from Europe.

In 1959, they were joined by Australia, Canada, South Africa and the USA.

The Federation developed gradually over the years into a truly global organisation with Member Associations representing countries from all regions of the world.

FIDIC also formed a partnership with the World Bank and other Multinational Development Banks working in the different geographical regions to support the financial requirements and international standards to deliver the infrastructure projects worldwide.

It was later referred to as the "International Federation of Consulting Engineers" and now has over 100 member countries.

FIDIC publishes various standard contracts to be used for infrastructure projects, construction works, consultancy services, major plant supplies, etc.

The aim is to have a standard form for international use which tries to achieve a balance between a comprehensive coverage of the circumstances, problems and obligations which are likely, but does not try to cover every conceivable problem, an impossible task which would result in an unmanageable document.

Written contracts are all about allocating risk and obligations and providing the appropriate machinery for resolving issues likely to be encountered.

Each of the FIDIC contracts applies to a specific type of work and/or procurement method and is characterised by an individual colour label e.g. Red Book, Yellow Book and Silver Book (referred to as "The Rainbow Suite), which form the subject area for this book, though other FIDIC contracts will be briefly reviewed.

The development of the FIDIC contracts

It is not proposed within this book to provide a full history, and to review all of the FIDIC contracts since its inception, but the more recent contracts, particularly the ones with which the author is familiar, are covered.

Note that the FIDIC contracts were originally modelled on the Institution of Civil Engineers (ICE) Contracts, which were in turn modelled on the Association of Consulting Engineers Contracts. The current day FIDIC contracts were modelled on the various contracts below.

Conditions of Contract for Works of Civil Engineering Construction: The Red Book (1987)

This contract is in two parts:

- Part I contains General Conditions, a Form of Tender with Appendix and a Form of Agreement. The Appendix is a part of the Form of Tender and makes specific provision for the insertion of details, which are essential and yet unique to each project.
  There are 72 clauses within the General Conditions.
- Part II contains Conditions of Particular Application with Guidelines for Preparation of Part II Clauses.

The Form of Tender and the Agreement provide that the Agreement should comprise the Letter of Acceptance, the Conditions of Contract, the Specification, the Drawings and the priced Bill of Quantities. If another document is to be given the same status this should be clearly stated in the Agreement.

The Priority of Documents is in the following order:

1 The Contract Agreement
2 The Letter of Acceptance
3 The Tender
4 Part II of the Conditions (The Particular Conditions)
5 Part I of the Conditions (The General Conditions)
6 Any other document forming part of the Contract

The Form of Tender provides for the undertaking to provide a Performance Security within 28 days of the Letter of Acceptance. However, the final decision as to whether to have a Performance Security is optional.

- The Contractor's obligation is to construct the Works according to the Employer's Consultant's design, although some design responsibility may be reallocated to the Contractor.
- "Necessary" levels of supervision must be provided by the Contractor, they should be suitably conversant with the nature of the works to be undertaken.
- The Contractor or a competent agent approved by the Engineer should be constantly kept on the works to receive directions and instructions on behalf of the Contractor.
- The Contractor is responsible for setting out and for the correctness of positioning levels, etc.
- The submission of a programme is an express requirement of the FIDIC Conditions (Clause 14).
- The Contractor becomes responsible for the care of the Works from the Works Commencement Date to the date of issue of a Taking-Over Certificate.
- An Engineer is needed who can act impartially between the parties in order to interpret obligations and responsibilities as circumstances change.
- The Engineer will normally appoint the Resident Engineer or Clerk of Works to be the Engineer's Representative.
- Contracts made under the FIDIC Conditions are remeasure and value contracts based on a Bill of Quantities. None of the documents state what the work will eventually cost. This is not a lump sum contract but one where the price is calculated on completion and remeasurement.
- The Contractor is deemed to have inspected and examined the site and its surroundings.
- Also, the Contractor is deemed to have satisfied himself as to the nature of the ground and sub-soil but only so far as is "<u>practicable</u>".
- Adverse Physical Conditions and Artificial Obstructions require the value of the Contract works to be ascertained and determined by the Engineer, and except as otherwise stated, to do so by a measurement.
- Variations are required to be valued by the Engineer in accordance with principles stated and after consultation with the Contractor, which require:
  - Works similar to that in the Bill of Quantities are priced at the same rates.
  - So far as reasonable Bill rates are to be used as the basis for pricing work.
  - Where neither of the above methods can be used, then a fair valuation shall be made.
  - Valuation on a daywork basis where the work is so ordered.
- Clause 60 sets out the scheme for payment which depends on the submission by the Contractor of monthly statements. The Contractor is required to set out the amounts under heads (a) to (e) of Clause 60.1. Other than for unfixed materials, it envisages that the amounts are estimated.
- The Contractor's rights to claim additional payment pursuant to the FIDIC Conditions are set out in various clauses. The categories for such payment include:
  - <u>Cost</u>-based claims (Cost only);
  - <u>Value</u>-based claims (Cost plus Profit);
  - Other claims as defined in the Conditions.

- If the Employer wishes the Contractor to complete certain parts of the works before the whole and to recover liquidated and ascertain damages for failure to achieve the same, then the Sectional Completion provision in the Appendix should be used. If it is used, then the Extension of Time mechanism in Clause 44 should be applied to each Section.
- Clause 44 deals with Extensions of Time applies to the Works and Sections and deals with time only, not money.
  - Variations;
  - Any cause of delay referred to in the Conditions;
  - Exceptionally adverse climactic conditions;
  - Any delay, impediment or prevention by Employer;
  - Special circumstances.
- There are provisions for liquidated damages for delay in Completion.
- There are also provisions for Resolution of Disputes and Arbitration.

The Orange Book: *Conditions of Contract for Design-Build and Turnkey* (1st Ed 1995)

The *Conditions of Contract for Design-Build and Turnkey* (The "Orange Book") is recommended for the provision of electrical and/or mechanical plant, and for the design and execution of building or Employer's Representative works. The Contractor designs and provides in accordance with the Employer's Requirements.

Note that the FIDIC 1999 Contracts were structured on the same basis as the FIDIC 1995 *Conditions of Contract for Design-Build and Turnkey* (The "Orange Book"), so it is worth considering this Contract in a little detail and readers will see how the FIDIC 1999 Yellow Book developed.

The Orange Book was published in 1995 due to a growing trend towards design and build projects, and at that time FIDIC recognised that the world of contract drafting was moving on.

The Orange Book contained Dispute Adjudication Board (DAB) provisions, whilst the Red Book had, in 1996, a DAB supplement published by FIDIC.

The structure of the Orange Book is as follows:

1 The Contract
2 The Employer
3 The Employer's Representative
4 The Contractor
5 Design
6 Staff and Labour
7 Plant, Materials and Workmanship
8 Commencement, Delays and Suspension
9 Tests on Completion
10 Employer's Taking Over
11 Tests after Completion
12 Defects Liability
13 Contract Price and Payment
14 Variations

15 Default of Contractor
16 Default of Employer
17 Risk and Responsibility
18 Insurance
19 Force Majeure
20 Claims, Disputes and Arbitration

The Priority of Documents under the Orange Book in case of ambiguities or inconsistencies is in the following order of importance:

a The Contract Agreement
b The Letter of Acceptance
c The Employer's Requirements
d The Tender
e The Conditions of Contract, Part II
f The Conditions of Contract, Part I
g The Schedules
h The Contractor's Proposal

Some relevant Clauses are as follows:

Clause 3: The employer's representative

The Employer's Representative is the person appointed by the Employer to act on his behalf and is named in the Appendix to Tender.

The Employer's Representative and his staff must be suitably qualified and competent to carry out their duties. If there is any limitation on the Employer's Representative's level of authority, for example, the need to obtain the Employer's approval on major changes then this must be included within the contract.

Clause 4: The contractor

Clause 4 covers many of the Contractor's general obligations.

Sub-Clause 4.1 states the Contractor's obligation to execute the Works in accidence with the Contract including compliance with the Employer's Requirements, Contractor's Proposal and Schedules.

Sub-Clause 4.3 describes the role of the Contractor's Representative who has all the authority necessary to act on the Contractor's behalf and his whole time to running the project.

Under Sub-Clause 4.9, the Employer is deemed to have made available all relevant data in his possession on sub-surface and hydrological conditions by the Base Date, which is 28 days before the date of submission of the Tender. Note that the Employer is required to make available any information he has, not to do an extensive site investigation himself!

The Contractor is responsible for interpreting the data and shall be deemed to have inspected and examined the Site, its surroundings and other available information, and satisfied himself on matters such as sub-surface conditions, hydrological and climactic conditions and the extent and nature of the work and works.

"Deemed" does not actually mean that the Contractor MUST visit the Site, but that he is "fixed with the knowledge" as if he had....there is a subtle difference!

Sub-Clause 4.11 defines "unforeseeable" as "not reasonably foreseeable by an experienced Contractor by the date for submission of the Tender".

If the Contractor encounters something which is unforeseeable he notifies the Employer's Representative who may award an extension of time and additional cost.

Sub-Clause 4.14 covers the provision of programmes showing how the Contractor proposes to execute the Works. Each such programme is to be supported by a report describing the methods that the Contractor proposes to adopt.

The Contractor is required to submit a programme to the Employer's Representative within the time stated in the Appendix to Tender.

The programme is required to show:

- a The order in which the Contractor proposes to carry out the Works (including design, procurement, manufacture, delivery, installation, testing and commissioning);
- b All major events and activities in the production of Construction Documents;
- c Periods for pre-construction reviews and for submissions, approvals and consents; and
- d The sequence of tests specified in the Contract;

The contract describes the content of the Contractor's programme which should include order and sequence, timing of tests, method statement and details of resource levels.

Monthly progress reports are to be prepared by the Contractor and six copies submitted to the Employer's Representative. The contract prescribes the content of each report which includes:

- a Photographs and detailed descriptions of progress including each stage of design, procurement, manufacture, delivery to Site, construction, erection, testing and commissioning;
- b Charts showing the status of Construction Documents, purchase orders, manufacture and construction;
- c For the manufacture of each main item of Plant and Materials, name of manufacturer, manufacture location, percentage progress and expected dates of commencement of manufacture, inspection, tests and delivery;
- d Records of personnel and Contractor's Equipment on Site;
- e Copies of quality assurance documents, test results, etc.;
- f Safety statistics including details of hazardous incidents and activities related to environmental aspects;
- g Comparisons of actual and planned progress with details of any aspects which may jeopardise completion.

Clause 5: Design

The Contractor is responsible for carrying out the design of the Works. He is required to submit the name of each designer, design Subcontractor to the Employer's Representative and gain prior consent.

The Contractor "holds himself" as having the experience and capability necessary for the design.

The Contractor is to prepare and update a complete set of "as built" records and keep them on Site. Two copies are to be submitted to the Employer's Representative prior to commencement of the Tests on Completion.

In addition, the Contractor must submit "as built" drawings of the Works which shall be submitted to the Employer's Representative for his inspection.

Prior to the issue of any Taking-Over Certificate the Contractor shall submit to the Employer's Representative one microfiche copy, one full size original copy and six printed copies of the relevant "as built" drawings, and any further Construction Docents stated in the Employer's Requirements. NB: The Works are not considered complete until this is done.

Clause 8: Commencement, delays and suspension

Sub-Clause 8.4 defines the Contractor's entitlement to extensions of time.

There are five grounds for extension of time:

- a A Variation (unless an adjustment to the Time for Completion is agreed under Sub-Clause 14.3);
- b A Force Majeure event (as defined in Sub-Clause 19.1);
- c A cause of delay giving rise to an entitlement under the contract;
- d Physical conditions or circumstances on the Site which are exceptionally adverse and not foreseeable by an experienced Contractor; or
- e Any delay, impediment or prevention by the Employer.

If the Contractor fails to comply with Sub-Clause 8.2 i.e. to complete within the Time for Completion, he must pay to the Employer the relevant sum stated in the Appendix to Tender for every day or part of a day elapsed between the relevant Time for Completion and the date stated in the Taking-Over Certificate. Except that the total payment may be limited within the Appendix to Tender.

In the event that an extension of time is granted under Sub-Clause 8.3, the amount shall be recalculated accordingly and any over payment refunded.

At any time after the Employer has become entitled to liquidated damages the Employer's Representative may give notice to the Contractor under Sub-Clause 15.2, requiring the Contractor to complete within a reasonable time for completion.

Clause 9: Tests on completion

The Contractor is required to carry out all Tests on Completion. He has to give the Employer's Representative not less than 21 days' notice prior to each Test on Completion. Unless otherwise agreed the Tests shall be carried out within 14 days after this date.

In considering the results of the Tests on Completion, the Employer's Representative makes allowances for the effect of any use of the Works by the Employer on the performance of the Works. As soon as the Works or a Section has passed the Tests on Completion the Contractor provides the Employer's Representative and the Employer with a certified report of the Test results.

Clause 10: Taking over of the works and sections

The Works are taken over by the Employer when they have been completed in accordance with the contract and have passed the Tests on Completion, and a Taking-Over Certificate has been issued. The Contractor may apply by notice to the Employer's Representative for a Taking-Over Certificate not earlier than 14 days prior to completion.

The Employer's Representative shall within 28 days after receiving the Contractor's application:

a Issue the Taking-Over Certificate to the Contractor stating the date on which the works were completed including passing the Tests on Completion.
b Reject the application giving his reasons and specifying the work required to be done by the Contractor to enable the Taking-Over Certificate to be issued, the Contractor then completes such work before issuing a further notice under this Sub-Clause.

If the Employer's Representative fails to either issue the Taking-Over Certificate or to reject the Contractor's application within 28 days, and the Works are substantially in accordance with the Contract, the Taking-Over Certificate shall be deemed to have been issued.

Clause 12: Defects liability

Defects Liability obligations are specified in Clause 12.

The issue of the Performance Certificate confirms that the Contractor has fulfilled all his obligations under the contract. The Certificate is issued by the date 28 days after the expiry of the Contract Period or as soon after such date as the Contractor has provided all the Construction Documents and completed and tested all the Works including remedying any defects. Only the Performance Certificate shall constitute approval of the Works.

Clause 13: The contract price

Unless otherwise stated in Part II:

a Payment for the Works shall be on a lump sum basis.
b The Contract Price shall not be adjusted for changes in the cost of labour, materials or other matters.
c The Contractor shall pay all duties and taxes, and the Contract Price shall not be adjusted for such costs.
d Any quantities which may be set out in a Schedule are only estimated quantities and not to be taken as actual and correct quantities of the Works to be executed by the Contractor.
e Any quantities, prices or rates of payment per unit quantity which may be set out in a Schedule are only to be used for the purposes stated in the Schedule.

If any part of the Works is to be paid according to quantity supplied or work done, the provisions for measurement and valuation shall be stated on Part II.

The Contractor is required to submit a Statement (six copies) to the Employer's Representative after the end of each month showing in detail the amounts to which he considers himself entitled together with supporting documents.

The Statement includes:

- a The estimated contract value of Construction Documents produced and the Works executed up to the end of the month;
- b Any amounts due to changes in legislation (Sub-Clause 13.16);
- c Any amount to be deducted for retention calculated by applying the retention percentage to the above amounts;
- d Any amounts to be added or deducted for advance payments;
- e Any amounts due for Plant and Materials intended for the Works (Sub-Clause 13.5);
- f Any other additions or deductions which may have become due other than under Sub-Clause 8.6;
- g Deduction of any amounts certified in previous certificates.

Provided the Employer has received and approved the Performance Security, the Employer's Representative certifies payments within 28 days of receiving each Contractor's Statement. A copy of the certificate is sent to the Employer and the Contractor.

Clause 14: Variations and adjustments

The Employer's Representative may vary the works at any time during the Contract Period either by instruction or by a request for the Contractor to submit a proposal.

If the Employer's Representative requests the Contractor to submit a proposal and subsequently elects not to proceed with the work, the Contractor is reimbursed for the Cost incurred.

Sub-Clause 14.2 allows the Contractor to submit a proposal to the Employer's Representative which could reduce the cost of constructing, maintaining or operating the Works, or improve the efficiency or value to the Employer of the completed Works. Any such proposal shall be prepared at the cost of the Contractor.

If the Employer's Representative requests a proposal, prior to instructing a Variation, the Contractor shall submit as soon as practicable:

- a A description of the proposed design and/or work to be performed and a programme for its execution;
- b The Contractor's Proposal for any necessary modification to the programme;
- c The Contractor's Proposal for adjustment to the Contract Price, Time for Completion and/or modifications to the Contract.

The Employer's Representative shall, as soon as practicable after receipt, respond with approval, rejection or comments.

If the Employer's Representative instructs a Variation, he shall proceed in accordance with Sub-Clause 3.5 to agree or determine adjustments to the Contract Price (including reasonable profit), Time for Completion and Schedule of Payments.

Clause 19: Force majeure

Clause 19 covers Force Majeure, which is defined in Sub-Clause 19.1 as an event which is beyond the control of the Employer and Contractor, which makes it impossible or illegal to perform. Examples include Acts of God, war, embargo nuclear contamination and riot, unless solely restricted to the employees of the Contractor or his Subcontractors.

Clause 20: Claims, disputes and arbitration

Clause 20 covers the submission of claims and the settlement of disputes.

If the Contractor considers himself entitled to additional payment, he gives notice to the Employer's Representative not later than 28 days of the start of the event giving rise to the claim.

The Contractor keeps contemporaneous records and upon receipt the Employer's Representative may instruct the keeping of further records.

Within 28 days of such notice, or some other time as may be agreed by the Employer's Representative, the Contractor sends to the Employer's Representative an account giving details and amount of the claim. Where the claim has a continuing effect, he sends an interim account with a final account within 28 days of the end of the effects of the event.

If the Contractor fails to comply with this Sub-Clause he shall not be entitled to additional payment.

The Contractor shall be entitled to have included in any Interim Payment Certificate such amount for any claim as the Employer's Representative considers due.

Unless the members of the DAB have been previously agreed upon between and by the parties, the parties shall within 28 days of the Effective Date ensure the appointment of a DAB.

The terms of appointment:

- a Incorporate the model terms published by the Federation Internationale des Ingenieurs-Conseils (FIDIC).
- b Require each member of the DAB to remain independent of the parties.
- c Require the DAB to act impartially and in accordance with the Contract.
- d Include undertakings by the parties that the members of the DAB shall not be liable for breach of duty or contract.

The Employer and the Contractor shall pay one half of the DAB's remuneration.

If any of the following conditions apply:

- a The parties fail to agree upon the sole member of a one person DAB within 28 days of the Effective Date.
- b Each party fails to nominate an acceptable member for the DAB of three members within 28 days of the Effective Date.
- c The parties fail to agree upon the appointment of the third member (to act as chairman) within 28 days of the Effective Date.
- d The parties fail to agree upon the appointment of a replacement member of the DAB within 28 days of the date on which a member of the DAB declines to act.

Both Party's attempt to settle disputes amicably before the commencement of arbitration. Arbitration commences within 56 days of a notice of dissatisfaction.

Any dispute in respect of which:

a The decision of the DAB has not become final and binding, and
b Amicable settlement has not been reached.

shall be settled by reference to arbitration.

The Harmonised Red Book (MDB Edition): *Conditions of Contract for Construction for Building and Engineering Works Designed by the Employer* (The Pink Book)

It is worth some consideration of this document, as an example of a collaborative exercise between major global financial institutions and the publishers of an international suite of contracts.

Several of the Multilateral Development Banks (MDBs) have, for many years, used the FIDIC Conditions of Contract for Construction as part of their standard bidding documents, which the MDBs require their borrowers to follow in financing construction projects.

In doing so, the MDBs introduced a number of additional clauses within the Particular Conditions.

These additional clauses, which were specific to the MDBs, had standard wording which had to be repeated whenever procurement documents were prepared for a new project.

Also, these additional clauses were not the same for each MDB, which created uncertainties amongst the users of the documents and increased the possibilities of disputes.

As a result the MDBs decided to harmonise their tender documents on an international basis, and resolved that there should be a modified form of the FIDIC Conditions of Contract for Construction, 1st Edition 1999, in which the General Conditions would contain the standard wording which previously had been incorporated into the Particular Conditions.

The first version of the MDB Harmonised Construction Contract was released in May 2005, with an amended second version in March 2006 and a third amended version in 2010.

"The Pink Book" was launched for use as part of the standard bidding documents by the MDBs.

The Islamic Development Bank and the World Bank worked with FIDIC in developing this contract, the MDBs being:

- The World Bank
- The European Bank for Reconstruction and Development
- African Development Bank
- Asian Development Bank
- Black Sea Trade and Development Bank (BSDB)
- Caribbean Development Bank (CDB)
- Council of Europe Development Bank (CEB)
- Inter-American Development Bank (IADB)

The MDBs represent lending agencies that fund projects on a global basis and as such play a crucial role in the development of major infrastructure projects.

Clause 8.1 in particular had been amended from the Red Book, the amended Sub-Clause requiring certain conditions to be met before the works can commence:

- The Contract must be signed by both parties.
- The Contractor must have received reasonable evidence of the Employer's financial arrangements.
- The Contractor must have been given possession of the Site by the Employer.
- There must be in place provision for any advance payment and corresponding guarantee/bond.

The Contract allowed the Contractor the option to terminate if the above matters were not fully in place within 180 days of the Letter of Acceptance, which actually introduced complexity and possible confusion prior to starting work; full possession of the Site may not always be necessary or even possible from the outset.

The following clauses are referred to in the Specification:

- Definition of "Employer's Equipment" (Clause 1.1.6.3)
- Permissions obtained by Employer (Clause 1.13(a))
- Opportunities for works by others (Clause 4.6
- Values for emissions and discharges (Clause 4.18)
- Details of Employer's Equipment and Free-Issue Materials (Clause 4.19)
- Criteria for designers (Yellow Book – Clause 5.1)
- Technical documents to be included in Contractor's Documents and Language for Contractor's Documents (Yellow Book – Clause 5.2)
- Contractor's Documents to be submitted for Review and/or Approval (Yellow Book – Clause 5.3)
- Other standards for compliance (Yellow Book – Clause 5.4)
- Training to be provided for Employer's personnel (Yellow Book – Clause 5.5)
- Numbers and types of copies of As-Built Drawings (Yellow Book – Clause 5.6)
- Operation and Maintenance manuals (Yellow Book – Clause 5.7)
- Arrangements for staff and labour (Clause 6.1)
- Facilities for staff and labour (Clause 6.6)
- Payment of royalties (Clause 7.8)

Those faced with a Pink Book will have often been used to the Red Book and therefore will be familiar with the general layout of the contract, the 20 clauses are still there but there have been amendments which some may think are for the better, and others otherwise.

Minor amendments have taken place to definitions of which some are worthy of being noted:

- Cost no longer referred to "reasonable" profit, but stated the profit percentage which was fixed at 5% unless stated otherwise in the Contract Data.
- Sub-Clause 1.5 was a new Sub-Clause that allowed the lender's representatives to inspect the Site and audit the Contractor's accounts and records relative to the project.

- Under Sub-Clause 2.1, access to the Site must be given such that the programme can proceed "without disruption".
- Under Sub-Clause 2.4, the Employer has to demonstrate its ability to pay the Contract Price before "the Commencement Date".
- Under Sub-Clause 2.5, the Employer had to give notice of its Claims within 28 days, which was quite onerous.
- Under Sub-Clause 3.1, the Engineer had to gain the Employer's approval before dealing with matters under Clauses in respect of Unforeseeable Physical Conditions and the issue of Variations.

  The Employer having to give approval to the Engineer in respect of Claims for quite major issues such as Unforeseeable Physical Conditions and Variations can be very dangerous and inadvisable, unless the Employer has a full understanding of the construction process, and the various actions required in administering that process. An ill-advised Employer (or possibly one that was advised but chose not to take that advice!) could cause significant delays and additional costs to the project.
- Under Sub-Clause 3.5, the Engineer had to give its determination "within 28 days from receipt of the corresponding Claim or request".
- Under Sub-Clause 6.2, there was an obligation upon the Contractor to inform its personnel of their liability to pay local income tax.
- Under Sub-Clause 8.1, the project cannot commence unless:
  - The contract agreement has been signed by both Parties.
  - The Contractor has reasonable proof that funding is in place.
  - The advance payment has been received by the Contractor.
- Under Sub-Clause 8.6, the Contractor could be paid for acceleration measures to overcome Employer delays.
- Under Sub-Clause 13.1, the Contractor is not bound to carry out a Variation if it would "trigger a substantial change in the sequence or progress of the works".
- Under Sub-Clause 15.5, whilst the Employer can terminate for convenience, it cannot terminate to pre-empt a rightful termination by the Contractor.
- Sub-Clause 15.6 is a new clause that attempts to deal with corrupt and/or fraudulent practices.
- Under Sub-Clause 16.2, the Contractor must demonstrate that the Employer's failures must "materially and adversely affect the economic balance of the contract and/or the ability of the Contractor to perform the contract" prior to termination.

  There are however two further grounds allowing the Contractor to terminate:
  - Failure of the funder to provide funds;
  - The absence of the Engineer's Instruction to commence work 180 days after the Letter of Acceptance.
- Under Sub-Clause 19.2, in order to claim Force Majeure, the claiming party must demonstrate that it has been <u>prevented</u> from performing "its substantial obligations".
- Under Sub-Clause 20.6, arbitration rules may differ according to the origin of the lending agency.

In February 2019, FIDIC entered into an agreement with the World Bank that will see the international funding organisation adopt the use of six FIDIC standard contracts for the next five years.

As part of the agreement, the World Bank is given a non-exclusive licence to refer to the six major FIDIC contracts for projects they finance and the documents will also be used as part of the World Bank's standard bidding documents.

The contracts are primarily the FIDIC 2017 contracts.

The agreement includes translation into five major languages: Arabic, Chinese, French, Spanish and Portuguese.

The six FIDIC contract documents covered by the FIDIC/World Bank agreement are:

- *Conditions of Contract for Construction for Building and Engineering Works Designed by the Employer* ("Red Book"), Second Edition 2017
- *Conditions of Contract for Plant & Design-Build for Electrical & Mechanical Plant & for Building & Engineering Works Designed by the Contractor* ("Yellow book"), Second Edition 2017
- *Conditions of Contract for EPC Turnkey Projects* ("Silver book"), Second Edition, 2017
- *Client/Consultant Model Services Agreement* ("White book"), Fifth Edition 2017
- *Conditions of Contract for Design, Build and Operate Projects* ("Gold book"), First Edition 2008
- *Short Form of Contract* ("Green book"), First Edition 1999

The Gold Book: DBO Contract – *Conditions for Design, Build and Operate Projects* (1st Ed 2008)

In September 2008, FIDIC published the Design, Build and Operate (DBO) contract, with many of its provisions now similar to the FIDIC 2017 Rainbow Suite, so it is again worth some consideration.

FIDIC had stated for some time that there was a growing need for a document which would combine a design and build obligation, with a long-term operation and maintenance commitment.

The question was whether it was best to have a design and build contract (such as the Yellow or Silver Book), with a separate contract to operate the facility over a period of time, the two contracts being completely separate or linked together, or alternatively to have a combined "all in one contract".

The Gold Book is the latter "all in one" combined contract, though it has been described as "a Yellow Book with an operate and maintenance contract bolted on", which makes it sound as if it is not truly fit for the purpose, but it actually works quite well.

In reality although the principle is of a Design and Build Contract with an extra part bolted on, it contains many differences to the Yellow Book in addition to the entirely new provisions which deal with the suggested period of 20 years for operation and maintenance of the facility.

Some said at the time that the Yellow Book should actually cover the Design and Build part as it is a familiar Contract, and have an annexe to cover the operation and maintenance part, but there was an opportunity to publish something new, and unique for this contractual relationship.

It is to be used in a "green field" situation where the Employer wishes to employ a single Contractor to design, build and subsequently operate and maintain the completed construction project for a period of 20 years.

It is likely that, although there is a single Contractor under the Gold Book, that Contractor will be a Joint Venture (JV) with one partner responsible for the "design and build" part, the other for the long-term "operation" part.

Note that the Gold Book is not a Design, Build, Finance, Operate (DBFO) contract, the Contractor has no responsibility either for financing the project or for its ultimate commercial success.

The Gold Book when launched in 2008 filled a gap within FIDIC's Contracts and has been widely used by various government departments, especially utilities providers.

The author has personal experience of the Gold Book being used in the Middle East by a petroleum company.

It is a DBO contract that the industry has needed for some time to reflect the ever-growing trend that Contractors no longer construct something then go away, but they also maintain and operate the facility for many years to come.

Clearly, there is a very long-term commitment by both Parties with a contract of this type, and any potential relationship issues or disputes between an Employer and a Contractor carrying out a design and build contract where standards have not been met would leave both Parties to have to live with each other for a very long period, but it does remedy the situation where one Contractor designs and builds a plant, and a totally separate entity maintains it.

When preparing the *Conditions of Contract for Design, Build and Operate Projects*, the drafting group attempted to include all conditions of a general nature, which were likely to apply to the majority of DBO contracts, into General Conditions.

The nature of DBO contracts and those who use them, whether they be public or private organisations, is that they often require special conditions of contract, or indeed particular procedures, which differ from those included in the General Conditions.

For this reason, the DBO document Particular Conditions Part B – Special Provisions which includes advice to drafters of contract documents who may wish to add Special Provisions to replace or supplement the clauses to be found in the General Conditions.

The Structure of the Contract is as follows:

1 General Provisions
2 The Employer
3 The Employer's Representative
4 The Contractor
5 Design
6 Staff and Labour
7 Plant, Materials and Workmanship
8 Commencement Date, Completion and Programme
9 Design-Build
10 Operation Service
11 Testing
12 Defects
13 Variations and Adjustments
14 Contract Price and Payment
15 Termination by Employer
16 Suspension and Termination by Contractor
17 Risk Allocation
18 Exceptional Risks
19 Insurance
20 Claims, Disputes and Arbitration

Whilst it is not intended to review the DBO Contract in detail within this book, certain provisions should be mentioned, as many of those provisions within the Gold Book have been adopted within the FIDIC 2017 Contracts:

- The Contract covers the design and build plus operation and maintenance for 20 years by the same Contractor on a green field site.
- The design and build phase is similar to the FIDIC Yellow Book with specific and challenging completion criteria, but also there is a cut-off date should the Contractor be 182 days late in completion of the phase, leading to termination if desired.
- Payment is on a lump sum basis but there is a defined "asset replacement fund" and schedule that notes the timing and cost of the replacement of certain assets.
- The cost of replacing Plant and Equipment outside the schedule will be the responsibility of the Contractor, as is a cost to the Contractor over that stated on the Schedule. Any surplus in the fund at the end of the 20 years is divided equally between the Parties.
- The Employer is entitled to deduct 5% from payments during the "Operation Service Period" (OSP) in case the Contractor does not fulfil its maintenance obligations. The fund is to be released, if not spent, within the Final Payment to the Contractor. The Contractor is responsible for correcting its own defects arising from the design and the construction in this period.
- An independent audit body is jointly appointed for the duration of the Operation Service Period to monitor the performance of the Contractor and the Employer. Whilst having no power, the Parties are intended to give "due regard" to matters raised by the audit body.
- A joint inspection is required at least two years before the end of the Operation Service Period. Any works identified must be carried out by the Contractor who will also face completion tests similar to those at the end of the design and build phase. Defaulting contractors risk losing the 5% Maintenance Retention Fund.
- A standing DAB is established from a set date for the design and build phase, and a new DAB every five years during the OPS.

The key to success rests with the Contractor, who must design and build a quality plant, ideally with low future operating and maintenance costs, but fit for purpose and built to last.

Some key features are as follows:

Priority of documents

a The Contract Agreement
b The Letter of Acceptance
c The Letter of Tender
d The Particular Conditions Part A – Contract Data
e The Particular Conditions Part B – Special Provisions
f The General Conditions
g The Employer's Requirements
h The Schedules
i The Contractor's Proposals and any other documents forming part of the contract:

- The Employer is represented by an Employer's Representative rather than an Engineer.
- The Contractor's general obligations include the obligation to design, execute and complete the Works and also to provide the Operation Service in accordance with the Contract.
- The Contractor is responsible for carrying out the design of the Works including the Contractor's Documents, as built records and operation and maintenance manuals.
- Clause 14 deals with the Contract Price, its payment, and also with the possibility of an advance payment. It also covers Application for Interim Payments. It also includes for an Asset Replacement Schedule.
- Clause 17 lists risks held by the Employer during the Design-Build and Operation periods and also the Contractor's risks during the Operation Service Period.
- The Employer gives the Contractor the Operating Licence to operate and maintain the Works during the Operation Service Period.
- The Employer's arrangements for financing the design, execution and operation of the Works including the Asset Replacement Fund are detailed in the Financial Memorandum.
- The Contractor shall, as specified in the Contract or as instructed by the Engineer, allow appropriate opportunities for carrying out the work to:
  - The Employer's Personnel;
  - Any other contractors employed by the Employer;
  - The personnel of any legally constituted public authorities.
- To the extent, which was practicable, the Contractor is deemed to have satisfied himself regarding
  - Form and nature of the Site;
  - Hydrological and climactic conditions;
  - The extent and nature of the works and goods necessary for the execution and completion of the Works;
  - Laws, procedures and labour practices;
  - Requirements for access, accommodation, facilities, personnel, etc.
- If the Contractor encounters Unforeseeable Physical Conditions, he notifies the Employer's Representative. The Contractor may be awarded an extension of time and/or Cost.
- Progress reports shall include:
  - Charts and detailed descriptions of progress;
  - Photographs showing status of manufacture and progress on Site;
  - The name of the manufacturer, location, progress for each item of Plant and Materials;
  - Records of Contractor's Personnel and Equipment;
  - Copies of quality assurance documents, test results, etc.;
  - Notices of Employer's Claims and Contractor's Claims;
  - Safety statistics;
  - Comparisons of actual and planned progress;
  - Order in which the Contractor intends to carry out the Works;
  - The period of Operation Service;
  - The period for reviews of Contractor's Documents, samples, approvals and consents;

  - Sequence and timing of inspections and tests;
  - Method statements and resource schedules.
- Each Party shall endeavour to advise the other Party in advance of any known or probable future event that could increase the Contract Price, or delay the execution of the Works or the Operation Period.
- Performance of the Contractor's obligations shall not be considered to have been completed until the Contract Completion Certificate has been signed by the Employer's Representative.
- The Contractor shall commence the design and execution of the Works within 28 days of the Commencement Date.
- Extension of Time causes:
  - i A Variation;
  - ii A cause of delay covered by a Sub-Clause of the Conditions;
  - iii Exceptionally adverse climactic conditions;
  - iv Unforeseeable shortages caused by epidemic or governmental actions;
  - v Any delay, impediment or prevention caused by the Employer or his personnel or other Contractors:
    - Contractor gives notice to the Employer's Representatives not later than 28 days after becoming aware of the event. Notice describes event.
    - Contractor keeps records.
    - Within 42 days after he becomes aware, Contractor submits detailed claim, further claims at monthly intervals.
    - Within 42 days after receiving the claim, Employer's Representative responds.
- The Contractor is obliged to comply with the Operation Management Requirements in the Contract including following the Operation and Maintenance Plan.

  This Clause also covers the provision of an Independent Compliance Audit to audit and monitor the performance of the Employer and the Contractor during the Operation Service in compliance with the Operation Management Requirements.
- The commencement of the Operation Service shall be from the date stated in the Commissioning Certificate.
- The Contractor provides the Operation Service in compliance with the Operation Management Requirements.
- If there any delays or interruptions during the Operation Service caused by the Contractor he shall compensate the Employer.
- If there any delays or interruptions during the Operation Service caused by the Employer he shall compensate the Contractor.
- If a failure to reach production outputs is caused by the Employer, the Employer notifies the Contractor of the steps he should take, the Contractor is paid Cost plus Profit.
- If a failure to reach production outputs is caused by the Contractor, the Contractor pays performance damages.
- Contractor may submit a written proposal which if adopted:
  - i Accelerates completion;
  - ii Reduces the cost of executing, operating or maintaining the Works;

iii Improves the efficiency or value of the Works;
iv Includes other benefit.
- Contractor is NOT paid a fee.

- Regarding claims, Contractor's and Employer's Claims are both dealt with under Clause 20, with strict provisions and time scales, which are common to the FIDIC 1999 and 2017 Contracts.

  There is also provision for avoidance of disputes, and also disputes arising during the Operation Service Period, the former being relevant to the standing nature of the DAB and the latter relating to post issue of the Commissioning Certificate.

  The initial Notice that has to be provided in submitting a claim is 28 days, and failure to issue the Notice renders the claim time barred.

  The FIDIC drafters, within the Gold Book, have perhaps considered those who believe that fatal notices are unfair as they have introduced a potential opportunity for the Contractor to be able to proceed with its claims despite not having submitted a Notice within the required time period.

  The solution which is similar to the FIDIC 2017 Contract is a referral to the DAB should the Contractor consider that there were circumstances to justify the late submission of the Notice. The DAB, which is a standing board, appointed at the commencement of the project has the authority to override the 28-day time limit if it considers it fair and reasonable to do so.

  Following submission of the Notice the Contractor must keep contemporaneous records to substantiate its claim.

  Within 42 days after the Contractor became aware (or should have become aware) of the event or circumstance giving rise to a claim for an entitlement the Contractor must send to the Employer's Representative a fully detailed claim which includes all supporting particulars of the contractual basis for the claim plus any delay analysis or quantum evidence as appropriate.

  There are relaxations to the 42 days if allowed by the DAB following an extension of the original 28-day Notice or if agreed between the Contractor and the Employer's Representative who may also ask for further supporting particulars.

  No matter how professional a Contractor may be it may often prove difficult to gather all relevant information within the 42-day period, especially if the Contractor has to rely upon a chain of third parties, such as Sub-Contractors and their suppliers, to provide basic information and potentially involve lawyers to advise upon the contractual or other basis of the claim.

The White Book: *Client/Consultant Model Services Agreement*

The FIDIC White Book is entitled the *Client/Consultant Model Services Agreement* and consists of:

- Agreement
- General Conditions
- Particular Conditions
- Appendices 1, 2, 3 and 4

The White Book has been drafted to create conditions of agreement that would span the life cycle of an Engineer's or other Consultant's involvement.

Accordingly, the document is suitable for use during:

- Pre-investment and feasibility studies;
- The design phase;
- The administration of a contract.

As with other FIDIC contracts there are both General Conditions and Particular Conditions of Contract which combine to set out the scope of the Consultant's work, payment terms and the like.

The White Book incorporates the same financial protection as afforded to Contractors, in that the consultant too can ask the Client (as opposed to the Employer) if it has the ability to pay the Consultant's fees.

In a similar vein, and maybe not surprising to some, the White Book limits the Consultant's responsibilities, and therefore liabilities, to "exercise reasonable skill, care and allegiance in the performance of his obligations under the Agreement".

This limitation is further qualified since nothing else in the agreement, or any legal requirement of the country or any other jurisdiction can impose a greater risk upon the consultant. Thus, the Consultant/Engineer has a limited risk that, it is suggested, is not in accord with the thoughts of employers and contractors alike.

The *Form of Contract for Dredging and Reclamation Works* (Blue Book)

The *Form of Contract for Dredging and Reclamation Works* (Blue Book) was originally launched the First Edition in 2006, then the Second Edition in 2016.

It is based on the FIDIC Short Contract (Green Book) and was originally termed the "Blue-Green Book", but has an Engineer working on behalf of the Employer, whereas the Green Book has an Employer's Representative (if required).

It was published to allow users to have a specialist marine contract, rather than having to use and adapt another member of the FIDIC contract family, in particular the FIDIC Conditions of Contract for Works of Civil Engineering Construction 1987.

Like the Green Book, it is abbreviated and flexible. In terms of being a smaller document the General Conditions are only 16 pages and 15 clauses long.

The format has also changed from the major forms with the Agreement and the Appendix to the contract being the first section.

Perhaps, it is the major forms that have the order incorrect since it is those particular terms that are the most important to recognise especially in such a flexible form as the Blue Book.

Similarities with the major forms are in the inclusion of standard forms, such as securities, a section on adjudication (a one- or three-person DAB) with rules and the adjudicator's agreement and the all-important guidance section.

The Engineer is still recognised within the contract; however, design responsibility can rest with either the Employer (and its Engineer) or the Contractor.

Payment terms are extremely flexible, and a list of options, such as lump sum, remeasurement and cost plus, are all noted within the Appendix.

It is perhaps apparent that a greater input of a Contractor's organisation has influenced some of the General Conditions, as has perhaps the use of other forms within the industry as a whole:

- Notices by the Contractor in respect of a claim must be given within 28 days but there is no fatal time bar.
- Claim items are listed as defined risks which may entitle the Contractor to monetary or time compensation.
- The defined risks, recognising the likely impacts of weather upon the Contractor's ability to make progress, potentially soften the usual clause wording, entitling the Contractor to make a claim if "any operation of the forces of nature affecting the Site/and or the Works", which was Unforeseeable or "against which an experienced Contractor could not reasonably have been expected to take precautions" but give the Employer (and Contractor) less room for debate by also defining the Employer's risk to be "climatic conditions more adverse than those specified in the Appendix".
- If disputes are not settled amicably they are to be settled by referral to adjudication by a DAB, and, if dissatisfied with the DAB's decision (or if no decision is made within the set timescale) the dispute can be referred to Arbitration.

The Blue Book is a model of a contract drafted by those with a particular section of the industry in mind and with the knowledge to incorporate the necessary variations to standard forms that may have been considered for use in the past.

The traditional method of appointing Contractors for a dredging and reclamation work is on a remeasurable basis with the Employer being responsible for design; the Employer also being responsible for obtaining planning permission, licences and permits with the Contractor responsible for customs clearances, and some licences and certificates. However, the Contract is flexible enough to allow for other payment methods.

In some cases, the Contractor may be responsible for design, and some of the Employer's obligations above.

Site Data is critical with a contract of this type, with the Contractor encountering different conditions to what he had expected (and priced and programmed for) and the contract specifically deals with issue in that respect.

There are also risks specific to work of this type, which again are covered by the Contract.

Many of the other provisions within the Blue Book are similar to the members of the Rainbow Suite (Red, Yellow and Silver Books), which are discussed in more detail in later chapters of this book, including similar terminology; the Employer having to provide sufficient evidence of funding, there is also the use of "Exceptional Events" rather than the more traditional "force majeure", in addition to limitations on liability.

Conditions of Contract for Underground Works **(2019 Emerald Book)**

In 2019 FIDIC launched a much anticipated new form of contract, the Emerald Book, for Underground Works.

Whilst some practitioners questioned the need for a separate contract for underground works, the nature of, for example, major tunnelling projects is that underground and sub-surface construction works are highly dependent on the hydrogeological, geological and geotechnical properties of the ground itself.

Underground Works are predominantly characterised by three unique and high risk features:

i The method of excavation and ground support, and the associated equipment and other resources to carry out the works are major factors for the successful outcome of the project.
ii Physical access to the Works is often limited to just a few locations or even a single location with difficult access, which places serious constraints on construction logistics, use of equipment and the environment.
iii The land above where the Works are to be carried out usually belongs to a number of third parties, many of whom have no knowledge of what is being carried out.

There are also other important features that are not unique to Underground Works:

i They require extensive investment in Contractor's Equipment.
ii Underground excavation and associated lining works are very time-consuming.

The Emerald Book is designed for use in connection with all sub-surface works, and as such it can be applied to all common excavation methods, including blasting and tunnelling using special tunnel boring machines (TBMs), and also may be appropriate for other types of work that include significant geotechnical uncertainty, for example, for caverns used in the energy sector.

The Emerald Book is based on the FIDIC Yellow Book 2017.

Particular features of the Emerald Book include:

- Disclosure of all available geological and geotechnical information;
- The inclusion of a contractual geotechnical baseline;
- The inclusion of a tailored "Unforeseeable Physical Conditions" clause;
- The provision of a flexible mechanism for remuneration according to ground conditions, both foreseen and unforeseen;
- Time for Completion being influenced by ground conditions;
- The implementation of a ground classification system and supporting particular conditions that reflect excavation and stabilisation activities.

The following contents of the Emerald Book are of particular relevance:

i The Geotechnical Baseline Report (GBR) is defined as the single contractual source of risk allocation related to subsurface physical conditions to the Parties. It addresses the reaction of the ground to excavation and support under the contractually agreed construction methodology. All subsurface physical conditions not addressed in the GBR are considered "Unforeseeable".
ii The risks arising out of the foreseen properties of the ground, obstacles and adverse reaction to the excavation and ground support processes are assigned to the Contractor, as well as the production rates and cost of performing the Works under the same conditions. Conversely, the risks arising out of unforeseen physical conditions of the ground, obstacles and adverse reaction to the excavation

and ground support processes are allocated to the Employer, with the Contractor being able to receive an extension of time and/or reimbursement of cost.

iii The Time for Completion of the Works, of a Section or of any Milestone may largely depend on the subsurface physical conditions that are encountered during excavation. As the risk related to these conditions is allocated to the Employer, Time for Completion should be adjusted by the variation of these conditions within the limits defined by the GBR, insofar as this variation affects the critical path of the Works, of a Section or of any Milestone. If the onerous conditions encountered are higher than described in the GBR, the Time for Completion is extended, but if less onerous, the Time for Completion is reduced.

iv The cost of the excavation and lining works will also largely depend on the subsurface physical conditions and/or ground reaction(s) to such works. The Conditions provide that the excavation and lining works are to be measured and paid for using rates and prices set out in the Bill of Quantities.

v Due to the typically long duration of underground works and to the high cost of investment in equipment, an important portion of the cost to the Contractor, such as, depreciation of and interest for equipment, and overhead charges, is time-related, whilst the remainder of the cost depends on the volume or quantity of the performed works.

2 "The old and the new" – FIDIC 1999 becomes FIDIC 2017

The FIDIC Rainbow Suite

This book is primarily based on the FIDIC Rainbow Suite, including the changes incorporated within the FIDIC 2017 Contracts.

The Rainbow Suite consists of:

- The Red Book, or to give it its full title *Conditions of Contract for Construction*, is primarily intended for building or engineering works where the Employer bears the design responsibility; however, the works may include some Contractor-designed civil, mechanical, electrical and/or construction works.

  The basis of payment to the Contractor is through rates and prices in the Bills of Quantities, though it can be amended to provide a lump sum price or as a cost reimbursable contract.
- The Yellow Book, full title "*Conditions of Contract for Plant and Design-Build*", is intended for contracts on electrical/mechanical installations, and for the design and execution of building or engineering works where the Contractor bears the design responsibility. The Contractor designs and provides the Works in accordance with the Employer's Requirements.
- The Silver Book, full title *Conditions of Contract for EPC/Turnkey Projects*, is suitable for the provision on a turnkey basis of a process or power plant, of a factory or similar facility, or of an infrastructure project or other type of development, where the Contractor takes total responsibility for the design and execution of the project, and a higher degree of certainty of final price and time is required.

More details will follow in terms of the specific use and content of each contract within the Rainbow Suite.

The launch of the FIDIC 2017 Contracts

New versions of the Red Book, Yellow Book and Silver Book were launched on 5 December 2017 (the FIDIC 2017 Contracts), and they constitute updates (with many amendments) of the former editions from 1999.

Note that the FIDIC 1999 Contracts can still be used, and probably will continue to be used for years to come.

During training and consultancy assignments around the world, the author found it surprising, even up to the present day, that Employers, and even entire countries, had not even adopted the FIDIC 1999 Contracts, preferring to use the older contracts (with "appropriate amendments"), in particular:

- *Conditions of Contract for Works of Civil Engineering Construction*: The Red Book (1987)
- *Conditions of Contract for Electrical and Mechanical Works including Erection on Site*: The Yellow Book (1987)
- *Conditions of Contract for Design-Build and Turnkey*: The Orange Book (1995)

The first and third of these Contracts were discussed in Chapter 1.

The FIDIC 1999 Contracts consist of:

- The General Conditions – Clauses 1–20
- Appendix
- General Conditions of Dispute Adjudication Agreement
- Guidance for the Preparation of Particular Conditions
- Annexes – Forms of Security
- Forms of Letter of Tender, Contract Agreement and Dispute Adjudication Agreement

The FIDIC 2017 Contracts consist of:

- The General Conditions – Clauses 1–21
- Appendix
- DAAB Procedural Rules
- Guidance for the Preparation of Particular Conditions
- Annexes – Forms of Security
- Forms of Letter of Tender, Contract Agreement and Dispute Adjudication Agreement

There is nothing particularly wrong with using an older version of any published Contract, provided there are "appropriate amendments" to incorporate current practice and laws, but the FIDIC 1999 Contracts brought in some welcome improvements in the structure and ease of use of the Contracts, and the FIDIC 2017 Contracts have continued with those improvements, so it is recommended that Employers should move to using the latest version of all published contracts, not just FIDIC, as soon as practicable (Figures 2.1 and 2.2).

FIDIC Conditions of Subcontract

Note that at the time of writing this book there are no Conditions of Subcontract published for use with the FIDIC 2017 Contracts.

The Conditions of Subcontract for Construction 2011 are for use in conjunction with the Conditions of Contract for Construction 1999, and the Multilateral Development Bank Harmonised Edition of the FIDIC Conditions of Contract for Construction (Pink Book).

These Subcontract Conditions supersede the previous Conditions of Subcontract for Works of Civil Engineering Construction 1994, which were for use in conjunction

FIDIC 1999 Contract Structure

Red Book

Clause 1: General Provisions

1.1 Definitions
1.2 Interpretation
1.3 Communications
1.4 Law and Language
1.5 Priority of Documents
1.6 Contract Agreement
1.7 Assignment
1.8 Care and Supply of Documents
1.9 Delayed Drawings or Instructions
1.10 Employer's Use of Contractor's Documents
1.11 Contractor's Use of Employer's Documents
1.12 Confidential Details
1.13 Compliance with Laws
1.14 Joint and Several Liability

Clause 2: The Employer

2.1 Right of Access to the Site
2.2 Permits, Licences or Approvals
2.3 Employer's Personnel
2.4 Employer's Financial Arrangements
2.5 Employer's Claims

Clause 3: The Engineer

3.1 Engineer's Duties and Authority
3.2 Delegation by the Engineer
3.3 Instructions of the Engineer
3.4 Replacement of the Engineer
3.5 Determinations

Yellow Book

Clause 1: General Provisions

1.1 Definitions
1.2 Interpretation
1.3 Communications
1.4 Law and Language
1.5 Priority of Documents
1.6 Contract Agreement
1.7 Assignment
1.8 Care and Supply of Documents
1.9 Errors in the Employer s Requirements
1.10 Employer's Use of Contractor's Documents
1.11 Contractor's Use of Employer's Documents
1.12 Confidential Details
1.13 Compliance with Laws
1.14 Joint and Several Liability

Clause 2: The Employer

2.1 Right of Access to the Site
2.2 Permits, Licences or Approvals
2.3 Employer's Personnel
2.4 Employer's Financial Arrangements
2.5 Employer's Claims

Clause 3: The Engineer

3.1 Engineer's Duties and Authority
3.2 Delegation by the Engineer
3.3 Instructions of the Engineer
3.4 Replacement of the Engineer
3.5 Determinations

Silver Book

Clause 1: General Provisions

1.1 Definitions
1.2 Interpretation
1.3 Communications
1.4 Law and Language
1.5 Priority of Documents
1.6 Contract Agreement
1.7 Assignment
1.8 Care and Supply of Documents
1.9 Confidentiality
1.10 Employer's Use of Contractor's Documents
1.11 Contractor's Use of Employer's Documents
1.12 Confidential Details
1.13 Compliance with Laws
1.14 Joint and Several Liability

Clause 2: The Employer

2.1 Right of Access to the Site
2.2 Permits, Licences or Approvals
2.3 Employer's Personnel
2.4 Employer's Financial Arrangements
2.5 Employer's Claims

Clause 3: The Employer's Administration

3.1 The Employer's Representative
3.2 Other Employer's Personnel
3.3 Delegated Persons
3.4 Instructions
3.5 Determinations

Figure 2.1 Structure of the FIDIC 1999 Contracts.

Clause 4: The Contractor

4.1 Contractor's General Obligations
4.2 Performance Security
4.3 Contractor's Representative
4.4 Subcontractors
4.5 Assignment of Benefit of Subcontract
4.6 Co-operation
4.7 Setting Out
4.8 Safety Procedures
4.9 Quality Assurance
4.10 Site Data
4.11 Sufficiency of the Accepted Contract Amount
4.12 Unforeseeable Physical Conditions
4.13 Rights of Way and Facilities
4.14 Avoidance of Interference
4.15 Access Route
4.16 Transport of Goods
4.17 Contractor's Equipment
4.18 Protection of the Environment
4.19 Electricity, Water and Gas
4.20 Employer's Equipment and Free-Issue Material
4.21 Progress Reports
4.22 Security of the Site
4.23 Contractor's Operations on Site
4.24 Fossils

Clause 5: Nominated Subcontractors

5.1 Definition of "nominated Subcontractor"
5.2 Objection to Nomination
5.3 Payments to nominated Subcontractors
5.4 Evidence of Payments

Clause 6: Staff and Labour

6.1 Engagement of Staff and Labour
6.2 Rates of Wages and Conditions of Labour
6.3 Persons in the Service of Employer
6.4 Labour Laws
6.5 Working Hours
6.6 Facilities for Staff and Labour
6.7 Health and Safety
6.8 Contractor's Superintendence
6.9 Contractor's Personnel
6.10 Records of Contractor's Personnel and Requirement
6.11 Disorderly Conduct

Clause 4: The Contractor

4.1 Contractor's General Obligations
4.2 Performance Security
4.3 Contractor's Representative
4.4 Subcontractors
4.5 Nominated Subcontractors
4.6 Co-operation
4.7 Setting Out
4.8 Safety Procedures
4.9 Quality Assurance
4.10 Site Data
4.11 Sufficiency of the Accepted Contract Amount
4.12 Unforeseeable Physical Conditions
4.13 Rights of Way and Facilities
4.14 Avoidance of Interference
4.15 Access Route
4.16 Transport of Goods
4.17 Contractor's Equipment
4.18 Protection of the Environment
4.19 Electricity, Water and Gas
4.20 Employer's Equipment and Free-Issue Material
4.21 Progress Reports
4.22 Security of the Site
4.23 Contractor's Operations on Site
4.24 Fossils

Clause 5: Design

5.1 General Design Obligations
5.2 Contractor's Documents
5.3 Contractor's Undertaking
5.4 Technical Standards and Regulations
5.5 Training
5.6 As-Built Documents
5.7 Operation and Maintenance Manuals
5.8 Design Error

Clause 6: Staff and Labour

6.1 Engagement of Staff and Labour
6.2 Rates of Wages and Conditions of Labour
6.3 Persons in the Service of Employer
6.4 Labour Laws
6.5 Working Hours
6.6 Facilities for Staff and Labour
6.7 Health and Safety
6.8 Contractor's Superintendence
6.9 Contractor's Personnel
6.10 Records of Contractor's Personnel and Requirement
6.11 Disorderly Conduct

Clause 4: The Contractor

4.1 Contractor's General Obligations
4.2 Performance Security
4.3 Contractor's Representative
4.4 Subcontractors
4.5 Nominated Subcontractors
4.6 Co-operation
4.7 Setting Out
4.8 Safety Procedures
4.9 Quality Assurance
4.10 Site Data
4.11 Sufficiency of the Accepted Contract Amount
4.12 Unforeseeable Difficulties
4.13 Rights of Way and Facilities
4.14 Avoidance of Interference
4.15 Access Route
4.16 Transport of Goods
4.17 Contractor's Equipment
4.18 Protection of the Environment
4.19 Electricity, Water and Gas
4.20 Employer's Equipment and Free-Issue Material
4.21 Progress Reports
4.22 Security of the Site
4.23 Contractor's Operations on Site
4.24 Fossils

Clause 5: Design

5.1 General Design Obligations
5.2 Contractor's Documents
5.3 Contractor's Undertaking
5.4 Technical Standards and Regulations
5.5 Training
5.6 As-Built Documents
5.7 Operation and Maintenance Manuals
5.8 Design Error

Clause 6: Staff and Labour

6.1 Engagement of Staff and Labour
6.2 Rates of Wages and Conditions of Labour
6.3 Persons in the Service of Employer
6.4 Labour Laws
6.5 Working Hours
6.6 Facilities for Staff and Labour
6.7 Health and Safety
6.8 Contractor's Superintendence
6.9 Contractor's Personnel
6.10 Records of Contractor's Personnel and Requirement
6.11 Disorderly Conduct

Figure 2.1 (Continued).

Clause 7: Plant, Materials and Workmanship

7.1 Manner of Execution
7.2 Samples
7.3 Inspection
7.4 Testing
7.5 Rejection
7.6 Remedial Work
7.7 Ownership of Plant and Materials
7.8 Royalties

Clause 8: Commencement, Delays and Suspension

8.1 Commencement of Works
8.2 Time for Completion
8.3 Programme
8.4 Extension of Time for Completion
8.5 Delays Caused by Authorities
8.6 Rate of Progress
8.7 Delay Damages
8.8 Suspension of Work
8.90 Consequences of Suspension
8.10 Payment for Plant and Materials in Event of Suspension
8.11 Prolonged Suspension
8.12 Resumption of Work

Clause 9: Tests on Completion

9.1 Contractor's Obligations
9.2 Delayed Tests
9.3 Retesting
9.4 Failure to Pass Tests on Completion

Clause 10: Employer's Taking Over

10.1 Taking Over of the Works and Sections
10.2 Taking Over Parts of the Works
10.3 Interference with Tests on Completion
10.4 Surfaces Requiring Reinstatement

Clause 11: Defects Liability

11.1 Completion of Outstanding Work and Remedying Defects
11.2 Cost of Remedying Defects
11.3 Extension of Defects Notification Period
11.4 Failure to Remedy Defects
11.5 Remedying of Defective Work
11.6 Further Tests
11.7 Right of Access
11.8 Contractor to Search
11.9 Performance Certificate
11.10 Unfulfilled Obligations
11.11 Clearance of Site

Clause 7: Plant, Materials and Workmanship

7.1 Manner of Execution
7.2 Samples
7.3 Inspection
7.4 Testing
7.5 Rejection
7.6 Remedial Work
7.7 Ownership of Plant and Materials
7.8 Royalties

Clause 8: Commencement, Delays and Suspension

8.1 Commencement of Works
8.2 Time for Completion
8.3 Programme
8.4 Extension of Time for Completion
8.5 Delays Caused by Authorities
8.6 Rate of Progress
8.7 Delay Damages
8.8 Suspension of Work
8.90 Consequences of Suspension
8.10 Payment for Plant and Materials in Event of Suspension
8.11 Prolonged Suspension
8.12 Resumption of Work

Clause 9: Tests on Completion

9.1 Contractor's Obligations
9.2 Delayed Tests
9.3 Retesting
9.4 Failure to Pass Tests on Completion

Clause 10: Employer's Taking Over

10.1 Taking Over of the Works and Sections
10.2 Taking Over Parts of the Works
10.3 Interference with Tests on Completion
10.4 Surfaces Requiring Reinstatement

Clause 11: Defects Liability

11.1 Completion of Outstanding Work and Remedying Defects
11.2 Cost of Remedying Defects
11.3 Extension of Defects Notification Period
11.4 Failure to Remedy Defects
11.5 Remedying of Defective Work
11.6 Further Tests
11.7 Right of Access
11.8 Contractor to Search
11.9 Performance Certificate
11.10 Unfulfilled Obligations
11.11 Clearance of Site

Clause 7: Plant, Materials and Workmanship

7.1 Manner of Execution
7.2 Samples
7.3 Inspection
7.4 Testing
7.5 Rejection
7.6 Remedial Work
7.7 Ownership of Plant and Materials
7.8 Royalties

Clause 8: Commencement, Delays and Suspension

8.1 Commencement of Works
8.2 Time for Completion
8.3 Programme
8.4 Extension of Time for Completion
8.5 Delays Caused by Authorities
8.6 Rate of Progress
8.7 Delay Damages
8.8 Suspension of Work
8.90 Consequences of Suspension
8.10 Payment for Plant and Materials in Event of Suspension
8.11 Prolonged Suspension
8.12 Resumption of Work

Clause 9: Tests on Completion

9.1 Contractor's Obligations
9.2 Delayed Tests
9.3 Retesting
9.4 Failure to Pass Tests on Completion

Clause 10: Employer's Taking Over

10.1 Taking Over of the Works and Sections
10.2 Taking Over Parts of the Works
10.3 Interference with Tests on Completion

Clause 11: Defects Liability

11.1 Completion of Outstanding Work and Remedying Defects
11.2 Cost of Remedying Defects
11.3 Extension of Defects Notification Period
11.4 Failure to Remedy Defects
11.5 Remedying of Defective Work
11.6 Further Tests
11.7 Right of Access
11.8 Contractor to Search
11.9 Performance Certificate
11.10 Unfulfilled Obligations
11.11 Clearance of Site

Figure 2.1 (Continued).

Clause 12: Measurement and Evaluation

12.1 Works to be Measured
12.2 Method of Measurement
12.3 Evaluation
12.4 Omissions

Clause 13: Variations and Adjustments

13.1 Right to Vary
13.2 Value Engineering
13.3 Variation Procedure
13.4 Payment in Applicable Currencies
13.5 Provisional Sums
13.6 Daywork
13.7 Adjustments for Changes in Legislation
13.8 Adjustments for Changes in Cost

Clause 14: Contract Price and Payment

14.1 The Contract Price
14.2 Advance Payment
14.3 Application for Interim Payment Certificates
14.4 Schedule of Payments
14.5 Plant and Materials intended for the Works
14.6 Issue of Interim Payment Certificates
14.7 Payment
14.8 Delayed Payment
14.9 Payment of Retention Money
14.10 Statement at Completion
14.11 Application for Final Payment Certificate
14.12 Discharge
14.13 Issue of Final Payment Certificate
14.14 Cessation of Employer's Liability
14.15 Currencies of Payment

Clause 15: Termination by Employer

15.1 Notice to Correct
15.2 Termination by Employer
15.3 Valuation at Date of Termination
15.4 Payment after Termination
15.5 Employer's Entitlement to Termination

Clause 16: Suspension and Termination by Contractor

16.1 Contractor's Entitlement to Suspend Work
16.2 Termination by Contractor
16.3 Cessation of Work and Removal of Contractor's Equipment
16.4 Payment on Termination

Clause 12: Tests after Completion

12.1 Procedure for Tests after Completion
12.2 Delayed Tests
12.3 Retesting
12.4 Failure to Pass Tests after Completion

Clause 13: Variations and Adjustments

13.1 Right to Vary
13.2 Value Engineering
13.3 Variation Procedure
13.4 Payment in Applicable Currencies
13.5 Provisional Sums
13.6 Daywork
13.7 Adjustments for Changes in Legislation
13.8 Adjustments for Changes in Cost

Clause 14: Contract Price and Payment

14.1 The Contract Price
14.2 Advance Payment
14.3 Application for Interim Payment Certificates
14.4 Schedule of Payments
14.5 Plant and Materials intended for the Works
14.6 Issue of Interim Payment Certificates
14.7 Payment
14.8 Delayed Payment
14.9 Payment of Retention Money
14.10 Statement at Completion
14.11 Application for Final Payment Certificate
14.12 Discharge
14.13 Issue of Final Payment Certificate
14.14 Cessation of Employer's Liability
14.15 Currencies of Payment

Clause 15: Termination by Employer

15.1 Notice to Correct
15.2 Termination by Employer
15.3 Valuation at Date of Termination
15.4 Payment after Termination
15.5 Employer's Entitlement to Termination

Clause 16: Suspension and Termination by Contractor

16.1 Contractor's Entitlement to Suspend Work
16.2 Termination by Contractor
16.3 Cessation of Work and Removal of Contractor's Equipment
16.4 Payment on Termination

Clause 12: Tests after Completion

12.1 Procedure of Tests after Completion
12.2 Delayed tests
12.3 Retesting
12.4 Failure to Pass Tests after Completion

Clause 13: Variations and Adjustments

13.1 Right to Vary
13.2 Value Engineering
13.3 Variation Procedure
13.4 Payment in Applicable Currencies
13.5 Provisions Sums
13.6 Daywork
13.7 Adjustments for Changes in Legislation
13.8 Adjustments for Changes in Cost

Clause 14: Contract Price and Payment

14.1 The Contract Price
14.2 Advance Payment
14.3 Application for Interim Payment Certificates
14.4 Schedule of Payments
14.5 Plant and Materials intended for the Works
14.6 Interim Payments
14.7 Timing of Payments
14.8 Delayed Payment
14.9 Payment of Retention Money
14.10 Statement at Completion
14.11 Application for Final Payment
14.12 Discharge
14.13 Final Payment
14.14 Cessation of Employer's Liability
14.15 Currencies of Payment

Clause 15: Termination by Employer

15.1 Notice to Correct
15.2 Termination by Employer
15.3 Valuation at Date of Termination
15.4 Payment after Termination
15.5 Employer's Entitlement to Termination

Clause 16: Suspension and Termination by Contractor

16.1 Contractor's Entitlement to Suspend Work
16.2 Termination by Contractor
16.3 Cessation of Work and Removal of Contractor's Equipment
16.4 Payment on Termination

Figure 2.1 (Continued).

Clause 17: Risk and Responsibility

17.1 Indemnities
17.2 Contractor's Care of the Works
17.3 Employer's Risks
17.4 Consequences of Employer's Risks
17.5 Intellectual and Industrial Property Rights
17.6 Limitation of Liability

Clause 18: Insurance

18.1 General Requirements for Insurances
18.2 Insurance for Works and Contractor's Equipment
18.3 Insurance against Injury to Persons and Damage to Property
18.4 Insurance for Contractor's Personnel

Clause 19: Force Majeure

19.1 Definition of Force Majeure
19.2 Notice of Force Majeure
19.3 Duty to Minimise Delay
19.4 Consequences of Force Majeure
19.5 Force Majeure Affecting Subcontractor
19.6 Optional Termination, Payment and Release
19.7 Release from Performance under the Law

Clause 20: Claims, Disputes and Arbitration

20.1 Contractor's Claims
20.2 Appointment of the Dispute Adjudication Board
20.3 Failure to Agree Dusted Adjudication Board
20.4 Obtaining Dispute Adjudication Board's Decisions
20.5 Amicable Settlement
20.6 Arbitration
20.7 Failure to Comply with Dispute Adjudication Board's Decision
20.8 Expiry of Dispute Adjudication Board's Appointment

Clause 17: Risk and Responsibility

17.1 Indemnities
17.2 Contractor's Care of the Works
17.3 Employer's Risks
17.4 Consequences of Employer's Risks
17.5 Intellectual and Industrial Property Rights
17.6 Limitation of Liability

Clause 18: Insurance

18.1 General Requirements for Insurances
18.2 Insurance for Works and Contractor's Equipment
18.3 Insurance against Injury to Persons and Damage to Property
18.4 Insurance for Contractor's Personnel

Clause 19: Force Majeure

19.1 Definition of Force Majeure
19.2 Notice of Force Majeure
19.3 Duty to Minimise Delay
19.4 Consequences of Force Majeure
19.5 Force Majeure Affecting Subcontractor
19.6 Optional Termination, Payment and Release
19.7 Release from Performance under the Law

Clause 20: Claims, Disputes and Arbitration

20.1 Contractor's Claims
20.2 Appointment of the Dispute Adjudication Board
20.3 Failure to Agree Dusted Adjudication Board
20.4 Obtaining Dusted Adjudication Board's Decisions
20.5 Amicable Settlement
20.6 Arbitration
20.7 Failure to Comply with Dispute Adjudication Board's Decision
20.8 Expiry of Dispute Adjudication Board's Appointment

Clause 17: Risk and Responsibility

17.1 Indemnities
17.2 Contractor's Care of the Works
17.3 Employer's Risks
17.4 Consequences of Employer's Risks
17.5 Intellectual and Industrial Property Rights
17.6 Limitation of Liability

Clause 18: Insurance

18.1 General Requirements for Insurances
18.2 Insurance for Works and Contractor's Equipment
18.3 Insurance against Injury to Persons and Damage to Property
18.4 Insurance for Contractor's Personnel

Clause 19: Force Majeure

19.1 Definition of Force Majeure
19.2 Notice of Force Majeure
19.3 Duty to Minimise Delay
19.4 Consequences of Force Majeure
19.5 Force Majeure Affecting Subcontractor
19.6 Optional Termination, Payment and Release
19.7 Release from Performance under the Law

Clause 20: Claims, Disputes and Arbitration

20.1 Contractor's Claims
20.2 Appointment of the Dispute Adjudication Board
20.3 Failure to Agree Dispute Adjudication Board
20.4 Obtaining Dispute Adjudication Board's Decisions
20.5 Amicable Settlement
20.6 Arbitration
20.7 Failure to Comply with Dispute Adjudication Board's Decision
20.8 Expiry of Dispute Adjudication Board's Appointment

Figure 2.1 (Continued).

FIDIC 2017 Contract Structure

Red Book

Clause 1: General Provisions

1.1 Definitions
1.2 Interpretation
1.3 Notices and Other Communications
1.4 Law and Language
1.5 Priority of Documents
1.6 Contract Agreement
1.7 Assignment
1.8 Care and Supply of Documents
1.9 Delayed Drawings or Instructions
1.10 Employer's Use of Contractor's Documents
1.11 Contractor's Use of Employer's Documents
1.12 Confidentiality
1.13 Compliance with Laws
1.14 Joint and Several Liability
1.15 Limitation of Liability
1.16 Contract Termination

Clause 2: The Employer

2.1 Right of Access to the Site
2.2 Assistance
2.3 Employer's Personnel and Other Contractors
2.4 Employer's Financial Arrangements
2.5 Site Data and Items of Reference
2.6 Employer Supplied Materials and Employer's Equipment

Clause 3: The Engineer

3.1 The Engineer
3.2 Engineer's Duties and Authority
3.3 The Engineer's Representative
3.4 Delegation by the Engineer
3.5 Engineer's Instructions
3.6 Replacement of the Engineer
3.7 Agreement or Determination
3.8 Meetings

Yellow Book

Clause 1: General Provisions

1.1 Definitions
1.2 Interpretation
1.3 Notices and other Communications
1.4 Law and Language
1.5 Priority of Documents
1.6 Contract Agreement
1.7 Assignment
1.8 Care and Supply of Documents
1.9 Errors in the Employer's Requirements
1.10 Employer's Use of Contractor's Documents
1.11 Contractor's Use of Employer's Documents
1.12 Confidentiality
1.13 Compliance with Laws
1.14 Joint and Several Liability
1.15 Limitation of Liability
1.16 Contract Termination

Clause 2: The Employer

2.1 Right of Access to the Site
2.2 Assistance
2.3 Employer's Personnel and Other Contractors
2.4 Employer's Financial Arrangements
2.5 Site Data and Items of Reference
2.6 Employer Supplied Materials and Employer's Equipment

Clause 3: The Engineer

3.1 The Engineer
3.2 Engineer's Duties and Authority
3.3 The Engineer's Representative
3.4 Delegation by the Engineer
3.5 Engineer's Instructions
3.6 Replacement of the Engineer
3.7 Agreement or Determination
3.8 Meetings

Silver Book

Clause 1: General Provisions

1.1 Definitions
1.2 Interpretation
1.3 Notices and other Communications
1.4 Law and Language
1.5 Priority of Documents
1.6 Contract Agreement
1.7 Assignment
1.8 Care and Supply of Documents
1.9 Employer's Use of Contractor's Documents
1.10 Contractor's Use of Employer's Documents
1.11 Confidentiality
1.12 Compliance with Laws
1.13 Joint and Several Liability
1.14 Limitation of Liability
1.15 Contract Termination

Clause 2: The Employer

2.1 Right of Access to the Site
2.2 Assistance
2.3 Employer's Personnel and Other Contractors
2.4 Employer's Financial Arrangements
2.5 Site Data and Items of Reference
2.6 Employer Supplied Materials and Employer's Equipment

Clause 3: The Employer's Administration

3.1 The Employer's Representative
3.2 Other Employer's Personnel
3.3 Delegated Persons
3.4 Instructions
3.5 Agreement or Determination
3.6 Meetings

Figure 2.2 Structure of the FIDIC 2017 Contracts.

Clause 4: The Contractor

4.1 Contractor's General Obligations
4.2 Performance Security
4.3 Contractor's Representative
4.4 Contractor's Documents
4.5 Training
4.6 Co-operation
4.7 Setting Out
4.8 Health and Safety Obligations
4.9 Quality Management and Compliance Verification Systems
4.10 Use of Site Data
4.11 Sufficiency of the Accepted Contract Amount
4.12 Unforeseeable Physical Conditions
4.13 Rights of Way and Facilities
4.14 Avoidance of Interference
4.15 Access Route
4.16 Transport of Goods
4.17 Contractor's Equipment
4.18 Protection of the Environment
4.19 Temporary Utilities
4.20 Progress Reports
4.21 Security of the Site
4.22 Contractor's Operations on Site
4.23 Archaeological and Geological Findings

Clause 5 : Subcontracting

5.1 Subcontractors
5.2 Nominated Subcontractors

Clause 6: Staff and Labour

6.1 Engagement of Staff and Labour
6.2 Rates of Wages and Conditions of Labour
6.3 Recruitment of Persons
6.4 Labour Laws
6.5 Working Hours
6.6 Facilities for Staff and Labour
6.7 Health and Safety of Personnel
6.8 Contractor's Superintendence
6.9 Contractor's Personnel
6.10 Contractor's Records
6.11 Disorderly Conduct
6.12 Key Personnel

Clause 4: The Contractor

4.1 Contractor's General Obligations
4.2 Performance Security
4.3 Contractor's Representative
4.4 Subcontractors
4.5 Nominated Subcontractors
4.6 Co-operation
4.7 Setting Out
4.8 Health and Safety Obligations
4.9 Quality Management and Compliance Verification Systems
4.10 Use of Site Data
4.11 Sufficiency of the Accepted Contract Amount
4.12 Unforeseeable Physical Conditions
4.13 Rights of Way and Facilities
4.14 Avoidance of Interference
4.15 Access Route
4.16 Transport of Goods
4.17 Contractor's Equipment
4.18 Protection of the Environment
4.19 Temporary Utilities
4.20 Progress Reports
4.21 Security of the Site
4.22 Contractor's Operations on Site
4.23 Archaeological and Geological Findings

Clause 5: Design

5.1 General Design Obligations
5.2 Contractor's Documents
5.3 Contractor's Undertaking
5.4 Technical Standards and Regulations
5.5 Training
5.6 As-Built Documents
5.7 Operation and Maintenance Manuals
5.8 Design Error

Clause 6: Staff and Labour

6.1 Engagement of Staff and Labour
6.2 Rates of Wages and Conditions of Labour
6.3 Recruitment of Persons
6.4 Labour Laws
6.5 Working Hours
6.6 Facilities for Staff and Labour
6.7 Health and Safety of Personnel
6.8 Contractor's Superintendence
6.9 Contractor's Personnel
6.10 Contractor's Records
6.11 Disorderly Conduct
6.12 Key Personnel

Clause 4: The Contractor

4.1 Contractor's General Obligations
4.2 Performance Security
4.3 Contractor's Representative
4.4 Subcontractors
4.5 Nominated Subcontractors
4.6 Co-operation
4.7 Setting Out
4.8 Health and Safety Obligations
4.9 Quality Management and Compliance Verification Systems
4.10 Use of Site Data
4.11 Sufficiency of the Contract Price
4.12 Unforeseeable Difficulties
4.13 Rights of Way and Facilities
4.14 Avoidance of Interference
4.15 Access Route
4.16 Transport of Goods
4.17 Contractor's Equipment
4.18 Protection of the Environment
4.19 Temporary Utilities
4.20 Progress Reports
4.21 Security of the Site
4.22 Contractor's Operations on Site
4.23 Archaeological and Geological Findings

Clause 5: Design

5.1 General Design Obligations
5.2 Contractor's Documents
5.3 Contractor's Undertaking
5.4 Technical Standards and Regulations
5.5 Training
5.6 As-Built Documents
5.7 Operation and Maintenance Manuals
5.8 Design Error

Clause 6: Staff and Labour

6.1 Engagement of Staff and Labour
6.2 Rates of Wages and Conditions of Labour
6.3 Recruitment of Persons
6.4 Labour Laws
6.5 Working Hours
6.6 Facilities for Staff and Labour
6.7 Health and Safety of Personnel
6.8 Contractor's Superintendence
6.9 Contractor's Personnel
6.10 Contractor's Records
6.11 Disorderly Conduct
6.12 Key Personnel

Figure 2.2 (Continued).

Clause 7: Plant, Materials and Workmanship

7.1 Manner of Execution
7.2 Samples
7.3 Inspection
7.4 Testing by the Contractor
7.5 Defects and Rejection
7.6 Remedial Work
7.7 Ownership of Plant and Materials
7.8 Royalties

Clause 8: Commencement, Delays and Suspension

8.1 Commencement of Works
8.2 Time for Completion
8.3 Programme
8.4 Advance Warning
8.5 Extension of Time for Completion
8.6 Delays Caused by Authorities
8.7 Rate of Progress
8.8 Delay Damages
8.9 Employer's Suspension
8.10 Consequences of Employer's Suspension
8.11 Payment for Plant and Materials after Employer's Suspension
8.12 Prolonged Suspension
8.13 Resumption of Work

Clause 9: Tests on Completion

9.1 Contractor's Obligations
9.2 Delayed Tests
9.3 Retesting
9.4 Failure to Pass Tests on Completion

Clause 7: Plant, Materials and Workmanship

7.1 Manner of Execution
7.2 Samples
7.3 Inspection
7.4 Testing by the Contractor
7.5 Defects and Rejection
7.6 Remedial Work
7.7 Ownership of Plant and Materials
7.8 Royalties

Clause 8: Commencement, Delays and Suspension

8.1 Commencement of Works
8.2 Time for Completion
8.3 Programme
8.4 Advance Warning
8.5 Extension of Time for Completion
8.6 Delays Caused by Authorities
8.7 Rate of Progress
8.8 Delay Damages
8.9 Employer's Suspension
8.10 Consequences of Employer's Suspension
8.11 Payment for Plant and Materials after Employer's Suspension
8.12 Prolonged Suspension
8.13 Resumption of Work

Clause 9: Tests on Completion

9.1 Contractor's Obligations
9.2 Delayed Tests
9.3 Retesting
9.4 Failure to Pass Tests on Completion

Clause 7: Plant, Materials and Workmanship

7.1 Manner of Execution
7.2 Samples
7.3 Inspection
7.4 Testing by the Contractor
7.5 Defects and Rejection
7.6 Remedial Work
7.7 Ownership of Plant and Materials
7.8 Royalties

Clause 8: Commencement, Delays and Suspension

8.1 Commencement of Works
8.2 Time for Completion
8.3 Programme
8.4 Advance Warning
8.5 Extension of Time for Completion
8.6 Delays Caused by Authorities
8.7 Rate of Progress
8.8 Delay Damages
8.9 Employer's Suspension
8.10 Consequences of Employer's Suspension
8.11 Payment for Plant and Materials after Employer's Suspension
8.12 Prolonged Suspension
8.13 Resumption of Work

Clause 9: Tests on Completion

9.1 Contractor's Obligations
9.2 Delayed Tests
9.3 Retesting
9.4 Failure to Pass Tests on Completion

Figure 2.2 (Continued).

Clause 10: Employer's Taking Over

10.1 Taking Over of the Works and Sections
10.2 Taking Over Parts
10.3 Interference with Tests on Completion
10.4 Surfaces Requirement Reinstatement

Clause 11: Defects After Taking Over

11.1 Completion of Outstanding Work and Remedying Defects
11.2 Cost of Remedying Defects
11.3 Extension of Defects Notification Period
11.4 Failure to Remedy Defects
11.5 Remedying of Defective Work off Site
11.6 Further Tests after Remedying Defects
11.7 Right of Access after Taking Over
11.8 Contractor to Search
11.9 Performance Certificate
11.10 Unfulfilled Obligations
11.11 Clearance of Site

Clause 12: Measurement and Valuation

12.1 Works to be Measured
12.2 Method of Measurement
12.3 Valuation of the Works
12.4 Omissions

Clause 13: Variations and Adjustments

13.1 Right to Vary
13.2 Value Engineering
13.3 Variation Procedure
13.4 Provisional Sums
13.5 Daywork
13.6 Adjustments for Changes in Laws
13.7 Adjustments for Changes in Cost

Clause 10: Employer's Taking Over

10.1 Taking Over of the Works and Sections
10.2 Taking Over Parts
10.3 Interference with Tests on Completion
10.4 Surfaces Requirement Reinstatement

Clause 11: Defects After Taking Over

11.1 Completion of Outstanding Work and Remedying Defects
11.2 Cost of Remedying Defects
11.3 Extension of Defects Notification Period
11.4 Failure to Remedy Defects
11.5 Remedying of Defective Work off Site
11.6 Further Tests after Remedying Defects
11.7 Right of Access after Taking Over
11.8 Contractor to Search
11.9 Performance Certificate
11.10 Unfulfilled Obligations
11.11 Clearance of Site

Clause 12: Tests after Completion

12.1 Procedure for Tests after Completion
12.2 Delayed Tests
12.3 Retesting
12.4 Failure to Pass Tests after Completion

Clause 13: Variations and Adjustments

13.1 Right to Vary
13.2 Value Engineering
13.3 Variation Procedure
13.4 Provisional Sums
13.5 Daywork
13.6 Adjustments for Changes in Laws
13.7 Adjustments for Changes in Cost

Clause 10: Employer's Taking Over

10.1 Taking Over the Works and Sections
10.2 Taking Over of Parts of the Works
10.3 Interference with Tests on Completion

Clause 11: Defects After Taking Over

11.1 Completion of Outstanding Work and Remedying Defects
11.2 Cost of Remedying Defects
11.3 Extension of Defects Notification Period
11.4 Failure to Remedy Defects
11.5 Remedying of Defective Work off Site
11.6 Further tests after Remedying Defects
11.7 Right of Access after Taking Over
11.8 Contractor to Search
11.9 Performance Certificate
11.10 Unfulfilled Obligations
11.11 Clearance of Site

Clause 12: Tests after Completion

12.1 Procedure for Tests after Completion
12.2 Delayed Tests
12.3 Retesting
12.4 Failure to Pass Tests after Completion

Clause 13: Variations and Adjustments

13.1 Right to Vary
13.2 Value Engineering
13.3 Variation Procedure
13.4 Provisional Sums
13.5 Daywork
13.6 Adjustments for Changes in Laws
13.7 Adjustments for Changes in Cost

Figure 2.2 (Continued).

Clause 14: Contract Price and Payment

14.1 The Contract Price
14.2 Advance Payment
14.3 Application for Interim Payment
14.4 Schedule of Payments
14.5 Plant and Materials intended for the Works
14.6 Issue of IPC
14.7 Payment
14.8 Delayed Payment
14.9 Release of Retention Money
14.10 Statement at Completion
14.11 Final Statement
14.12 Discharge
14.13 Issue of FPC
14.14 Cessation of Employer's Liability
14.15 Currencies of Payment

Clause 15: Termination by Employer

15.1 Notice to Correct
15.2 Termination for Contractor's Default
15.3 Valuation after Termination for Contractor's Default
15.4 Payment after Termination for Contractor's Default
15.5 Termination for Employer's Convenience
15.6 Valuation after Termination for Employer's Convenience
15.7 Payment after Termination for Employer's Convenience

Clause 16: Suspension and Termination by Contractor

16.1 Suspension by Contractor
16.2 Termination by Contractor
16.3 Contractor's Obligations after Termination
16.4 Payment after Termination by Contractor

Clause 17: Care of the Works and Indemnities

17.1 Responsibility for Care of the Works
17.2 Liability for Care of the Works
17.3 Intellectual and Industrial Property Rights
17.4 Indemities by Contractor
17.5 Indemities by Employer
17.6 Shared Indemnities

Clause 14: Contract Price and Payment

14.1 The Contract Price
14.2 Advance Payment
14.3 Application for Interim Payment
14.4 Schedule of Payments
14.5 Plant and Materials intended for the Works
14.6 Issue of IPC
14.7 Payment
14.8 Delayed Payment
14.9 Release of Retention Money
14.10 Statement at Completion
14.11 Final Statement
14.12 Discharge
14.13 Issue of FPC
14.14 Cessation of Employer's Liability
14.15 Currencies of Payment

Clause 15: Termination by Employer

15.1 Notice to Correct
15.2 Termination for Contractor's Default
15.3 Valuation after Termination for Contractor's Default
15.4 Payment after Termination for Contractor's Default
15.5 Termination for Employer's Convenience
15.6 Valuation after Termination for Employer's Convenience
15.7 Payment after Termination for Employer's Convenience

Clause 16: Suspension and Termination by Contractor

16.1 Suspension by Contractor
16.2 Termination by Contractor
16.3 Contractor's Obligations after Termination
16.4 Payment after Termination by Contractor

Clause 17: Care of the Works and Indemnities

17.1 Responsibility for Care of the Works
17.2 Liability for Care of the Works
17.3 Intellectual and Industrial Property Rights
17.4 Indemnities by Contractor
17.5 Indemnities by Employer
17.6 Shared Indemnities

Clause 14: Contract Price and Payment

14.1 The Contract Price
14.2 Advance Payment
14.3 Application for Interim Payment
14.4 Schedule of Payments
14.5 Plant and Materials intended for the Works
14.6 Interim Payments
14.7 Payment
14.8 Delayed Payment
14.9 Release of Retention Money
14.10 Statement at Completion
14.11 Final Statement
14.12 Discharge
14.13 Final Payment
14.14 Cessation of Employer's Liability
14.15 Currencies of Payment

Clause 15: Termination by Employer

15.1 Notice to Correct
15.2 Termination for Contractor's Default
15.3 Valuation after Termination for Contractor's Default
15.4 Payment after Termination for Contractor's Default
15.5 Termination for Employer's Convenience
15.6 Valuation after Termination for Employer's Convenience
15.7 Payment after Termination for Employer's Convenience

Clause 16: Suspension and Termination by Contractor

16.1 Suspension by Contractor
16.2 Termination by Contractor
16.3 Contractor's Obligations after Termination
16.4 Payment after Termination by Contractor

Clause 17: Care of the Works and Indemnities

17.1 Responsibility for Care of the Works
17.2 Liability for Care of the Works
17.3 Intellectual and Industrial Property Rights
17.4 Indemnities by Contractor
17.5 Indemnities by Employer
17.6 Shared Indemnities

Figure 2.2 (Continued).

Clause 18: Exceptional Events	**Clause 18: Exceptional Events**	**Clause 18: Exceptional Events**
18.1 Exceptional Events	18.1 Exceptional Events	18.1 Exceptional Events
18.2 Notice of an Exceptional Event	18.2 Notice of an Exceptional Event	18.2 Notice of an Exceptional Event
18.3 Duty to Mainimise Delay	18.3 Duty to Minimise Delay	18.3 Duty to Minimise Delay
18.4 Consequences of an Exceptional Event	18.4 Consequences of an Exceptional Event	18.4 Consequences of an Exceptional Event
18.5 Optional Termination	18.5 Optional Termination	18.5 Optional Termination
18.6 Release from Performance under the Law	18.6 Release from Performance under the Law	18.6 Release from Performance under the Law
Clause 19: Insurance	**Clause 19: Insurance**	**Clause 19: Insurance**
19.1 General Requirements	19.1 General Requirements	19.1 General Requirements
19.2 Insurance to be provided by the Contractor	19.2 Insurance to be provided by the Contractor	19.2 Insurance to be provided by the Contractor
Clause 20: Employer's and Contractor's Claims	**Clause 20: Employer's and Contractor's Claims**	**Clause 20: Employer's and Contractor's Claims**
20.1 Claims	20.1 Claims	20.1 Claims
20.2 Claims for Payment and/or EOT	20.2 Claims for Payment and/or EOT	20.2 Claims for Payment and/or EOT
Clause 21: Disputes and Arbitration	**Clause 21: Disputes and Arbitration**	**Clause 21: Disputes and Arbitration**
21.1 Constitution of the DAAB	21.1 Constitution of the DAAB	21.1 Constitution of the DAAB
21.2 Failure to appoint DAAB Members	21.2 Failure to appoint DAAB Members	21.2 Failure to appoint DAAB Members
21.3 Avoidance of Disputes	21.3 Avoidance of Disputes	21.3 Avoidance of Disputes
21.4 Obtaining DAAB's Decision	21.4 Obtaining DAAB's Decision	21.4 Obtaining DAAB's Decision
21.5 Amicable Settlement	21.5 Amicable Settlement	21.5 Amicable Settlement
21.6 Arbitration	21.6 Arbitration	21.6 Arbitration
21.7 Failure to Comply with DAAB's Decision	21.7 Failure to Comply with DAAB's Decision	21.7 Failure to Comply with DAAB's Decision
21.8 No DAAB in Place	21.8 No DAAB in Place	21.8 No DAAB in Place

Figure 2.2 (Continued).

with the FIDIC Conditions of Contract for Works of Civil Engineering Construction 1987.

The Subcontract is similar to the Conditions of Contract for Construction 1999, and many of the clauses are "back to back" with the Main Contract, but some are unique to the Subcontractor.

Again, as with the Main Contract, there are:

- General Conditions
- Particular Conditions

There are also, similar to the main contract, various sample forms for use with the Subcontract.

The Structure of the General Conditions is:

1 Definitions and Interpretation
2 The Main Contract
3 The Contractor
4 The Subcontractor
5 Assignment of the Subcontract and Subcontracting
6 Co-operation, Staff and Labour
7 Equipment, Temporary Works, Other Facilities, Plant and Materials
8 Commencement and Completion
9 Tests on Completion
10 Completion and Taking-Over the Subcontract Works
11 Defects Liability
12 Measurement and Evaluation
13 Subcontract Variations and Adjustments
14 Subcontract Price and Payment
15 Termination of the Main Contract and Termination of the Subcontract by the Contractor
16 Suspension and Termination by the Subcontractor
17 Risk and Indemnities
18 Subcontract Insurances
19 Subcontract Force Majeure
20 Notices, Subcontractor's Claims and Disputes

In 2019, FIDIC launched the *Conditions of Subcontract for Plant and Design-Build*, but curiously this is only for use with the FIDIC 1999 Yellow Book.

This subcontract is structured as follows:

1 Definitions and Interpretation
2 The Main Contract
3 The Contractor
4 The Subcontractor
5 Subcontract Design
6 Co-operation, Staff and Labour
7 Equipment, Temporary Works, Other Facilities, Plant and Materials
8 Commencement and Completion

9 Tests on Completion
10 Completion and Taking-Over the Subcontract Works
11 Defects Liability
12 Subcontract Tests after Completion
13 Subcontract Variations and Adjustments
14 Subcontract Price and Payment
15 Termination of the Main Contract and Termination of the Subcontract by the Contractor
16 Suspension and Termination by the Subcontractor
17 Risk and Indemnities
18 Subcontract Insurances
19 Subcontract Force Majeure
20 Notices, Subcontractor's Claims and Disputes

This subcontract is for use between a Contractor and a Subcontractor where the main contract is the *Conditions of Contract for Plant and Design-Build* (Yellow Book) 1999 Edition (not the 2017 version) and is to be used where the subcontractor is both designing and building part of the main works. Note that it is not used where the Subcontractor is only building/supplying part of the works.

The *Conditions of Subcontract for Plant and Design-Build* are based on the *Conditions of Subcontract for Construction*, 2011 (commonly referred to as the "Red Book (1999) Subcontract"), but have been amended to be used with the FIDIC 1999 Yellow Book.

Basic differences between the FIDIC 1999 and 2017 Contracts

The FIDIC 2017 Contracts are structurally similar to the FIDIC 1999 Contracts, although they are considerably larger, with far more pages than the previous editions.

For example, the General Conditions within the Red Book now comprise 106 pages, whereas the FIDIC 1999 version was only 62 pages.

The FIDIC 2017 Red, Yellow and Silver Books all have the same structure, each having 21 Clauses with mostly the same titles, with the following exceptions:

- The Red and Yellow Books both have an Engineer to act for the Employer, but the Silver Book has an Employer's Representative, so Clause 3 of the Red and Yellow Books is entitled "The Engineer", whilst Clause 3 of the Silver Book is entitled "The Employer's Administration".
- Because the Yellow and Silver Books are Design and Build Contracts Clause 5 of those books is entitled "Design", whilst Contractor's design (if applicable) is incorporated within Clause 4.1.
- The Red Book is based on measurement of the Works for valuation and payment purposes, so Clause 12 of the Red Book is entitled "Measurement and Valuation", whilst Clause 12 of the Yellow and Silver Books is entitled "Tests after Completion".

There is also a significant increase in administrative requirements within the FIDIC 2017 Contracts, mostly to achieve greater certainty of project outcome over a wider range of projects in a wider range of locations, though many users may consider them as over wordy and prescriptive.

Some Sub-Clauses have been moved to what the drafters believe are more suitable locations within the General Conditions of Contract. For example:

- Limitation of liability has been moved from Clause 17 of the FIDIC 1999 Contracts to Clause 1 in FIDIC's 2017 contract updates. This was done to align with the changes to Clause 17 (please see below).
- The Employer's obligation to provide Site Data, and the provisions relating to materials and/or equipment to be supplied by the Employer for the Contractor's use have been moved from Clause 4 to new Sub-Clauses under Clause 2.
- Currency of payment for Variations has been moved from Sub-Clause 13.4 in the FIDIC 1999 contracts to Sub-Clause 14.15 in the FIDIC 2017 Contracts.

The Contracts are also more balanced in the rights and obligations of the Parties, and a number of the contractual provisions that applied to just one Party under the FIDIC 1999 forms of contract are now reciprocal between the Employer and the Contractor.

For example:

- Confidentiality of contract documents;
- Assistance to obtain the necessary permits/ permissions/licences/approvals for the Works;
- The right to notify a claim if the other Party does not obtain the necessary permits/permissions/ licences/approvals or comply with one that has been obtained;
- The right of one Party to require the other Party to remove from his/her personnel any person who is found to have engaged in corrupt, fraudulent or coercive practice;
- The prohibition of recruitment by one Party from the other Party's personnel;
- The claims procedure to be followed is the same for Contractor's and Employer's claims.

Where relevant, in later chapters, as the various parts of the FIDIC 2017 Contracts are reviewed, any changes in the contents of Clauses and Sub-Clauses between the FIDIC 1999 Contracts and the FIDIC 2017 Contracts are mentioned.

There are many improvements in the FIDIC 2017 Contract forms, though there are a number of examples of possibly unnecessary changes, bearing in mind that people do prefer some familiarity, so it is expected that there will be challenges as the use of the FIDIC contracts grows.

It would appear that many knowledgeable and experienced practitioners were consulted in the drafting of the FIDIC 2017 as there have been quite substantial changes across the spectrum in terms of communications, programme, quality, claims and dispute avoidance and resolution.

In the drafting process FIDIC's 2017 Contract updates were subjected to a number of systematic reviews and controls by over 50 eminent and experienced professionals, specially chosen and invited by FIDIC from across the spectrum of interested users (including employers, contractors, engineers, lawyers and bilateral/multilateral development banks from around the world).

In addition to the Yellow, Red and Silver Books, there is an updated FIDIC White Book, which is intended for consultancy/advisory services, and which was published in the spring of 2017.

With the aim of increasing clarity and reducing potential misunderstandings on interpretation, which is in itself to be welcomed, FIDIC has improved the contract provisions in the 2017 Contracts by making them more prescriptive and introducing structured project management and procedural mechanisms.

These set out exactly what is expected from the Employer, the Contractor, the Engineer and the Employer's Representative (where appropriate) during the performance of the Contract.

In updating the three forms of contract, FIDIC has also attempted to use simple language since many users' first language is not English, and again that must be welcomed. This approach is also intended to facilitate the translation of the documents into other languages.

As recognised international standard forms of construction contract, FIDIC's aim is to improve and strengthen the application of these contract forms in all jurisdictions, and in all situations, and there are references to relevant legislation and in many cases recommendations that the Parties should seek legal advice where appropriate, particularly regarding issues of termination, specifically where it is actioned for the Employer's convenience, and also references to terms such as "gross negligence" which can have different meanings under differing legislations.

All three contracts have been prepared for projects where the contract is subject to a tendering process, but they may also, with a few adjustments, be applied without a tender, for example direct negotiation between the Parties.

In addition to the Red, Yellow and Silver Books, an updated FIDIC White Book for consultancy/advisory services was published in the spring of 2017, though it is not the intention of the author to review the White Book in any detail within this book.

FIDIC also offers a wide variety of other standard contracts.

Of all the contracts in use today, FIDIC is the nearest to a truly "international" form of contract for construction works, large-scale machinery supplies, infrastructure projects, consultancy services, etc.

The aim is to have a standard form for international use which tries to achieve a balance between comprehensive coverage of the circumstances, problems and obligations which are likely, but does not try to cover every conceivable problem, an impossible task which would result in an unmanageable document.

Written contracts are all about allocating risk and obligations and providing the appropriate machinery for resolving issues likely to be encountered.

The previous launch of the FIDIC 1999 Contracts

Note that when the FIDIC 1999 Contracts were originally launched they were not "revisions" of the previous FIDIC contracts, they were "brand new" versions with many structural changes, and hence, they were referred to as "First Edition" Contracts.

The general layout of the 1999 Contracts is as follows:

- General Conditions, including an Appendix entitled "General Conditions of Dispute Adjudication Agreement" which includes the Procedural Rules for a Dispute Adjudication Board (DAB);
- A section giving guidance for the preparation of any Particular Conditions, this section also includes examples of guarantees, securities and bonds that are commonplace on international projects;

- A section entitled "Forms", here FIDIC provides examples of:
  - Letter of Tender with supporting Appendix to Tender;
  - Contract Agreement;
  - DAB Agreements (either a one-person DAB or three-person DAB).

Within the very useful Forward to the Contract, there are graphics indicating timelines relative to:

1 Principal events from invitation to tender to return of the Performance Security;
2 Payment procedures under Clause 14 (Contract Price and Payment);
3 The sequence of events under Clause 20 (Claims, Disputes and Arbitration) following either Party giving notice of its intention to refer a dispute to a DAB.

As stated above, the main contracts (referred to as the "Rainbow Suite"), which were originally published in September 1999, are discussed below.

Conditions of Contract for Construction (The "Red Book")

The Red Book is recommended for building or engineering works where most of the design is provided by the Employer or his Representative. The Contractor undertakes the construction work in accordance with the Employer's design; however, the works may include some Contractor-designed civil, mechanical, electrical and/or construction works.

The basis of payment to the Contractor is through rates and prices in the Bills of Quantities, though it can be amended to provide a lump sum price or as a cost reimbursable contract.

Other features of the Red Book are:

- The Contract is administered on behalf of the Employer by the Engineer, its actions including approval of work, certification of payments and determination of entitlement to Extensions of Time.
- The Employer bears the risks of "adverse physical conditions" and the "operation of the forces of nature" that are considered to be Unforeseeable.
- Claims by both Parties have to follow procedures, with various time scales attached, some of which contain time bars.
- The Contractor has some financial protection, in that it can request evidence from the Employer that it has the finances to pay the estimated Contract Price.
- Materials can be paid for both on and off Site if strict criteria are followed, including the listing of materials for which payment may be sought within the Contractor's tender.

Conditions of Contract for Plant and Design-Build (The "Yellow Book")

The Yellow Book is recommended for the provision of electrical and/or mechanical plant, and for the design and execution of building or engineering works. The Contractor designs the Works, in accordance with the Employer's Requirements.

Design and build as a procurement strategy is explored in more detail within Chapter 5 ("Design").

Payment to the Contractor is based upon a lump sum price and which may be assessed a milestone schedule to be achieved by the Contractor. The Yellow Book will most likely be used for process plants and the like where a high degree of offsite manufacture of plant and equipment is foreseen and payment terms can be drafted to recognise this situation subject to the listing of such plant and equipment within the Contractor's tender as within the Red Book.

Other features of the Yellow Book are:

- The Contract is administered on behalf of the Employer by the Engineer, its actions including approval of work, certification of payments and determination of entitlement to Extensions of Time.
- Testing procedures leading to completion are likely to be more complicated than within the Red Book, again reflecting the likely nature of the project.
- The Employer bears the risks of "adverse physical conditions" and the "operation of the forces of nature" that are considered to be Unforeseeable.
- Claims by both Parties have to follow procedures, with various time scales attached, some of which contain time bars.
- The Contractor has some financial protection, in that it can request evidence from the Employer that it has the finances to pay the estimated Contract Price.
- Materials can be paid for both on and off Site if strict criteria are followed, including the listing of materials for which payment may be sought within the Contractor's tender.

Conditions of Contract for EPC/Turnkey Project (The "Silver Book")

The Silver Book is suitable for the provision on a turnkey basis of a process or power plant, of a factory or similar facility or of an infrastructure project or other type of development, where

- A higher degree of certainty of final price and time is required, and
- The Contractor takes total responsibility for the design and execution of the project.

Short Form of Contract (The "Green Book")

The final contract to be issued in the FIDIC 1999 Contracts was the Green Book or *Short Form of Contract*, and is worth a mention.

This Contract recognised a need for a much simpler and shorter FIDIC contract, to suit projects with a relatively low Contract Price and a short time duration.

The Green Book is suitable for a small contract (say, under £250,000 or equivalent local currency) if the construction time is short (say, less than six months), or for relatively simple or repetitive work, irrespective of whether the design is provided by the Employer or the Contractor, and of whether the project involves civil, electrical, mechanical and/or construction works.

The Contract itself is very familiar as a FIDIC contract in terms of its structure and format, and any reader will recognise the Contract as being from the same family as the Red, Yellow and Silver Books, albeit it has only 15 clauses and a total of 10 pages but is designed to be very flexible.

The Clauses are short and easily understood, and whilst design can be carried out by either Party, an Engineer is not foreseen; however, the Employer may appoint a Representative.

The Price, and the payments to the Contractor, can be set as a lump sum, by remeasurement or cost reimbursable.

As with the major forms the Contract includes guidance notes (noted as not forming part of the Contract) as well as an Agreement together with its Appendix and Rules for Adjudication.

The noticeable absentee being the Particular Conditions section, in this respect FIDIC considers that the Green Book can work without such conditions; however, a cautionary note is provided should an Employer deem it necessary to amend the drafted Contract.

Structure of Contract

1 General Provisions
2 The Employer
3 Employer's Representatives
4 The Contractor
5 Design by Contractor
6 Employer's Liabilities
7 Time for Completion
8 Taking-Over
9 Remedying Defects
10 Variations and Claims
11 Contract Price and Payment
12 Default
13 Risk and Responsibility
14 Insurance
15 Resolution of Disputes

A brief summary of the FIDIC *Short Form of Contract* (Green Book)

Clause 1: General Provisions

- There is a very short list of definitions used in the Conditions of Contract and is divided into "The Contract", "Persons", "Dates, Tests and Periods", "Money and Payments" and "Other Definitions".
- The Clause then covers interpretation such as gender, singular/plural, in addition to marginal words and other headings not forming the Conditions.
- All the documents comprising the contract are mutually explanatory of one another i.e. so there is no priority of documents.
- The Law of the Contract is stated in the Appendix.
- All communications are to be written in the language stated in the Appendix.
- The Contractor is required to comply with the Laws of the countries where activities are performed.

Clause 2: The Employer

- The Employer is required to give the Contractor the right of access at the times stated in the Appendix.
- The Employer is required to "assist" the Contractor where possible in obtaining copies of the Laws of the Country and any necessary permits to work in that country. It is fundamentally the responsibility of the Contractor to obtain these.

- The Contractor is required to comply with all instructions given by the Employer.
- No approval or consent by the Employer affects the Contractor's obligations.

Clause 3: Employer's Representative

- One of the Employer's personnel shall have authority to act for him. This person is stated in the Appendix.
- The Employer may appoint a firm or individual to act for him. This person is stated in the Appendix.

Clause 4: The Contractor

- Sub-Clause 4.1 states the Contractor's obligation to execute the Works.
- Sub-Clause 4.3 requires the Contractor to name and receive consent to representative.
- Sub-Clause 4.4 covers Subcontractors, who are subject to the Employer's consent.
- Sub-Clause 4.2 requires the Contractor to provide a Performance Security if stated in the Appendix.

Clause 5: Design By Contractor

The Contractor carries out any design as required by the Appendix. All designs are to be submitted to the Employer who may comment within 14 days of receipt.

The Contractor shall be responsible for his design including infringing any patents, copyrights, etc.

Clause 6: Employer's Liabilities

- There are 16 Employer's Liabilities listed in the contract, any of which can give rise to an Extension of Time under Clause 7.3.

Clause 7: Time for Completion

- The Contractor is required to complete the Works within the Time for Completion.
- The Contractor is required to submit a programme to the Employer in the form stated in the Appendix.
- There are 16 grounds for Extension of Time, basically the Employer's Liabilities under Clause 6.1.
- The Contractor is liable to the Employer and pays the amounts stated in the Appendix.

Clause 8: Taking Over

- The Contractor is to notify the Employer when he considers that the Works are complete.
- The Employer notifies the Contractor when he considers the Contractor has completed the Works.

Clause 9: Remedying Defects

- The Employer may notify the Contractor at any time of a Defect. The Contractor remedies at no cost to the Employer.
- The Employer may give instructions to the Contractor to uncover or test any work. The Contractor is paid if this shows that the work was not defective.

Clause 10: Variations and Claims

- The Employer may instruct Variations.
- Variations are valued at lump sums or rates agreed by the Parties.
- Either Party is required to notify the other as soon as it is aware of anything that could delay or disrupt the Works or give rise to a claim for additional payment. The Contractor's rights can be affected by the timing or lack of an Early Warning.
- The Contractor can recover costs as a result of Employer's Liabilities.
- The Contractor is required to submit an itemised breakdown of the value of Variations and claims within 28 days of an instruction or event giving rise to a claim.

Clause 11: Contract Price and Payment

- The Works are valued in accordance with the Appendix.
- The Contractor is required to submit a Statement to the Employer showing the amounts to which he considers himself entitled.
- Within 28 days of delivery of each Statement, the Employer pays the Contractor the amount due.
- One half of the retention is released within 14 days of a Take-Over Notice.
- The remaining retention is released within 14 days of the expiry of the Defects Notification Period.
- Within 42 days of the expiry of the Defects Notification Period, the Contractor is to submit a final account.
- Payments are in the currency of stated in the Appendix.
- The Contractor is entitled to be paid interest at the rate stated in the Appendix.

Clause 12: Default

- If the Contractor defaults by abandoning the Works, refusing to comply with an instruction or fails to proceed with the Works, the Employer can issue a notice. If the Contractor has not remedied within 14 days the Employer may give notice to terminate the Contract.
- If the Employer fails to pay, or is in breach of the Contract the Contractor may give notice. If the matter is not remedied within 7 days the Contractor may suspend the Works, if not remedied within 28 days, the Contractor can terminate.
- Either Party may terminate immediately due to insolvency.
- Following termination, the Contractor is entitled to be paid any outstanding amounts and if the Contractor has terminated, loss of profit.

Clause 13: Risk and Responsibility

- This Sub-Clause defines the Contractor's responsibility for care of the Works.
- Sub-Clause 17.3 lists the risks which, if they occur, entitle the Contractor to compensation in respect of loss or damage to the Works.

Clause 14: Insurance

- The Contractor is required to take out and maintain insurance for loss or damage to the Works, third-party liability and Employer's Liability.
- All instances are required to conform with any requirements in the Appendix.
- If the Contractor fails to insure, the Employer may take cover and deduct the premiums from the Contractor.

Clause 15: Resolution Of Disputes

- Any dispute may be referred to the Adjudicator named in the Appendix.
- If a Party is dissatisfied with an Adjudicator's decision, the Party can give notice within 28 days of receipt of the decision, otherwise the adjudicator's decision is binding.
- Any dispute which has been the subject of a notice of dissatisfaction can be settled by reference to arbitration.

What is the difference between the FIDIC *Conditions of Contract for Plant and Design-Build* (The "Yellow Book") and the *Conditions of Contract for EPC/Turnkey Project* (The "Silver Book")?

FIDIC 1999 and 2017 Contract users are often confused about whether to use the *Conditions of Contract for Plant and Design-Build* (The "Yellow Book") or the *Conditions of Contract for EPC/Turnkey Project* (The "Silver Book").

They are both design and build contracts, but there are some significant differences between them.

Let us first consider the primary features of both contracts:

Conditions of Contract for Plant and Design-Build *(The "Yellow Book")*

- The FIDIC Red and Yellow Books are fairly simple, in that they share a fairly balanced view of risk sharing between the two Parties, so that anything for which the Contractor should logically receive an Extension of Time and/or recovery of cost (possibly with Profit) will probably be contained within the provisions, with the Employer only giving additional time and/or paying the Contractor for the risk if it occurs. The Contractor effectively does not have to include within its tender for risks that are particularly difficult to value, as relief is available through the Contract.
- The period for pricing tenders does not have to be particularly lengthy (certainly not in comparison with the Silver Book) and where risks such as hydrological, sub-surface and other matters affecting the Site are not readily available, those risks remain with the Employer.
- It is not possible for the tenderers to properly inspect the Site, or the amount of underground work is so significant or complex that for the Contractor to price all risks under the Silver Book would be both inequitable and cost prohibitive.

- Similar to the *Conditions of Contract for Construction* (The "Red Book") the Yellow Book is administered on behalf of the Employer by the Engineer.
- The Yellow Book is structured similar to the Red Book, the primary difference being the Contractor's Design Clauses, where under the Red Book, the Contractor may have design responsibility, whereas in the Yellow Book, the Contractor has full design responsibility.
- The Employer via its Engineer requires a close relationship with the Contractor throughout the duration of the project including the more likely potential to cause change to design.

Conditions of Contract for EPC/Turnkey Project *(The "Silver Book")*

Let us first consider Engineering, Procurement and Construction (EPC) Contracts generally, which are a common form of contract used to undertake large-scale construction works, including complex oil and gas projects.

Under an EPC Contract a Contractor is obliged to deliver a complete facility to a developer who needs only to "turn a key" to start operating the completed facility; hence, EPC Contracts are sometimes called "turnkey contracts".

In addition to delivering a complete facility, the Contractor must deliver that facility for a guaranteed price by a guaranteed date and it must perform to the specified level.

Failure to comply with any requirements will usually result in the Contractor incurring significant financial liabilities.

EPC Contracts present a high risk to Contractors for the following reasons:

Single point of responsibility

The Contractor is responsible for all design, engineering, procurement, construction, commissioning and testing activities. Therefore, if any problems occur, the Contractor is required to remedy and/or compensate the Employer.

With many large EPC Contracts, the Contractor consists of a Joint Venture (JV) between various entities. It is critical that the relationship between those entities is known, and also the EPC Contract must state that those entities are jointly and severally liable to the Employer.

Fixed Contract Price

Risk of cost overruns and the benefit of any cost savings are the Contractor's liability. The Contractor has a very limited ability to claim additional money which is limited to circumstances where the Employer has directly delayed the Contractor or has ordered Variations to the Works.

Fixed Time for Completion

EPC Contracts include a guaranteed completion date that is either a fixed date or a fixed period after the commencement of the EPC Contract (FIDIC contracts have a Commencement Date and a Time for Completion).

If the time scale is not met, the Contractor is liable for delay/liquidated damages. These damages are designed to compensate the Employer for loss and damage suffered as a result of late completion of the facility, as if the Works had not been completed late.

In addition, the EPC Contract must provide for the Contractor to be granted an Extension of Time when it is delayed by the acts or omissions of the Employer to prevent time becoming at large.

Performance specification

Unlike a traditional construction contract, an EPC Contract usually contains a performance specification.

The performance specification details the performance criteria that the Contractor must meet but does not dictate how they must be met. This is left to the Contractor to determine. A delicate balance must be maintained. The specification must be detailed enough to ensure the Employer knows what it is contracting to receive but not so detailed that if problems arise the Contractor can argue they are not its responsibility.

Whilst there are, as described above, numerous advantages to using an EPC Contract, there are some disadvantages. These include the fact that it can result in a higher contract price than alternative contractual structures. This higher price is a result of a number for factors not least of which is the allocation of almost all the construction risk to the Contractor.

This has a number of consequences, one of which is that the Contractor will have to factor into its price the cost of absorbing those risks. This will result in the Contractor building contingencies into the contract price for events that are unforeseeable and/or unlikely to occur. If those contingencies were not included the contract price would be lower. However, the Employer would bear more of the risk of those unlikely or unforeseeable events. Sponsors have to determine, in the context of their particular project, whether the increased price is worth paying.

As a result, sponsors and their advisers must critically examine the risk allocation on every project. Risk allocation should not be an automatic process.

Instead, the Employer should allocate risk in a sophisticated way that delivers the most efficient result.

For example, if a project is being undertaken in an area with unknown geology and without the time to undertake a proper geotechnical survey, the Employer may be best served by bearing the site condition risk itself as it will mean the Contractor does not have to price a contingency it has no way of quantifying. This approach can lower the risk premium paid by the Employer.

Alternatively, the opposite may be true. The Employer may wish to pay for the contingency in return for passing on the risk which quantifies and caps its exposure.

This type of analysis must be undertaken on all major risks prior to going out to tender. Another consequence of the risk allocation is the fact that there are relatively few engineering and construction companies that can and are willing to enter into EPC Contracts. As mentioned in the Introduction some bad publicity and a tightening insurance market have further reduced the pool of potential EPC Contractors. The scarcity of EPC Contractors can also result in relatively high contract prices.

Performance Security

It is standard for the Contractor to be required to provide a Performance Security to protect the Employer should the Contractor not comply with its obligations under the EPC Contract. The security takes a number of forms including a bank guarantee or bond for a percentage, normally in the range of 5%–15%, of the Contract Price.

Variations

The Employer has the right to order Variations and to agree to Variations suggested by the Contractor. If the Employer wants the right to omit works either in their entirety or to be able to engage a different Contractor this must be stated specifically. In addition, a properly drafted variations clause should make provision for how the price of a variation is to be determined.

In the event the Parties do not reach agreement on the price of a variation, the Employer or its representative should be able to determine the price. This determination is subject to the dispute resolution provisions. In addition, the variations clause should detail how the impact, if any, on the performance guarantees is to be treated. For some larger Variations the Employer may also wish to receive additional security. If so, this must also be dealt with in the Variations clause.

Defects Liability

The Contractor is usually obliged to repair defects that occur in the 12 to 24 months, following completion of the performance testing.

Defects Liability clauses can be tiered. That is the clause can provide for one period for the entire facility and a second, extended period, for more critical items.

Termination

This sets out the contractual termination rights of both Parties. The Contractor usually has very limited contractual termination rights. These rights are limited to the right to terminate for non-payment or for prolonged suspension or prolonged Force Majeure.

The Employer will have more extensive contractual termination rights. They will usually include the ability to terminate immediately for certain major breaches or where the Contractor becomes insolvent and the right to terminate after a certain period for other breaches.

In addition, the Employer may have a right to terminate for convenience. It is likely the Employer's ability to exercise its termination rights will also be limited by the terms of the financing agreements.

Another major disadvantage of an EPC Contract becomes evident when problems occur during construction. In return for receiving a guaranteed price and a guaranteed completion date, the Employer concedes most of the day-to-day control over the construction.

Therefore, Employers have limited ability to intervene when problems occur during construction. The more an Employer interferes during the construction, the greater the likelihood of the Contractor claiming additional time and costs. In addition, interference by the Employer will make it substantially easier for Contractors to defeat claims for liquidated damages and defective works.

Obviously, ensuring the project is completed satisfactorily is usually more important than protecting the integrity of the contractual structure.

However, if an Employer interferes with the execution of the works they will, in most circumstances, have the worst of both worlds.

They will have a contract that exposes them to liability for time and costs incurred as a result of their interference without any corresponding ability to hold the Contractor liable for delays in completion or defective performance. The same problems occur even where the EPC Contract is drafted to give the Employer the ability to intervene. In many circumstances, regardless of the actual drafting, if the Employer becomes involved in determining how the Contractor executes the works then the Contractor will be able to argue that it is not liable for either delayed or defective performance.

As a result, it is vitally important that great care is taken in selecting the Contractor and in ensuring the Contractor has sufficient knowledge and expertise to execute the works. Given the significant monetary value of EPC Contracts, and the potential adverse consequences if problems occur during construction, the lowest price should not be the only factor used when selecting Contractors.

The FIDIC Silver Book reflects a market desire for certainty of cost and time perhaps by a totally risk adverse Employer willing to "pay the price" for guaranteed delivery to a price.

The Contractor is required to allow within its tender for a wide range of risks relative to cost and time (see below), such risks most likely including all ground conditions (potentially in a country of which the Contractor will have little knowledge) and the completion of the Works will be based upon a strict but often brief performance-related specification.

The Employer will still bear some risks such as those related to war, terrorism and Force Majeure but the unbalanced risk profile of this Contract will undoubtedly be a higher price, a factor that Employer's must accept.

The main features of the Silver Book can be summarised as:

- Design liability rests solely with the Contractor, the Employer will provide its requirements, but these are often in the form of a brief performance specification.
- The Contractor carries out all EPC often including performance tests after completion; a "turn-key" project allowing operation of the facility upon completion.
- There is not an Engineer within the Contract, the Silver Book is administered by the Employer (who appoints an Employer's Representative).
- It is a lump sum contract with payment terms similar to those envisaged under the Yellow Book.
- The Silver Book is structured similar to the Red and Yellow Book in terms of programme, claims, disputes, etc. The Contractor's Design Clauses are the same as the Yellow Book, where the Contractor has full design responsibility.
- The Silver Book allocates significant risks to the Contractor including hydrological, sub-surface and other conditions which may affect the Site and/or the Works. The Contractor also holds some risks in respect of the Employer's Requirements, specifically for any omissions.

  Risk allocation is the key difference between the Yellow Book and the Silver Book. Employers must appreciate that the allocation of risk to the Contractor within the Silver Book comes with a price which can be significantly above the price under the Yellow Book if one were to compare similar tenders under the two books. However, whilst under the Silver Book the Contractor prices the risk at tender stage, the Contract Price under the Yellow Book could rise considerably if certain risks held by the Employer materialise during the progress of the works.

Detailed comparison of the FIDIC 2017* Conditions of Contract for Plant and Design-Build *(The "Yellow Book") and* the Conditions of Contract for EPC/ Turnkey Project *(The "Silver Book")

The structure of the two contracts would at first glance look similar, this is a conscious effort by FIDIC to make all of the Rainbow Suite (Red, Yellow and Silver Books) have the same "look and feel", but there are some notable differences between the Yellow and Silver Books.

Clause 1 – General Provisions

Priority of Documents – Sub-Clause 1.5

Yellow Book

a The Contract Agreement
b The Letter of Acceptance
c The Letter of Tender
d The Particular Conditions Part A – Contract Data
e The Particular Conditions Part B – Special Provisions
f The General Conditions
g The Employer's Requirements
h The Schedules
i The Contractor's Proposal
j The JV Undertaking (if the Contractor is a JV)
k Any other documents forming part of the Contract

Silver Book

a The Contract Agreement
b The Particular Conditions Part A – Contract Data
c The Particular Conditions Part B – Special Provisions
d The General Conditions
e The Employer's Requirements
f The Schedules
g The Tender
h The JV Undertaking (if the Contractor is a JV)
i Any other documents forming part of the Contract

A critical difference is that Sub-Clause 1.9 [Errors in the Employer's Requirements] is included within the Yellow Book, but does not exist within the Silver Book, and this is a major risk for the Contractor.

Under Sub-Clause 1.9 of the Yellow Book, if the Contractor finds an error, fault or defect within the Employer's Requirements as a result of scrutinising them under Sub-Clause 5.1 (i.e. promptly after receiving the Notice from the Engineer stating the Commencement Date (Clause 8.1)), the Contractor gives a Notice to the Engineer, and the Engineer agrees or determines whether or not there is an error and whether an experienced Contractor would have discovered the error, fault or defect before submitting the tender, and if an experienced Contractor would not have discovered the error, fault or defect prior to tender, it is judged to be a Variation, and the Contractor, if he suffers delay and/or incurs Cost as a result, is entitled to an Extension of Time and/or payment of Cost plus Profit.

Sub-Clause 1.9 is not included within the Silver Book, so the Contractor is responsible (and liable) for any error, fault or defect within the Employer's Requirements, and as stated above, this is a major risk for the Contractor.

Clause 2 – The Employer

Site Data and Items of Reference – Sub-Clause 2.5

Whilst Sub-Clause 2.5 looks very similar under the Yellow and Silver Books, the final paragraph in the Silver Book states that "*the Employer shall have no responsibility for the accuracy, sufficiency or completeness of such data and/or items of reference except as stated in Sub-Clause 5.1 [General Design Obligations]*" (see also previous item).

Clause 3 – The Employer

In the Yellow Book, the Employer appoints the Engineer, and in the Silver Book, the Employer appoints the Employer's Representative. (See Chapter 2 for the actions of the Engineer/Employers Representative.)

Clause 4 – The Contractor

Sub-Clause 4.7 – Setting Out

Under the Yellow Book <u>and</u> the Silver Book, the Contractor is responsible and liable for setting out the Works, with the Employer responsible for items of reference e.g. levels or datums, but the Contractor must verify their accuracy.

The Contractor is required to set out the Works in relation to the items of reference under Sub-Clause 2.5 [Site Data and Items of Reference].

Under the Yellow Book, any delay and/or cost caused by the Employer providing incorrect items of reference, e.g. incorrect finished floor levels, is recoverable by the Contractor.

Under the Yellow Book, the Contractor is required to:

a Verify the accuracy of all these items of reference before they are used for the Works;
b Promptly deliver the results of each verification to the Engineer;
c Rectify any error in the positions, levels, dimensions or alignment of the Works; and
d Be responsible for the correct positioning of all parts of the Works.

If the Contractor finds an error in any items of reference, he gives a Notice to the Engineer describing it, and if there is an error in those items of reference, whether or not an experienced contractor exercising due care would have discovered such an error when examining the Site, the Drawings and the Specification before submitting the Tender, or if the items of reference are specified on the Drawings and/or in the Specification and the Contractor's Notice is given after the expiry of the period stated in sub-paragraph (a) of Sub-Clause 4.7.2, and also, if an experienced contractor would not have discovered the error, and he suffers delay and/or incurs Cost as a result of the error, the Contractor is entitled to an Extension of Time and/or payment of such Cost Plus Profit.

Under the Silver Book the Contractor is required to comply with (a), (c) and (d) above, but there is no entitlement to an Extension of Time and/or payment of Cost Plus Profit.

Clearly, there is a significant risk held by the Contractor under the Silver Book, and in order to make a judgement on his liability (if it is possible to be judged) it is clear that he needs time to make and value that consideration, and this is one of the reasons (amongst many others) why Contractor tendering under the Silver Book must be given a significantly longer tender period than under the Yellow Book.

Sub-Clause 4.10 – Site Data

Sub-Clause 4.10 picks up from Sub-Clause 4.7 above.

Under the Yellow Book, the Contractor is responsible for interpreting all data referred to under Sub-Clause 2.5, and to that extent the Contractor is deemed to have taken into account all necessary information as to risks and is deemed to inspect the Site, and to have been satisfied before submitting the tender on the form and nature of the Site including sub-surface conditions, the hydrological and climactic conditions, the extent and nature of the works and Goods necessary for the Works, the Laws, procedures and the Contractor's requirements for access, accommodation, resources, etc.

The Silver Book has always been very brief in that "*the Contractor shall be responsible for verifying and interpreting all data made available by the Employer under Sub-Clause* 2.5".

Sub-Clause 4.12 – Unforeseeable Physical Conditions (Yellow Book)

Sub-Clause 4.12 – Unforeseeable Difficulties (Silver Book)

Under the Yellow Book, Sub-Clause 4.12 defines "physical conditions" and covers those which are "Unforeseeable".

"Physical Conditions" is defined under Sub-Clause 4.12 of the Yellow Book as

> natural physical conditions and physical obstructions (natural or man-made) and pollutants, which the Contractor encounters at the Site during execution of the Works, including sub-surface and hydrological conditions but excluding climatic conditions at the Site and the effects of those climatic conditions.

"Unforeseeable" is defined under Sub-Clause 1.1.87 of the Yellow Book as *"not reasonably foreseeable by an experienced contractor by the Base Date". "Base Date" is defined as "the date 28 days before the latest date for submission of the Tender".*

If the Contractor encounters physical conditions which it considers to have been Unforeseeable and that the finding will have an adverse effect on the progress and/or increase the Cost of the execution of the Works, the Contractor is required, as soon as possible, to give a Notice to the Engineer, describing the physical conditions, so that they can be inspected by the Engineer, setting out the reasons why the Contractor considers the physical conditions to be Unforeseeable, and describe the manner in which the physical conditions will have an adverse effect on the progress and/or increase the Cost of the execution of the Works.

The Engineer is then required to inspect and investigate the physical conditions within seven days (or a longer period agreed with the Contractor) after receiving the Contractor's Notice.

The Contractor is then required to comply with any instructions which the Engineer may give for dealing with the physical conditions, which may be a Variation.

If the Contractor suffers delay and/or incurs Cost due to these physical conditions, he is entitled to Extension of Time (EOT) and/or payment of such Cost.

Under the Silver Book, "Unforeseeable" is defined the same as under the Yellow Book i.e. under Sub-Clause 1.1.77 as "*not reasonably foreseeable by an experienced contractor by the Base Date*". "Base Date" is defined as "*the date 28 days before the latest date for submission of the Tender*".

Under the Silver Book, Sub-Clause 4.12 [Unforeseeable Difficulties] simply states that except as stated in the Particular Conditions the Contractor is deemed to have obtained all necessary information as to risks, contingencies and circumstances which may influence or affect the Works, and accepts total responsibility for having foreseen all difficulties and costs of successfully completing the Works, and the Contract Price is not be adjusted to take account of any Unforeseeable difficulties.

The only possibility for the Contractor to perhaps recover Cost (under certain circumstances) or gain an Extension of Time is if the event is of such magnitude that it can be considered to fall within the definition of an Exceptional Event under Clause 18, but that is unlikely.

Sub-Clause 5.1 – Design

After receiving a Notice to commence the Works, the Contractor is required to scrutinise the Employer's Requirements, and if he discovers any error, fault or other defect Sub-Clause 1.9 [Errors in the Employer's Requirements] shall apply.

Under Sub-Clause 1.9 (see also Chapter 2) if the Contractor finds an error, fault or defect in the Employer's Requirements he gives a Notice to the Engineer within the period stated in the Contract Data (if not stated, 42 days) calculated from the Commencement Date. The Contractor may notify later than the period stated see (b) below.

The Engineer then determines whether or not there is an error, fault or defect in the Employer's Requirements, and also whether or not an experienced contractor exercising due care would have discovered the error, fault or other defect:

- a When examining the Site and the Employer's Requirements <u>before</u> submitting the Tender; or
- b If the Contractor's Notice is given after the period stated in the Contract Data, when scrutinising the Employer's Requirements under Sub-Clause 5.1 [General Design Obligations]; and
- c What measures (if any) the Contractor is required to take to rectify the error, fault or defect.

If, under sub-paragraph (b) above, an experienced contractor would not have discovered the error, fault or other defect:

- i Sub-Clause 13.3.1 [Variation by Instruction] applies to the measures that the Contractor is required to take (if any), and
- ii If the Contractor suffers delay and/or incurs Cost as a result of the error, fault or defect, the Contractor is entitled to an Extension of Time and/or payment of Cost plus Profit.

<u>Note that Sub-Clause 1.9 is not included within the Silver Book</u>

The Silver Book states that the Employer is not responsible for any error, inaccuracy or omission with the Employer's Requirements, and any data or information received by the Contractor from the Employer does not relieve the Contractor of any obligation to provide the Works.

The only exclusions are:

a Data and information which is specifically stated within the Contract as being the responsibility of the Employer;
b Definitions of intended purpose of the Works;
c Criteria for testing and performance of the completed Works;
d Data and information which cannot be verified by the Contractor.

In both the Yellow and Silver Books, the Contractor is responsible for designing the Works in accordance with the Employer's Requirements.

The difference in the Contractor's liability for the correctness of those Employer's Requirements has been discussed previously.

As stated previously, under the Yellow Book, if the Contractor, after scrutinising the Employer's Requirements, finds errors, those errors can be accepted and remedied by the Engineer in the form of a Variation. The Contractor would also get an Extension of Time and/or payment of Cost plus Profit.

The critical difference within the Silver Book is that the Contractor takes responsibility for the Employer's Requirements including design criteria and calculations, with only certain exceptions for which the Employer is responsible for those being:

a Portions, data and information which are stated in the Contract as being the responsibility of the Employer;
b Definitions of intended purposes of the Works, or any parts of the Works;
c Criteria for the testing and performance of the completed Works;
d Portions, data and information which cannot be verified by the Contractor, except as otherwise stated in the Contract.

Whilst readers may conclude that Employers should use the Silver Book because it apportions significant risks to the Contractor, it must be remembered that:

i Contractors who understand the risks within the Silver Book will price those risks, and therefore, the tender prices could be significantly greater than under the Yellow Book, possibly artificially higher dependent on the tenderers appetite for risks. Obviously, the Contractors who do not understand the risks with the Silver Book (or any Contract) may gain a perceived advantage at tender stage, in that their tender price will be lower, but that is likely lead to problems later on.

   It is important at tender stage, and prior to Contract award that Clients question Contractors as to their knowledge of the Silver Book, possibly with a questionnaire at "expressions of interest" stage.
ii Clearly, the tendering contractors need significant additional time to price those risks; therefore, tender periods should be longer for the Silver Book.

As stated above, Employers who choose to employ the Silver Book should do so after careful consideration and careful selection of Contractors to submit bids, ensuring that tenderers have correctly interpreted and anticipated (and priced and programmed for) the risks with the Contract. The Silver Book is "different" to other standard forms of Contract.

Employers using the Silver Book should also recognise the significant costs for a Contractor to produce a tender and limit the number of tenderers accordingly.

Employers should also allow Contractors the freedom to carry out the works in the manner and the sequence they wish to use to achieve the required objectives. The Contract is restrictive enough without Employers adding further restrictions!

If Employers are not prepared to take the above into account when using the Silver Book, it begs the question, is the Silver Book the correct choice? Many Employers may be unable or unwilling to comply with the above advice, and in hindsight find that they probably should have used the Yellow Book, or even the Red Book.

In the author's opinion, and from the Contractor's point of view, Employers who use the Silver Book "ring alarm bells" in the minds of tenderers as to why they are using it, and it is usually because of unknown risks.

Above all both Parties should know the Yellow and Silver Books, and recognise and manage the risks to avoid disputes!

The author was involved some years ago with a Silver Book Contract, due to non-disclosure provisions it is not possible to name the Employer or the project, but the Employer, who was engaged in the "new technology" business, knew exactly what he wanted from the engineering project, and was able to properly define the performance criteria for the completed facility within the Contract.

The Employer was not from a construction-related industry, nor did he intend to be, and was happy, having chosen what he was advised was the best Contractor, and not the cheapest, fully trusted the Contractor to deliver, and he did, thereby creating a model of the type of Employer, and Contractor, who should be using the Silver Book.

The structure of the FIDIC 1999 Contracts

The Red, Yellow and Silver Books have 20 clauses, covering similar subject matter in all three Books.

Wherever possible, similar wording is used in all three Books in the equivalent Clauses and Sub-Clauses.

If we take the *Conditions of Contract for Construction* (Red Book) as an example:

1 General Provisions
2 The Employer
3 The Engineer (The Employer's Administration – Silver Book)
4 The Contractor
5 Nominated Subcontractors (Design – Yellow and Silver Books)
6 Staff and Labour
7 Plant, Materials and Workmanship
8 Commencement, Delays and Suspension
9 Tests on Completion
10 Employer's Taking Over
11 Defects Liability
12 Measurement and Evaluation (Tests after Completion – Yellow and Silver Books)
13 Variations and Adjustments
14 Contract Price and Payment
15 Termination by Employer
16 Suspension and Termination by Contractor

17 Risk and Responsibility
18 Insurance
19 Force Majeure
20 Claims, Disputes and Arbitration

Each book has its own set of tender documents:
Conditions of Contract for Construction (The "Red Book"):

- Letter of Tender
- Conditions of Contract
- Specification
- Drawings
- Schedules

Conditions of Contract for Plant and Design-Build (The "Yellow Book"):

- Letter of Tender
- Conditions of Contract
- Employer's Requirements
- Schedules
- Contractor's Proposal

Conditions of Contract for EPC/Turnkey Project (The "Silver Book"):

- Conditions of Contract
- Employer's Requirements
- Tender

A Construction or Plant and Design-Build Contract will also include a Letter of Acceptance and/or a Contract Agreement, and either document may bring the Contract into force. Typically, the Contract Agreement will bring an EPC Contract into force.

The Conditions of Contract comprise the "General Conditions", which form part of the "Conditions of Contract for...", First Edition 1999, published by the Federation International des Ingenieurs-Conseils (FIDIC), and the "Particular Conditions", which include amendments and additions to such General Conditions.

A principle that was applied throughout the drafting has been to facilitate the preparation of the tender documents, and in particular the preparation of the Conditions of Contract. Unlike most other published form of Conditions, those published by FIDIC recognise the reality that the tender documents for a particular project typically have to include provisions which are not appropriate for other projects. Preparation of the tender documents has been facilitated by anticipating alternative arrangements (for example, a lump sum contract under the Construction Book), by stating in the General Conditions which provisions are subject to what is stated in the Particular Conditions (because they may not be universally applicable), by maximising the General Conditions so as to minimise Particular Conditions and by identifying one location for essential contract-specific data.

In all three Contracts the General Conditions were drafted on the principle that users would find it more convenient if any provisions which they did not wish to apply could simply be deleted or not invoked, than if additional text had to be written in the Particular Conditions (because the General Conditions did not cover their requirements). If this criterion was considered inapplicable for a Sub-Clause, it is intended to contain provisions that were considered applicable to most contracts.

It should therefore be noted that some of the provisions contained in the General Conditions may be inappropriate for an apparently typical contract. The basic concept was to provide maximum convenience for users, particularly those who prepare the tender documents, for whom a degree of competence is necessary.

Thus, as much text as possible is incorporated into the General Conditions, including some from the old Part II. Some provisions cannot apply unless invoked/enabled by the insertion of the required data, and provisions can readily be disabled by omission of the data or by deletion.

For Contractor-designed Works, the Employer's Requirements should include all relevant criteria, including quality, performance criteria and testing, but need not specify any matters which would be imposed on the Works by the applicable law. Quality should be specified in terms that are not so detailed as to reduce the Contractor's design responsibilities, not so imprecise as to be difficult to enforce and not subject to anybody's future opinions which tenderers may consider impossible to forecast.

Detailed overview of the FIDIC 1999 Contracts

In considering the FIDIC contracts a short review of each Clause within the FIDIC 1999 Contracts follows, with a more detailed analysis of each of the FIDIC 2017 Clauses within Chapters 4–12, with references to the changes from FIDIC 1999 to FIDIC 2017 Contracts.

Note that this is not a precise review of the FIDIC 1999 Contracts but intended as an indication of the structure and content of the Red, Yellow and Silver Books.

This short review covers:

- *Conditions of Contract for Construction* (The "Red Book")
- *Conditions of Contract for Plant and Design-Build* (The "Yellow Book")
- *Conditions of Contract for EPC/Turnkey Projects* (The "Silver Book")

Clause 1: General Provisions

This Clause includes the meaning of words or expressions used in the Conditions of Contract and is divided into "The Contract"; "Parties and Persons"; "Dates, Tests, Periods and Completion"; "Money and Payments"; "Works and Goods"; and "Other Definitions".

It also covers interpretation such as gender, singular/plural, in addition to marginal words and other headings not forming the Conditions. This is a common provision within most contracts.

All communications including approvals, certificates, consents, determinations, notices and requests are to be in writing, and sent to the address stated in the Appendix to Tender. Electronic transmissions e.g. emails, intranets, etc., are as stated in the Appendix to Tender.

There is also a requirement to copy other Parties into certificates and notices.

FIDIC is an international family of contracts so the applicable laws are critical, the Contract being governed by the law of the country stated in the Appendix to Tender. Normally, it is the law of the country in which the project is based. The ruling language and the language for communications are also stated in the Appendix to Tender.

There is a priority of documents in case of ambiguities or inconsistencies, with the following descending order of importance:

Red Book

a The Contract Agreement (if any)
b The Letter of Acceptance
c The Letter of Tender
d The Particular Conditions
e The General Conditions
f The Specification
g The Drawings
h The Schedules and any other documents forming part of the contract

The Yellow Book

a The Contract Agreement
b The Letter of Acceptance
c The Letter of Tender
d The Particular Conditions
e The General Conditions
f The Employer's Requirements
g The Schedules
h The Contractor's Proposal

The Silver Book

a The Contract Agreement
b The Particular Conditions
c The General Conditions
d The Employer's Requirements
e The Tender and any other documents forming part of the Contract

The Specification and Drawings remain in the custody of the Employer, two copies of each and every subsequent document is supplied to the Contractor.

Any documents supplied by the Contractor e.g. calculations, software, etc., remain in the custody of the Contractor.

Any delay in issue of drawings or instructions will give rise to an Extension of Time and/or recovery of Costs plus Profit.

NB: In the Yellow Book this is replaced with "Errors in the Employer's Requirements".

The Contractor retains copyright but gives the Employer a license to copy, use and communicate the Contractor's Documents. Note that this applies for the actual or intended working life of the relevant parts of the Works. Similarly, the Employer retains copyright but gives the Contractor a license to use the Employer's Documents.

Clause 2: The Employer

Clause 2 covers the role of the Employer, commencing with his obligation to give right of access and possession of the Site as stated in the Appendix to Tender. This right may not be exclusive. Rights may be withheld until the Performance Security has been received.

If this is delayed, the Contractor can give notice with the ensuing rights to Extension of Time and Cost plus Profit.

The Employer is required to "assist" the Contractor where possible in obtaining copies of the Laws of the Country and any necessary permits to work in that country. It is fundamentally the responsibility of the Contractor to obtain these.

Clause 2.4 introduced under the FIDIC 1999 Contracts, required the Employer, if requested by the Contractor, to provide evidence that he has, or will have the finance available to pay for the project, the evidence having to be provided within 28 days of any request by the Contractor.

Note that, under Clause 16.1, if the Employer fails to comply, the Contractor can give 21 days' notice to suspend work. If the Employer felt it necessary to constrain his obligation, he would have to include an amendment (in the Particular Conditions) saying what if any evidence he would provide, and tenderers would be able to take account of such an amendment when tendering. Sadly, this Clause is often deleted from the Contract.

The next Sub-Clause 2.5 sets out the procedure for the Employer to follow in the event of a claim against the Contractor. Similar procedures apply to claims by either Party.

Clause 3: The Engineer

The Engineer is the person appointed by the Employer to act on his behalf and is named in the Appendix to Tender. The Engineer and his staff must be suitably qualified and competent to carry out their duties. If there is any limitation on the Engineer's level of authority, for example, the need to obtain the Employer's approval on major changes then this must be included within the contract.

If the Engineer is to be changed, the Employer must comply with Clause 3.4.

The Engineer may delegate any of his duties at any time by notifying the Contractor in writing which may later be revoked, again in writing. It is important to note that the authority of the delegated representative is limited to the authority contained within the delegation notice.

The Contractor is required to comply with any instruction given by the Engineer. Instructions should only be taken from that person or any previously notified delegate. Wherever possible this should be in writing, though there is a fall back situation where the instruction is oral, it can be confirmed by the Contractor within two working days and if not dissented from within two further working days it becomes an instruction.

The Engineer is required to notify the Contractor at least 42 days prior to that replacement. The Contractor may make reasonable objection to a proposed replacement Engineer.

Sub-Clause 3.5 sets out the procedure for agreeing or determining various matters including Employer's claims, Contractor's claims and the evaluation of Variations and of the Works. The Engineer is required to consult with each Party in an attempt to reach agreement. In the absence of an agreement, the Engineer makes a fair determination.

NB: Clause 3 in the Silver Book is entitled "The Employer's Administration", and is similar to Clause 3 under the Red and Yellow Books, but refers to the Employer and the Employer's Representative rather than the Engineer.

Clause 4: The Contractor

Clause 4 covers many of the Contractor's general obligations.

Sub-Clause 4.1 states the Contractor's obligation to execute the Works, and requires any Contractor-designed Works to be (when completed) fit for the intended purpose defined in the Contract. Note that most design and build contracts provide for the design to be to reasonable skill and care.

The Contractor is only responsible for the design of the Permanent Works to the extent specified in the Contract. Sub-paragraphs (a) to (d) specify general requirements in respect of such designs, unless they are defined otherwise in the Particular Conditions. These general requirements are also stated in 13.3 (b) as being applicable to any subsequent Contractor-designed "value engineering" proposals, unless otherwise agreed by both Parties: see below.

Sub-Clause 4.2 requires the Contractor to provide a Performance Security at his own cost in the amount specified in the Appendix to Tender and in accordance with the Particular Conditions.

Sub-Clause 4.3 describes the role of the Contractor's Representative who has all the authority necessary to act on the Contractor's behalf and his whole time to running the project.

Unless the Representative is named in the Contract, the Contractor has to submit the name of his Representative for the Engineer to give his consent.

In the case of absence, any replacement must be subject to the Engineer's consent.

In the case of any delegation by the Contractor's Representative this must be notified to the Engineer before the delegation takes place.

Sub-Clause 4.4 covers Subcontractors, who are subject to the Engineer's consent. The Contractor does not have to have the consent to suppliers or previously named Subcontractors in the Contract.

Sub-Clause 4.6 is an unusual clause as it requires the Contractor to allow "appropriate opportunities" for carrying out work to various Parties employed by the Employer, or of any public authority. This is presumably to encourage the use of local labour. The Contractor is able to recover the cost as "Unforeseeable Cost".

Any delay and/or cost occasioned by the Employer providing incorrect items of reference, e.g. finished floor levels, is recoverable by the Contractor.

Sub-Clause 4.8 covers the Contractor's responsibility to comply with all safety procedures including local regulations during the Works. This includes provision of all necessary fencing, lighting and guarding.

The Contractor must institute a quality assurance system which is subject to the Engineer's audit.

Under Sub-Clause 4.10, the Employer is deemed to have made available all relevant data in his possession on sub-surface and hydrological conditions by the Base Date, which is 28 days before the date of submission of the Tender. The Contractor is responsible for interpreting the data and shall be deemed to have inspected and examined the Site. This does not actually mean that he MUST visit the Site, but that he is "fixed with the knowledge" as if he had....there is a subtle difference!

"Unforeseeable" is defined as "not reasonably foreseeable by an experienced Contractor by the date for submission of the Tender", allowing nearly 28 days after the Base Date for tenderers to assess the date. The Employer is also required to make available the same types of date that come into his possession thereafter.

The Contractor is deemed to have satisfied himself as to the correctness and sufficiency of the Accepted Contract Amount.

Sub-Clause 4.12 defines "physical conditions" and covers those which are "Unforeseeable". Under a new provision, the Engineer may also review whether other physical conditions in similar parts of the Works were more favourable than could reasonably have been foreseen when the Contractor submitted the Tender.

This provision may be particularly relevant for a tunnel, where there may be adverse conditions, in respect of similar parts of the Works, will not result in a net reduction in the Contract Price (otherwise, an Employer might have been tempted to conceal favourable information from tenderers).

The Contractor is required to bear all costs and charges in respect of rights of way including access to Site. He may also obtain at his own risk and cost any other facilities.

The Contractor is deemed (whether he has or has not) to have satisfied himself as to suitability and availability of access routes onto the Site.

The Contractor is to give the Engineer 21 days' notice prior to the date on which "any Plant or major item of other goods will be delivered to Site". The Contractor is responsible for all associated costs in connection with transporting, unloading, distribution and protection.

Equipment is what is normally known as "Plant" in other contracts, the Contractor being responsible for provision and care. The Contractor has to have the Engineer's consent to remove such Equipment.

The Contractor is required to take all reasonable steps to protect the environment. The Specification and the Law prescribe limits of any discharges.

Sub-Clauses 4.19 and 4.20, if enabled by details in the Specification, cover the arrangements for the Employer's provision of power, other services, "Employer's Equipment" and free-issue material.

Clause 4.30 covers any Equipment to be provided by the Employer. The Contractor is required to pay at the rates stated in the Specification.

Monthly progress reports are to be prepared by the Contractor and six copies submitted to the Engineer. The contract prescribes the content of each report which includes progress to date, photographs, critical documents and health and safety records.

The Contractor is responsible for keeping unauthorised personnel off the Site.

The Contractor is required to confine his operations to the Site and to keep all Equipment and Personnel within the Site.

On the issue of a Taking-Over Certificate, the Contractor is required to properly clear away.

All fossils, coins, archaeological finds, etc., are to be placed in the care of the Employer. The Contractor is required to inform the Engineer upon finding such articles and is able to recover any delay and/or cost through the contract.

Clause 5: Nominated Subcontractors (Red Book)

Clause 5 covers the appointment of nominated Subcontractors, because they are sometimes required, not because their use is recommended.

Sub-Clause 5.1 defines a nominated Subcontractor as one stated in the Contract, whom the Engineer under Clause 13 instructs the Contractor to employ as a Subcontractor.

The Contractor has a right to notify the Engineer of "reasonable objection" for various reasons e.g. the Subcontractor has insufficient competence or resources, or some other reason which will place the Contractor at a higher risk.

The Engineer certifies amounts due to the nominated Subcontractor under Clause 14. The Contractor is also paid for overheads and profit as stated in the Appendix to Tender.

The Contractor is required to prove discharge of amounts due to the nominated Subcontractor under a previous Payment Certificate.

NB: The Yellow and Silver Books do not have provision for Nominated Subcontractors, Clause 5 deals with the Contractor's design obligations.

The Contractor is responsible for carrying out and the design of the Works. He is required to submit the name of each designer and design Subcontractor to the Engineer. Once the Contractor is given notice to commence under Clause 8.1, he gives notice to the Engineer of any errors in the Employer's Requirements.

The Contractor's Documents comprise the technical documents specified in the Employer's Requirements, documents to satisfy regulatory approvals and as-built documents.

There is a 21-day review period from submission of a Contractor's Document during which the Engineer can give notice to the Contractor that a Contractor's Document fails to comply with the contract.

The Contractor is to prepare and update a complete set of "as-built" records and to issue them to the Engineer.

The Contractor is to supply prior to commencement of Tests on Completion operation and maintenance manuals. The Works are not deemed to be complete for taking-over until the Engineer has received them.

Clause 5: Design (Yellow and Silver Books)

The Contractor is responsible for carrying out and the design of the Works. He is required to submit the name of each designer and design Subcontractor to the Engineer. Once the Contractor is given notice to commence under Clause 8.1 he gives notice to the Engineer of any errors in the Employer's Requirements.

The Contractor's Documents comprise the technical documents specified in the Employer's Requirements, documents to satisfy regulatory approvals and as-built documents.

The Contractor prepares all the documents, and the Employer's Personnel may inspect them during their preparation.

There is a 21-day review period from the submission of a Contractor's Document during which the Engineer can give notice to the Contractor that a Contractor's Document fails to comply with the contract.

The Contractor undertakes that the design shall be in accordance with the Laws of the Country and the documents forming the contract.

The design, the Contractor's Documents and the execution of the work are to be in accordance with the contract and the applicable Laws.

The Contractor is to provide training of the Employer's Personnel in the operation and maintenance of the Works.

The Contractor is to prepare and update a complete set of "as-built" records and to issue them to the Engineer.

The Contractor is to supply prior to commencement of Tests on Completion operation and maintenance manuals. The Works are not deemed to be complete for taking-over until the Engineer has received them.

Design errors are to be corrected at the Contractor's cost.

Clause 6: Staff and Labour

Sub-Clause 6.1 requires the Contractor to make all necessary arrangements for his staff and labour.

The Contractor is required to pay rates of wages and observe labour conditions not lower than those established for the industry where the work is carried out.

The Contractor cannot recruit staff and labour from the Employer's personnel.

The Contractor is required to take all reasonable precautions to maintain the health and safety of his personnel including the appointment of a qualified accident prevention officer.

The Contractor provides all necessary superintendence includes the quantity and quality of personnel to plan, arrange, direct, manage, inspect and test the work.

The Contractor's personnel shall be appropriately qualified. The Engineer has the right to require the Contractor to remove any person for one of four reasons: misconduct, incompetence/negligence, failure to comply with the Contract or health and safety/protection of the environment.

The Contractor is required to prevent any disorderly conduct by his personnel. This also extends to preserve "peace and protection" of persons or property near the Site.

Clause 7: Plant, Materials and Workmanship

This Clause generally covers the Contractor's obligations regarding the quality of the work and associated testing.

The need for samples should be clear, and the Contractor has an obligation to submit the necessary samples, either at the Contractor's cost if specified in the Contract or as a Variation if the Engineer instructs additional samples.

The Contractor is to provide all apparatus, assistance, documents, etc., to carry out the specified tests. The Engineer is required to give the Contractor not less than 24 hours' notice before carrying out such tests. If despite giving the notice the Engineer does not attend, the tests will be carried out as if the Engineer had attended.

If work fails a test, the Contractor is required to make good the defect and repeat the test.

The Engineer may instruct the Contractor to remedy any defects within a reasonable time, or immediately if urgency requires.

Sub-Clause 7.7 defines the date when each item of Plant and Materials becomes the property of the Employer. In the event of suspension, property passes when conditions 8.10 (a) and (b) have been satisfied, without further pre-conditions relation to payment.

Unless the Specification states otherwise, the Contractor is to pay all royalties, rents and other payments in connection with the Works.

Clause 8: Commencement, Delays and Suspension

Clause 8 covers three time-related subjects:

i The start and duration of the project;
ii Programme, delays and extensions of time;
iii Suspension of work by the Engineer.

The Employer is required to give the Contractor not less than 7 days' notice of the Commencement Date, this normally being within 42 days after the Contractor receives the Letter of Acceptance.

The Contractor is required to complete all the works for take over including the passing of tests. Any Sections (Sectional Completion) must be described in the Appendix to the Tender.

The Contractor is required to submit a programme to the Engineer within 28 days of receiving the notice under Clause 8.1.

Sub-Clause 8.3 covers the provision of programmes showing how the Contractor proposes to execute the Works. Each such programme is to be supported by a report describing the methods that the Contractor proposes to adopt. Programmes are not "approved": the Contractor proceeds in accordance with the Programme, except to the extent that (i) he has been advised of its non-compliance, or (ii) it does not actually comply with the Contract. Unrealistic programmes do not provide a sound basis for claims.

The contract describes the content of the Contractor's programme which should include order and sequence, timing of tests, method statement and details of resource levels.

There are five grounds for Extension of Time:

a Variations;
b A cause of delay giving rise to an entitlement under the contract;
c Exceptionally adverse climactic conditions;
d Unforeseeable shortages in availability of personnel or goods caused by epidemic or government action;
e Any delay, impediment or prevention caused by the Employer, the Employer's Personnel or the Employer's other contractors on the Site.

Sub-Clause 8.5 gives the Contractor a right to an Extension of Time for delays caused by public authorities as long as the Contractor has diligently followed procedures.

If progress is too slow to complete within the Time for Completion or progress has fallen behind the current programme, the Engineer can instruct the Contractor to submit a revised programme and proposals to rectify matters.

Sub-Clause 8.7 covers delay damages, previously called liquidated damages (a phrase which is also applicable to some of the possible types of non-performance damages).

The Engineer may instruct the Contractor to suspend the progress of all or part of the Works during which time the Contractor must protect, save and store. The Engineer may, but does not have to, notify the Contractor of the reason for the suspension.

If the Contractor suffers delay and/or cost resulting from suspension he can claim it through the contract.

If the suspension has continued for more than 84 days the Contractor can request the Engineer's permission to proceed. If the Engineer does not give such permission the Contractor could terminate.

After permission to proceed is given, the Contractor and Engineer jointly inspect the Works and the Contractor makes good any deterioration or defect.

Clause 9: Tests on Completion

The Contractor is required to carry out all tests on Completion. He has to give the Engineer not less than 21 days' notice prior to each test.

If the Tests on Completion are delayed by the Employer Clause 10.3 applies giving the Contractor right to recover delays and/or costs as a result of any delay caused. If the Tests are delayed by the Contractor, the Engineer can give the Contractor 21 days to carry out the Tests.

If the Works fails the Tests on Completion, the Engineer is entitled to order further repetition of tests, reject the Works and issue a Taking-Over Certificate if the Employer requests.

Clause 10: Employer's Taking Over

Clause 10 specifies the conventional procedures for certification of completion and for the Employer's Taking Over.

The Works are taken over by the Employer when they have been completed in accordance with the contract and a Taking-Over Certificate has been issued. The Contractor may apply for a Taking-Over Certificate not earlier than 14 days prior to completion.

The Contractor is able to recover any additional costs associated with this action.

If the Contractor is prevented for more than 14 days from carrying out Tests on Completion, the Employer is deemed to have taken over the Works and the Engineer issues a Taking-Over Certificate. Any delay and/or cost incurred by the Contractor as a result is paid by the Employer.

Clause 11: Defects Liability

Defects Liability obligations are specified in Clause 11.

The Contractor is required to complete any work which is outstanding on the dates stated in a Taking-Over Certificate including rectifying defects or damage.

Sub-Clause 11.3 entitles the Employer to claim (under Sub-Clause 2.5) an extension of the Defects Notification Period if the Works, a Section or a major item of Plant cannot be used due to a defect or damage.

The Contractor is required to remedy defects within a reasonable time, failing which the Employer can give the Contractor a specific date to remedy, failing which the employer may carry out the work himself and charge the Contractor, or require the Engineer to determine a reduction in the Contract Price or terminate the contract as a whole.

Further tests may be required on the same basis as the original tests.

The Contractor has the right of reasonable access to the Works until the Performance Certificate has been issued.

The Engineer has a right to issue instructions to search for a defect. If the search shows that work was in accordance with the contract, the Contractor is entitled to be paid the Cost plus reasonable Profit.

The issue of the Performance Certificate confirms that the Contractor has fulfilled all his obligations under the contract. The Certificate is issued within 28 days after the latest of the expiry dates of the Defects Notification Periods.

Although this seems a contradiction in terms as the Contractor has fulfilled his obligations, there may be outstanding items such as clearance of the Site, Final Payment Certificate and the possibility of latent defects.

The Contractor leaves Site and removes any remaining Equipment, surplus material, etc., upon receiving the Performance Certificate. If he does not, the Employer may sell or dispose of remaining items.

Clause 12: Measurement and Evaluation (Red Book)

The Works are to be measured and valued for payment in accordance with the Contract. The Engineer decides what is to be measured and when, the Contractor then attends by giving him reasonable notice. If the contractor fails to attend, the Engineer's measurement is deemed to be accepted as accurate.

The method of measurement is in accordance with the Bills of Quantities or other Schedules.

However, Clause 12 covers the measurement and valuation of the Works. Irrespective of a "Variation", a Bill rate or price for an item of work may be amended if:

- i The measured quantity of the item is changed by more than 10% from the quantity of this item in the Bill of Quantities of other Schedule;
- ii This change in quantity multiplied by such specified rate for this item exceeds 0.01% of the Accepted Contract Amount;
- iii This change in quantity directly changes the Cost per unit quantity of this item by more than 1%;
- iv This item is not specified in the Contract as a "fixed rate item".

The cost which has been incurred and is not covered within the omission of any work is determined by the Engineer.

NB: Clause 12 under the Yellow and Silver Books is entitled "Tests after Completion".

If Tests after Completion are specified in the contract, the Clause applies. The Employer provides all electricity, equipment, suitably qualified staff, etc. the tests are required as soon as possible after the Works or a Section have been taken over by the Employer.

If the Employer delays tests, the Contractor gives notice and can recover Costs plus reasonable Profit.

If the Works fail to pass Tests after Completion, the tests are repeated.

Clause 12: Tests after Completion (Yellow and Silver Books)

If Tests after Completion are specified in the contract, the Clause applies. The Employer provides all electricity, equipment, suitably qualified staff, etc., the tests are required as soon as possible after the Works or a Section have been taken over by the Employer.

If the Employer delays tests, the Contractor gives notice and can recover Costs plus reasonable Profit.

If the Works fail to pass Tests after Completion, the tests are repeated.

If the Works fails the Tests after Completion, the Engineer is entitled to order further repetition of tests, reject the Works and issue a Taking-Over Certificate if the Employer requests.

Clause 13: Variations and Adjustments

Clause 13 covers Variations and Adjustments for changes in legislation and in cost. Variations are not binding if the Contractor promptly notifies that he cannot readily obtain the required "Goods".

The Engineer may vary the works, and the Contractor being bound to carry out the Works subject to the above exceptions.

Sub-Clause 13.2 allows the Contractor to submit a proposal to the Engineer which could accelerate completion, reduce cost, improve the efficiency or value of the Works or provide some other benefit. The Contractor shares the benefit of any betterment to the Employer as a fee.

The general procedure for variations is described in Sub-Clause 13.3. The Engineer may request a proposal for the Contractor, which may have resulted from his initiative under Sub-Clause 13.2 or from the request mentioned in the first sentence of Sub-Clauses 13.1 and 13.3.

The Engineer instructs regarding Provisional Sums with the Contract Price being adjusted accordingly.

Works of a minor or incidental nature which cannot be valued by measurement may be executed on a time and materials basis. This Clause contains the details of the resources to be recorded.

Increases or decreases in cost due to changes in legislation after the Base Date are paid to the Contractor.

Sub-Clause 13.8 specifies the adjustments for changes in cost, so as to avoid users having to write the necessary text in the Particular Conditions.

Clause 14: Contract Price and Payment

Clause 14 describes the procedures for payments to the Contractor. Payment procedures commence with the Contractor's submission of a (monthly or other) Statement, with supporting documents which are to include the progress report under Sub-Clause 4.21.

Sub-Clause 14.2 provides for the possibility of an advance payment, and Sub-Clause 14.5 provides for the possibility of making payment for Plant and Materials which have been shipped to the Country and/or have arrived at the Site.

The Contractor is required to submit a Statement (six copies) to the Engineer showing in detail the amounts to which he considers himself entitled.

The Statement includes:

a The estimated value of Works executed;
b Any amounts due to changes in legislation;

c Any amount to be deducted for retention;
d Any amounts to be included for advance payments;
e Any amounts due for Plant and Materials intended for the Works;
f Any amounts due in respect of claims, disputes and arbitration;
g Deduction of any amounts certified in previous Certificates.

This Sub-Clause provides for listed Plant and Materials listed in the contract and delivered to Site for incorporation into the Works to be paid.

Provided the Employer has received and approved the Performance Security, the Engineer certifies payments within 28 days of receiving each Contractor's Statement. There is provision in the Contract for a minimum amount of Payment Certificate to be stated in the Appendix to Tender to prevent the issuing of Certificates for minor amounts.

The Employer is required to pay the advanced payment within 56 days of Letter of Acceptance or 21 days of the Performance Security and Interim Payments within 56 days of the Engineer receiving the Contractor's Statement.

The entitlement to financing charges under Sub-Clause 14.8 is based upon when the Engineer received the Statement, progress report and other supporting documents, irrespective of the date on which the interim Payment Certificate (if any) was issued. Such certification date could, for example, be after the date defined in Sub-Clause 14.7 (b) by which payment should have been made.

Normal rules of retention apply. The first half of retention money is certified by the Engineer upon the issue of the Taking-Over Certificate, the outstanding balance being certified upon the expiry of the Defects Notification Periods.

Sub-Clause 14.10 provides for a Statement at completion, which is to include all amounts to which the Contractor considers he has or will become entitled. Under Sub-Clause 14.14, the Employer's Liability is limited to amounts included in this Statement and amounts arising after the issue of the Taking-Over Certificate for the Works.

Within 56 days of receiving the Performance Certificate, the Contractor submits a draft Final Statement (six copies). Once agreed this is called the Final Statement.

The Contractor when submitting the Final Statement confirms the amount constitutes full and final settlement of all monies due.

Within 28 days after receiving the Final Statement and discharge from the Contractor, the Engineer issues a Final Payment Certificate.

The Employer is not liable to the Contractor for any matter not included within the Final Statement.

The currency which constitutes the Contract price is named in the Appendix to Tender.

Clause 15: Termination by Employer

The Engineer may notify the Contractor of any failure under the contract requiring him to remedy within a reasonable period of time.

There are six grounds for the Employer to terminate the Contractor's employment. Subject to giving the Contractor 14 days' notice to remedy, the Employer may terminate.

As soon as possible after termination, the Engineer determines the value of Works completed by the Contractor.

Following termination, the Employer may withhold further payments from the Contractor till the cost of execution, defects, etc., has been established. The Employer may recover any damages incurred as a result of the termination.

The Employer may terminate at will, but he cannot terminate to complete the Works himself or to arrange for it to be completed by another Contractor.

Clause 16: Suspension and termination by Contractor

Under Sub-Clause 16.1, the Contractor is entitled to give notice of suspension in the event of specified defaults, which include the Engineer's failure to issue a Payment Certificate.

There are seven grounds for the Contractor to terminate his own employment. Subject to giving the Employer 14 days' notice to remedy, the Contractor may terminate.

Once a notice of termination has taken effect, the Contractor leaves the Site.

Following termination, the Employer returns the Performance Security to the Contractor and pays any losses resulting from the termination.

Clause 17: Risk and Responsibility

Sub-Clause 17.1 defines the extent to which the Employer and the Contractor indemnifies the other in respect of claims for personal injury or damage to property which occurs as a result of the actions of personnel or other for which they are responsible during the "design, execution and completion of the Works". Property excludes the Works itself which is dealt with separately at Sub-Clause 17.2.

As a general statement, unless risks are specifically allocated to the Employer and those defined as being under its responsibility, events are the risk and responsibility of the Contractor.

Depending upon the matters stated in Clause 19, some of the risks listed in Sub-Clause 17.3 may be so exceptional as to be Force Majeure.

This Sub-Clause defines the Contractor's responsibility for care of the Works.

Sub-Clause 17.2 provides for the Contractor's care of the Works, for which it is fully responsible until such time as the Taking-Over Certificate (or Taking-Over Certificates in the case of Sectional Completion) is issued or deemed to be issued.

Whilst this Sub-Clause specifically deals with the care of the Works until a Taking-Over Certificate has been issued or earlier termination, but it does not mention suspension by either the Contractor or the Employer possibly leading to termination.

The obligations of the Contractor to care for the Works should remain throughout a suspension, irrespective of responsibility for the events leading to the suspension, until such time as work recommences and a Taking-Over Certificate is issued or Termination takes place and the Contractor is released from its obligations in this respect.

Sub-Clause 17.3 covers Employer's Risks, which are the risks that the Contractor has no control over and if they occur, entitle the Contractor to compensation in respect of loss or damage to the Works, Goods or Contractor's Documents.

If any of the risks in Sub-Clause 17.3 occur, the Contractor is entitled to an Extension of Time and/or payment of Cost.

Sub-Clause 17.5 specifies general provisions for intellectual and industrial property rights, defining the extent of each Party's indemnity and describing procedures in the event of claims for infringement. Sub-Clause 17.6 specifies limits to each Party's liability.

Neither Party is liable for loss of use, loss of profit or any indirect or consequential loss other than that covered by the Contract.

Clause 18: Insurance

Clause 18 covers insurances, most of which are required to be arranged by the "insuring Party".

The insuring Party is subsequently named as the Contractor unless otherwise stated in the Particular Conditions (in this case, the second paragraph makes Clause 18 subject to any insurance terms agreed between the Parties). The provisions thus provide for the possibility that, for some of the insurances (typically those for the Works), the Employer may wish to effect insurance: by defining himself (in the Particular Conditions) as the insuring Party in respect of the relevant insurances (in this case, the third paragraph makes Clause 18 subject to the details annexed to the Particular Conditions).

The Works are to be insured in the joint names of the Contractor and Employer. The insurer should make each claims payment jointly to both the Parties, although this seldom happens.

If a Subcontractor is named in the insurance policy as additional insured, the insurer would thus make each claims payment jointly to each such additional insured as well as the two Parties: which could require a physically large cheque to make the payment. The situation is covered in the fourth paragraph of Sub-Clause 18.1.

Some Employers require payments to be made wholly to them, with the Engineer determining how much the Contractor should receive (often allocating blame), which is considered to be unwise and possibly wrong. Insurers only compensate a Party for actual loss, so the Party who bore the loss claimed should then be reimbursed, even if he appears blameworthy.

If the insuring Party fails (or is unable) to effect and keep in force its insurance, the other Party may effect the insurance at the cost of the defaulting Party. For various reasons, such other Party may not do so, possibly because insurance is no longer available at commercially reasonable terms. Sub-Clause 18.1 therefore concludes with provisions that address these circumstances.

Under Sub-Clause 18.2 (d), the insuring Party is required to effect insurance in respect of some of the Employer's Risks, excluding (in each case) risks which are not insurable at commercially reasonable terms and the exclusions described in sub-paragraph (e). If this cover ceases to be available, the last paragraph of Sub-Clause 18.2 comes into effect.

The insuring Party insures the Works, Plant and Materials for not less than full reinstatement costs including professional fees and profit.

The insuring Party insures against each Party's liabilities with respect to any loss, death or bodily injury (Public Liability), the indemnity limit set within the Appendix to Tender.

This covers the Contractor's liability for his own workforce (Employer's Liability).

Clause 19: Force Majeure

Clause 19 covers Force Majeure, which is defined in Sub-Clause 19.1 as an event which a Party could not have provided against, and could not control, avoid or overcome. The Clause includes general protection to the Contractor in respect of legal or physical impossibility.

Either Party may give a notice to the other, within 14 days of becoming aware, specifying the obligations which cannot be performed.

Each Party is obliged to use all reasonable endeavours to minimise any delay caused by a Force Majeure.

The Contractor is entitled to an Extension of Time and/or recovery of costs as a result of a Force Majeure.

Any subcontract should provide relief form Force Majeure. NB: The FIDIC subcontracts do not provide for Force Majeure.

If the execution of the Work is prevented for a period of 84 days (continuous) then either Party may terminate.

Any matter, in addition to Force Majeure, can give rise to release from performance.

Clause 20: Claims, Disputes and Arbitration

Clause 20 covers the submission of claims and the settlement of disputes.

If the Contractor considers himself entitled to an Extension of Time and/or additional payment, he gives notice to the Engineer not later than 28 days of becoming aware (this 28 day period has a time bar). The Contractor is required to give the Engineer full details, and within 42 days of receiving it the Engineer responds giving the extension of time/award of costs as appropriate.

Below is a list of the Sub-Clauses which entitle the Contractor to claim an Extension of Time and possible additional payment, including in some cases the Contractor is also entitled to a "reasonable profit".

Sub-Clause 1.9 – Delayed Drawings or Instructions (Red Book only)

If delay or disruption is caused or likely to be caused as a result of late drawings or instructions the Contractor is entitled to claim:

- Extension of Time;
- Cost plus reasonable Profit.

Sub-Clause 1.9 – Errors in the Employer's Requirements (Yellow Book only)

If delay is caused or Cost is incurred as a result of Errors in the Employer's Requirements which were not previously discoverable the Contractor is entitled to claim:

- Extension of Time;
- Cost plus reasonable Profit.

Sub-Clause 2.1 – Right to Assess to the Site

If delay is caused or Cost incurred as a result of the Employer failing to give the Contractor access to the Site at the prescribed time the Contractor is entitled to claim:

- Extension of Time;
- Cost plus reasonable Profit.

Sub-Clause 4.7 – Setting Out

If delay is caused or Cost incurred as a result of errors in the original setting out points and levels of reference notified by the Engineer the Contractor is entitled to claim:

- Extension of Time;
- Cost plus reasonable Profit.

Sub-Clause 4.12 – Unforeseeable Physical Conditions

If delay is caused or Cost incurred as a result of the Contractor encountering physical conditions which are Unforeseeable the Contractor is entitled to claim:

- Extension of Time;
- Cost only.

It is worth noting that Unforeseeable is a defined term meaning "not reasonably foreseeable by an experienced contractor by the date for the submission of the Tender".

Sub-Clause 4.24 – Fossils

If delay is caused or Cost incurred as a result of the Contractor's compliance with instructions issued by the Engineer to deal with the discovery of fossils and the like the Contractor is entitled to claim:

- Extension of Time;
- Cost (only).

Sub-Clause 7.4 – Testing

If delay is caused or Cost incurred as a result of testing being delayed by the Employer or on behalf of the Employer the Contractor is entitled to Claim:

- Extension of Time;
- Cost plus reasonable Profit.

Sub-Clause 8.5 – Delays caused by Authorities

If delay or disruption is caused or Cost incurred as a result of the actions or non-actions of Authorities the Contractor is entitled to claim:

- Extension of Time;
- Cost, with or without Profit, appears not to have been specifically considered.

Sub-Clause 8.9 – Consequences of Suspension

If delay is caused or is likely to be caused or Cost incurred as a result of the Engineer's instructions to suspend work the Contractor is entitled to claim:

- Extension of Time;
- Cost (only).

Sub-Clause 10.2 – Taking Over of Parts of the Works

If the Contractor incurs Cost as a result of the Employer taking over or using a part of the Works the Contractor is entitled to claim:

- Cost plus reasonable Profit.

Sub-Clause 10.3 – Interference with Tests on Completion

If delay is caused or Cost incurred as a result of tests being delayed by a reason for which the Employer is responsible the Contractor, amongst other remedies, is entitled to claim:

- Extension of Time;
- Cost plus reasonable Profit.

Sub-Clause 11.8 – Contractor to Search

If the Contractor incurs Cost as a result of searching for a defect for which it was not liable the Contractor is entitled to claim:

- Cost plus reasonable Profit.

Sub-Clause 12.2 – Delayed Tests (Yellow Book only)

If the Contractor incurs Cost as a result of carrying out Tests delayed by the Employer until after Completion the Contractor is entitled to claim:

- Cost plus reasonable Profit.

Sub-Clause 12.4 – Failure to Pass Tests after Completion (Yellow Book only)

If the Contractor incurs Cost as a result of the Employer delaying access to allow Tests to be carried out the Contractor is entitled to claim:

- Cost plus reasonable Profit.

Sub-Clause 13.7 – Adjustments for Changes in Legislation

If delay is caused or is likely to be caused or Cost incurred or likely to be incurred as a result of changes in the Laws of the Country the Contractor is entitled to claim:

- Extension of Time;
- Cost (only).

It is worth noting that Country is a defined term meaning the Country in which the Site (or most of it) is located, where the Permanent Works are to be executed.

Thus, this definition is very limited in the field of international contracting where Contractors, Suppliers and Sub-Contractors may all have originated from Countries other than where the project is being carried out and may suffer as a result of changes in legislation.

Sub-Clause 16.1 – Contractor's Entitlement to Suspend Work

If delay is caused or Cost incurred as a result of the Contractor properly suspending work (or reducing the rate of work) the Contractor is entitled to claim:

- Extension of Time;
- Cost plus reasonable Profit.

Sub-Clause 19.4 – Consequences of Force Majeure

If delay is caused or Cost incurred as a result of Force Majeure events the Contractor is entitled to claim:

- Extension of Time;
- Cost (only) in respect of the events listed at Sub-Clauses 19.1 (ii), 19.1 (iii) and 19.1 (iv)

In any event at some point the Contractor and the Engineer, and sometimes also the DAB or Arbitrator if it becomes a dispute, have to consider the causes of delay, establish periods of Extension of Time against those causes and also establish if there are any associated payment entitlements.

In respect of the Contractor declaring its required profit, it is suggested that instead of waiting until such time as a claim exists and the Employer and its Engineer/ Representative negotiating a percentage, the Parties follow the lead given by FIDIC within the MDB Harmonised Edition (the Pink Book) and agree a fixed percentage applicable to <u>all</u> claims.

The Pink Book states "in these Conditions, provisions including the expression 'Cost plus profit' require this profit to be one twentieth (5%) of this Cost unless otherwise indicated in the Contract Data".

Contractors, when drafting Sub-Contracts, should also take note of the differing provisions within the entitlement clauses such that there is a consistent approach "downstream" as well as "upstream".

There is no definition of Cost relative to the Employer's claims towards the Contractor, and therefore, it must be left to the Engineer or the Employer's Representative, to determine if profit should be passed on as a legitimate claim item.

Disputes may be referred to the DAB jointly appointed by the Parties.

If the Parties cannot agree on a member of the DAB, the Party named in the Particular Conditions resolves the matter.

Any dispute may be referred to the DAB which responds within 84 days.

Both Parties attempt to settle disputes amicably before the commencement of arbitration. Arbitration commences within 56 days of a notice of dissatisfaction.

Any dispute can be settled by reference to arbitration.

Any failure to comply can be referred to arbitration.

If a dispute arises and there is no DAB in place, the dispute can be directly referred to arbitration.

Which FIDIC contract for which project?

From the contracts described above and on Chapter 1, you can see that over the years FIDIC has developed very specialist contracts, but if we can consider the three most commonly used:

- Red Book = Construction Contract
- Yellow Book = Plant and Design-Build Contract
- Silver Book = EPC/Turnkey Contract

FIDIC provides guidance in terms of which of these FIDIC contracts should be used for which projects, and further, more detailed guidance being contained later in this book.

Is the project a larger or more complex project, or does the project have a relatively small value, a short construction time, or does it involve simple or repetitive work?

For larger or more complex projects:

- Red Book
- Yellow Book
- Silver Book

For relatively small value, short construction time or involving simple or repetitive projects:

- Green Book
- The Green Book is suitable for a small contract (say, under £250,000 or equivalent local currency) if the construction time is short (say, less than six months), or for relatively simple or repetitive work, irrespective of whether the design is provided by the Employer or the Contractor, and of whether the project involves civil, electrical, mechanical and/or construction works.
- Again, as with the Red Book, the Price can be set as a lump sum, by remeasurement or cost reimbursable.
- It does not matter whether the design is provided by the Employer (or his Engineer/Architect if he has one) or by the Contractor.
- It does not matter whether the project involves construction, electrical, mechanical, or other engineering work.

Let us examine larger or more complex projects in a little more detail:

As stated above, for these projects, the Employer has a choice of the Red Book, the Yellow Book or the Silver Book.

The next question to ask is:

Who is going to be responsible for all or most of the design of the project?

If the Employer is responsible for most or all of the design, use the Red Book (though the Contractor can be made responsible for some design), or if the Contractor is responsible for all of the design, use the Yellow Book or the Silver Book.

As stated previously the Red Book is recommended for building or engineering works where most of the design is provided by or on behalf of the Employer however, the Works may include some Contractor-designed civil, mechanical, electrical and/or construction works.

The Yellow Book or the Silver Book should be used where the Contractor will be responsible for all or most of the design.

In many projects involving large-scale electrical and mechanical works, the Contractor does the majority of the design, e.g. the detailed design of the plant or equipment, so that the plant meets the outline or performance specification prepared by the Employer, and in the relatively more recent design-build and turnkey-type projects, the Contractor also does the majority of the design, not only of plant projects but also of various infrastructure and other types of projects, and the project is required to fulfil the "Employer's Requirements", i.e. an outline or performance specification prepared by the Employer

- And an Engineer is required to administer the Contract, monitor the manufacture and erection on Site or construction work and certify payment,
- And with payment according to achieved milestones generally on a lump sum basis.

The Yellow Book is probably the best choice.

The Silver Book is suitable for the provision on a turnkey basis of a process or power plant, of a factory or similar facility or of an infrastructure project or other type of development, where

- A higher degree of certainty of final price and time is required, and
- The Contractor takes total responsibility for the design and execution of the project.

However, Employers should consider using the Yellow Book, rather than the Silver Book where:

- There is insufficient time, or insufficient information at tender stage, for tenderers to properly scrutinise and check the Employer's Requirements, or for them to carry out their designs, risk assessment studies and estimating.
- Under the Silver Book, the Contractor is responsible for the consequences of encountering unforeseeable ground conditions, and also he is responsible for certain aspects of the Employer's Requirements, such as the applicability of ISO standards, so tenderers need information on the matters related to such risks and they need time to assess it and to evaluate all risks.

These matters will be discussed in detail later in the book:

- Construction will involve substantial work underground or work in other areas which tenderers cannot inspect for these types of works, and the risks of encountering unforeseen conditions may be so great that the lowest tender is the one submitted by the least knowledgeable tenderer or most reckless gambler.
- The Employer intends to supervise closely or control the Contractor's work, or to review most of the construction drawings: with the more Contractor's risks, he needs to have greater freedom of action and less interference by the Employer.

- The amount of each interim payment is to be determined by an official or other intermediary. The Silver Book does not provide for an "Engineer" to administer the Contract and determine the amount of each monthly (or other) interim payment, which needs to be determined in advance and defined in a Schedule of Payments.

Is it the sort of project where the Concessionaire takes total responsibility for the financing, construction and operation of the Project?

In this case, the Concessionaire (the "Employer") probably requires an EPC (Engineer, Procure, Construct) Contract with the construction Contractor, where the Contractor takes <u>total</u> responsibility for the design and construction of the facility, and where there is a higher degree of certainty that the agreed Contract Price and time will not be exceeded.

Also, that the Employer does not wish to be involved in the day-to-day progress of the work, provided the end result meets the performance criteria he has specified.

And also, the Parties concerned, e.g. sponsors, lenders and the Employer, are willing to pay more for the construction of the Project in return for the Contractor bearing the extra risks associated with enhanced certainty of final price and time.

If this is what is wanted choose the *Conditions of Contract for EPC/Turnkey Projects* (EPC/Turnkey Contract) (Silver Book).

Is it a Process or Power Plant or Infrastructure Project (e.g. highways, rail, water or drainage/sewage treatment plant, electrical power line) where the Employer wishes the Project to be delivered on a Fixed-Price Turnkey Basis?

Note if the Employer wishes the Contractor to take <u>total</u> responsibility for the design and construction of the process or power facility and hand it over ready to operate "at the turn of a key" it is normally referred to as a Turnkey Project.

The Employer:

- Wishes to have a higher degree of certainty that the agreed Contract Price and time will not be exceeded;
- Wishes the Project to be organised on a strictly two-party approach i.e. without an "Engineer" being involved;
- Does not wish to be involved in the day-to-day progress of the construction work, provided the end result meets the performance criteria he has specified;
- Is prepared to pay more for the construction of his Project (compared to a simple Design and Build Contract) in return for the Contractor bearing the extra risks associated with enhanced certainty of final price and time.

If this is what is wanted, the Employer should choose the *Conditions of Contract for EPC/Turnkey Projects* (Silver Book).

Is it a Building Project where the Employer wishes to have his building(s) constructed on a Fixed-Price Turnkey Basis complete with all furniture, fittings and equipment already installed?

In the case of a building, the Employer or his Architect may have done some or most of the design prior to appointing the Contractor, but with suitable modification regarding design responsibility the Silver Book may be chosen.

3 Obligations and responsibilities of the Employer, the Employer's Representative, the Engineer and the Contractor

This chapter will review the roles and responsibilities of the various parties under the FIDIC Red, Yellow and Silver Books:

- The Employer;
- The Engineer (Red and Yellow Books)/The Employer's Representative (Silver Book);
- The Contractor.

The Employer

Clause 2 in the FIDIC 2017 Contracts contains the primary roles and responsibilities of the Employer under the FIDIC 2017 Contracts.

The Employer and the Contractor are the contracting parties (referred to in the Contract as a "Party" or the "Parties").

Although the Employer is a contracting party, like other construction contracts, the Employer has very few direct actions stated under the FIDIC 2017 Contracts.

The Employer's fundamental responsibilities under any construction contract are:

- Giving the Contractor right of access to, and/or possession of, parts of the Site as required;
- Making payment to the Contractor as certified by the Engineer/Employer's Representative;
- Not preventing or hindering the Contractor in meeting his contractual obligations;
- Not interfering with the duties of the Engineer/Employer's Representative or with certification.

Note, however, that the Silver Book refers frequently to "the Employer", but the actions under which the Employer is named will usually be carried out by the Employer's Representative (see below).

In writing this book, for clarity, the author has consistently used the word "Employer" where the Contract states that though the inference is that the Employer's Representative may actually carry out the actions on behalf of the Employer.

Right of access

Under Sub-Clause 2.1, the Employer is required give the Contractor right of access to, and possession of, all parts of the Site within the times stated in the Contract Data. In reality the Employer may not be able to give access and possession to all parts of the Site, and the Contractor might not need it all at once either.

NB: "Contract Data" is in Part A of the Particular Conditions, formerly referred to in the FIDIC 1999 Contracts as the "Appendix to Tender".

Note that this right of access and possession of the Site may not be exclusive to the Contractor, and particularly with a large project, he may have to share the Site with other contractors and other occupants of the Site.

This is an important point, as the Contractor may need to co-ordinate its activities with other contractors or occupants and is deemed to have priced and programmed the Works accordingly.

Note also that the Employer may withhold any such right or possession until the Performance Security has been received. Most Employers tend to insist on the Performance Security having been received before the Contractor starts work.

If no such time is stated in the Contract Data, the Employer is required to give the Contractor right of access to, and possession of, those parts of the Site within such times as may be required to enable the Contractor to proceed in accordance with the Contractor's Programme submitted under Clause 8.3.

Giving access and possession of the Site is a fundamental obligation upon the Employer. Late possession is an Employer Risk Event, and if the Employer does not give access and possession, and the Contractor suffers delay and/or incurs cost as a result of the failure, the Contractor is entitled to an Extension of Time and/or payment of Cost plus Profit.

Note that the Contractor is not entitled to an Extension of Time and/or payment of Cost plus Profit, purely because the Employer is late in giving possession and the amount of that delay e.g. 1 week delay in giving possession = 1 week Extension of Time with all Costs, the Contractor has to prove that he suffers delay and/or incurs Cost, to which Profit is added.

Under the FIDIC 2017 Contracts, the Employer has some additional duties and responsibilities, which are quite wide ranging and broadly covered by Clause 2 of the Contract.

Employer's use of Contractor's Documents

The Employer may use the Contractor's Documents under a non-terminable transferable non-exclusive royalty-free licence to copy, use and communicate as required under the Contract, but the Contractor retains the copyright and any intellectual property rights in these documents. The licence applies throughout the operational life of the relevant parts of the Works and extends to anyone who is in possession of the relevant part of the Works.

In the event of termination of the Contract by the Employer, the Employer may copy, use and communicate the Contractor's Documents for which the Contractor has received payment for the purpose of completing the Works and/or arranging for any other entities to do so.

The Contractor's Documents must not, without the Contractor's prior consent, be communicated to a third party by other than those permitted under the Contract.

Contractor's use of Employer's Documents

Similarly, the Contractor may at his own cost, copy, use and communicate these documents for the purposes of the Contract.

The Employer's Documents must not, without the Employer's prior consent, be copied, used or communicated to a third party by the Contractor, except as necessary for the purposes of the Contract.

The Employer retains copyright and other intellectual property rights in the Specification and Drawings and other documents made by (or on behalf of) the Employer.

Assistance

Under Sub-Clause 2.2, if the Contractor requests, the Employer is required to provide reasonable assistance to allow the Contractor to obtain:

- a Copies of the Laws of the Country which may not be readily available. They should be available, and it is the Contractor's responsibility to obtain them, but the Employer, particularly if in the public domain e.g. a government department, may be in the position to assist and expedite that availability.
- b Any permits, permissions, licences or approvals required by the Laws of the Country, which the Contractor is required to obtain for the delivery of Goods and/or for the export of Contractor's Equipment when it is removed from the Site. Again, it is the Contractor's responsibility to obtain them, the Employer is only assisting.

As stated, the Employer is only giving assistance if requested, it is fundamentally the Contractor's obligation to obtain them and to be fully aware of their content.

Employer's Personnel and other Contractors

Sub-Clause 2.3 requires that the Employer is responsible for ensuring that the Employer's Personnel and the Employer's other contractors (if any) on or near the Site co-operate with the Contractor's efforts under Sub-Clause 4.6 [Co-operation], and comply with the same obligations which the Contractor is required to comply with in terms of health and safety and/or environmental issues.

The Contractor may also require the Employer to remove any of the Employer's Personnel or of the Employer's other contractors who is found to have engaged in corrupt, fraudulent, collusive or coercive practice. Terms such as "corrupt, fraudulent, collusive or coercive practice" have legal connotations so the Parties should seek legal advice in this respect.

Employer's financial arrangements

The FIDIC 1999 Contracts introduced a new provision that, if requested by the Contractor, the Employer must provide evidence that he has, or will have the finance available to pay for the project, the evidence having to be provided within 28 days of any request by the Contractor. Also, if the Employer makes any material change to his financial arrangements he had to notify the Contractor with details.

Under Clause 16.1, if the Employer failed to comply, the Contractor can give 21 days' notice to suspend work. Also, if the Contractor did not receive reasonable evidence within 42 days of giving a notice to suspend work, the Contractor could terminate the Contract.

Under FIDIC 2017, Sub-Clause 2.4 has been expanded in that the Employer is required to describe his financial arrangements within the Contract Data and must notify the Contractor if there is a material change.

Also, if the Contractor requests that financial arrangements are in place for payment of Variations if they exceed the stated thresholds i.e. 10% of the Accepted Contract Amount, or the accumulated total of Variations exceeds 30% of the Accepted Contract Amount, or if the Contractor does not receive payment in accordance with Sub-Clause 14.7, or if the Contractor becomes aware of a material change in the Employer's financial arrangements of which the Contractor has not received a Notice giving details.

For any of the above issues the Contractor may request and the Employer within 28 days after receiving this request, must provide reasonable evidence that financial arrangements have been made and are being maintained which will enable the Employer to pay the part of the Contract Price remaining to be paid at that time.

Note that many Employers delete this provision from their contracts.

Site Data and items of reference

Under Sub-Clause 2.5, the Employer must have made available to the Contractor for his information, before the Base Date (which is 28 days before the tender submission date), all relevant data in the Employer's possession on the topography (the general arrangement, levels and physical features) of the Site and on sub-surface, hydrological, climatic and environmental conditions at the Site.

The Employer is also required to promptly make available to the Contractor all such data which comes into his possession after the Base Date.

Note that this provision only requires the Employer to share the information he already has in his own possession, and he does not have to obtain any additional information to fulfil the requirement.

Items of reference such as levels will normally be shown on the Drawings.

Note that, whilst Sub-Clause 2.5 looks very similar under the Yellow and Silver Books, the final paragraph which has been added in the Silver Book states that "*the Employer shall have no responsibility for the accuracy, sufficiency or completeness of such data and/or items of reference except as stated in Sub-Clause 5.1 [General Design Obligations]*". This additional paragraph can present a high risk to the Contractor!

Employer-Supplied Materials and Employer's Equipment

If there are Employer-Supplied Materials and/or Employer's Equipment which are listed in the Specification for the use of the Contractor in carrying out the Works (Sub-Clause 2.6), the Employer must make them available to the Contractor in accordance with the details, times, arrangements, rates and prices as stated in the Specification.

The Contractor is responsible for each item of Employer's Equipment and for any loss or damage occurring whilst any of the Contractor's Personnel is operating it, driving it, directing it, using it or in control of it.

The Engineer/Employer's Representative

See also specific roles of the Engineer or Employer's Representative in later chapters.

Note as stated above, the Silver Book refers frequently to "the Employer", but the actions under which the Employer is named will usually be carried out by the Employer's Representative. For clarity as stated previously, the author has consistently used the word "Employer" where the Contract states that.

The roles are essentially the same, but the Engineer has certain qualifications and experience as required by the Contract, which the Employer's Representative may not have. Note that an Engineer by profession may act as the Employer's Representative, but he is not acting as "the Engineer" under the Contract. An Employer's Representative cannot act as the Engineer, unless he actually is an Engineer.

One question to be considered is when should the Engineer (Red and Yellow Books) or Employer's Representative (Silver Book) be appointed?

Whilst the Contract can only refer to the Engineer/Employer's Representative acting from the point when the Contractor is appointed i.e. when the FIDIC contract comes into existence, they are not necessarily appointed at that stage, though their role is fundamentally different if they are not.

Concept/design stage

This is not strictly the stage at which an Engineer/Employer's Representative is expected to be appointed, it is more often the role of Designers formulating design alternatives and Quantity Surveyors/Cost Consultants producing budgets and cost plans, but if it is the case, it may require the Engineer/Employer's Representative to be involved in the design process and various decisions related to it.

The Engineer/Employer's Representative would be appointed in advance of the design team in most instances, and would assist the Employer in the selection and appointment of the design team, the Designer(s) and the Contractor.

Tender stage

The Engineer/Employer's Representative may be appointed at tender stage when the Employer may have developed the design, or at least the requirements, and wishes the Engineer/Employer's Representative to manage the tender process and following appointment of the Contractor into the construction phase, though often a Quantity Surveyor is appointed to support the Employer and the Engineer/Employer's Representative.

The Engineer/Employer's Representative would provide assistance in selection of the Contractor and continue with the contract role of the Engineer/Employer's Representative for the remainder of the contract.

Contract stage

In some instances, the Employer will progress the design, or at least the requirements, and appoint a Contractor, but require an Engineer/Employer's Representative to undertake the contract role of the Engineer/Employer's Representative for the remainder of the process, and in accordance with the FIDIC contract.

The roles and responsibilities of the Engineer/Employer's Representative

Before discussing specific issues and differences regarding the FIDIC 2017 Red, Yellow and Silver Books it is worth noting that FIDIC has amended the role of the Engineer in the Red and Yellow Books (the Silver Book has an Employer's Representative) from the impartial, quasi-arbitral role of previous editions.

Note that under the FIDIC 2017 Silver Book the Employer is obliged to appoint an "Employer's Representative", whereas under FIDIC 1999 it was just an option.

It is now an obligation, so that the Contractor and the Contractor's Representative have a single point of contact acting for the Employer, rather than a corporate entity called the "Employer", which is particularly important for giving of instructions, valuing the Works, making decisions on quality, etc., also there is a focal point for dealing with the Parties' claims, whether they be from the Contractor or the Employer.

The duties of the Employer's Representative are similar to that of an Engineer, but because the Contractor is responsible for all of the design it is felt that there is no need for a qualified Engineer to act for the Employer.

If the Engineer/Employer's Representative is a legal entity i.e. a company, then a person employed by the Engineer/Employer's Representative acts on its behalf.

It is very important that when carrying out his duties under the Contract, the Engineer/Employer's Representative acts neutrally between the Parties and not to act solely for the Employer.

There is no problem with the Engineer/Employer's Representative being from the same organisation as the Employer, so there is no requirement to be truly independent, but he must be impartial on all matters.

The Engineer/Employer's Representative must have the necessary authority to act for the Employer on all matters under the Contract.

The Contract specifically provides that the Engineer is required to be:

a A professional Engineer having suitable qualifications experiences and competences to act as the Engineer under the Contract. This is a critical requirement on that someone who is not a qualified Engineer cannot take the title "Engineer" under the contract. The Engineer is not specifically required to be based on Site, unlike an Engineer's Representative who is required to be based on Site;
b Fluent in the ruling language defined in the Contract.

If the Engineer is to be changed, the Employer must comply with Sub-Clause 3.4.

The Engineer may delegate any of his duties at any time by notifying the Contractor in writing which may later be revoked, again in writing. It is important to note that the authority of the delegated representative is limited to the authority contained within the delegation notice.

The Contractor is required to comply with any instruction given by the Engineer. Instructions should only be taken from that person or any previously notified delegate. Wherever possible this should be in writing, though there is a fallback situation where the instruction is oral, it can be confirmed by the Contractor within two working days and if not dissented from within two further working days it becomes an instruction.

The Engineer has no authority to amend the Contract.

If there is any requirement for the Engineer to obtain the consent of the Employer on any matter this must be stated in the Particular Conditions.

There is a new provision that if an Engineer exercises a specified authority for which the Employer's consent is required, then consent is deemed to have been given.

Any acceptance, agreement, approval, check, certificate, Notice, No-objection, etc., by the Engineer, the Engineer's Representative or any assistant does not relieve the Contractor from any duty, obligation or responsibility he has under the Contract.

Notices

One of the key changes within the FIDIC 2017 Contracts is the issue of Notices, of which there are far more than in the FIDIC 1999 Contracts, some with time limits for giving, or not giving, Notices.

The FIDIC 2017 Contracts define Notices under Sub-Clause 1.1.56 (Silver Book Sub-Clause 1.1.48) as a written communication identified as a Notice and issued in accordance with Sub-Clause 1.3 [Notices and Other Communications].

So, the Notice must be in writing, and clearly identifiable as a "Notice", not just a communication between the Parties.

For example, under new Sub-Clause 3.5 (Yellow/Red) and Sub-Clause 3.4 (Silver), if an instruction is a Variation the Engineer (Yellow/Red) or Employer (or Employer's Representative) (Silver) should state explicitly that it is a Variation.

If it is not stated as a Variation, but the Contractor thinks that the instruction constitutes a Variation, then the Contractor must immediately give a Notice to the Engineer/Employer's Representative to that effect, and give reasons.

If the Engineer (Employer under the Silver Book) does not respond within seven days, by giving another Notice (confirming, reversing or varying the instruction), he is deemed to have revoked the instruction.

Sub-Clause 1.3 (Notices and Other Communications) also reinforces the Contract requirements in terms of Notices.

- Sub-Clause 1.3(a) requires a Notice to be a paper-original, signed by the Contractor's Representative, the Engineer or the authorised representative of the Employer, or an electronic original generated from any of the systems of electronic transmission stated in the Contract Data.
- Sub-Clause 1.3(b) also states that a Notice should be identified as a Notice as opposed to another form of communication. A Notice need not refer to the clause under which it is issued, whereas another form of communication must do so.

However, it is important to bear in mind that a Notice given under certain clauses must state that it is given under the relevant clause, for example, Sub-Clauses 15.2.1 [Termination for Contractor's Default: Notice], 15.5 [Termination for Employer's Convenience], 16.1 [Suspension by Contractor] and 16.2.1 [Termination by Contractor Notice].

Also, all notices and other types of communication shall "not be unreasonably withheld or delayed".

If a Notice is issued by a Party or the Engineer, the Notice must also be issued to the other Party and/or the Engineer as appropriate.

There are certain actions required of the Engineer/Employer's Representative in FIDIC 2017 Contracts.

The FIDIC 2017 Contracts introduce a new term "No-objection", which is defined in Sub-Clause 1.1.55 (Silver Book Sub-Clause 1.1.48) as meaning that "*the Engineer/Employer has no objection to the Contractor's Documents, or other documents submitted by the Contractor under these Conditions, and such Contractor's Documents or other documents may be used for the Works*".

No-objection is a term frequently experienced by the author when working on projects in the Middle East, where the Engineer or Employer's Representative, having reviewed the work or document(s) submitted, does not actually "approve" or "accept", but does not object to it.

Cynics may say that the Engineer/Employer's Representative is trying to absolve himself of liability, by only giving "No-objection" rather than formally approving or accepting or even rejecting, but what he is doing by giving No-objection is not absolving the Contractor of his liability, and this is how the FIDIC 2017 No-objection Notice should be viewed and applied.

"Review" is defined as an "*examination and consideration by the Engineer/Employer's Representative of a Contractor's submission in order to assess whether (and to what extent) it complies with the Contract and/or with the Contractor's obligations under or in connection with the Contract*".

The Engineer's Representative

The Engineer may appoint an Engineer's Representative and delegate to him the authority necessary to act on the Engineer's behalf at the Site, except to replace the Engineer's Representative.

The Engineer's Representative (if appointed) is required to be based at the Site for the whole time that the Works are being executed at the Site.

The role of Engineer's Representative existed in older versions of the FIDIC contracts, but this was not included within the FIDIC 1999 Contracts, and it has been reintroduced into the FIDIC 2017 Contracts.

The Engineer's Representative is not required to be an Engineer but must be based at the Site for the whole time the Works are being executed, and therefore, the Engineer should consider this requirement in appointing an Engineer's Representative.

It is important to recognise that in delegating, the Engineer is sharing an authority with the delegate, be they an Engineer's Representative or an assistant, who will then represent him, it is not passing ultimate responsibility on to them. The authority is shared, and the Engineer may still take an action which they have delegated, but ultimately responsibility will remain with the Engineer as the principal.

It is also important to note that one can only delegate outwards from the principal i.e. no-one can assume authority possibly because, within an organisation, they are senior to the Engineer.

Delegation by the Engineer

There is only one Engineer for the Contract, but he may delegate authority to assistants by giving a Notice to the Parties, describing the duties and the delegated authority of each assistant, for example designers, inspectors, etc. The delegation must not take effect until this Notice has been received by both Parties.

However, there are two matters which the Engineer cannot delegate to anyone else:

a Acting under Sub-Clause 3.7 [Agreement or Determination]; and/or
b Issuing a Notice to Correct under Sub-Clause 15.1 [Notice to Correct].

Assistants shall be suitably qualified natural persons, who are experienced and competent to carry out these duties and exercise this authority, and who are fluent in the language defined in the Contract.

Engineer's instructions

The Engineer (Employer under the Silver Book), or his delegated representative, may issue instructions to the Contractor, which may be necessary for the execution of the Works.

It is critical that the Contractor only takes instructions from the Engineer (Employer under the Silver Book), or from the Engineer's Representative, or an assistant to whom the authority has been delegated to give instructions.

If the instruction constitutes a Variation, Sub-Clause 13.3.1 applies and it is dealt with as a Variation in the normal way.

If the Engineer (Employer under the Silver Book) does not state that the instruction constitutes a Variation, and this could be quite a common occurrence, and the Contractor considers that the instruction constitutes a Variation, or does not comply with applicable Laws, or will reduce the safety of the Works, or is technically impossible issues a Notice to the Engineer (Employer under the Silver Book) giving reasons.

If the Engineer (Employer under the Silver Book) does not respond to the Notice within seven days after receiving it, the Engineer is deemed to have revoked the instruction. If the Engineer responds, the Contractor is required to comply with the response.

This is a change from FIDIC 1999, and it is hoped that it will bring matters to a head quicker and will give the matter of Engineer's (Employer under the Silver Book) Instructions, and their status, some clarity, which is an advantage to the Contractor, the Engineer (Employer under the Silver Book) and also to the Employer who has to plan and budget for the changes.

Note that if a Contractor believes that an instruction from the Engineer (Employer under the Silver Book) constitutes a Variation, but the Engineer (Employer under the Silver Book) does not, although the Contract does not expressly deal with the issue, the Contractor should not refuse to comply with the instruction unless or until the Engineer confirms that it constitutes a Variation. The Contractor should comply with the instruction and raise the question of whether it is a Variation as a Claim and/or Dispute.

Replacement of the Engineer

The Engineer is required to notify the Contractor at least 42 days prior to replacing the Engineer including the name, address and relevant experience of the intended replacement Engineer.

The Contractor is given a right of objection, but if he does not respond within 14 days after receiving the Notice, by stating any objection with reasons, the Contractor is deemed to have accepted the replacement. The Employer must not replace the Engineer with a person against whom the Contractor has raised reasonable objection by a Notice.

Often this Sub-Clause is amended as the Employer would prefer the Contractor to have no right of objection, or to be seen as having any choice as to who should be the replacement Engineer, but deletion of this provision would give tenders reasonable concern regarding the Employer's intentions.

If the Engineer is unable to act as a result of death, illness, disability or resignation, the Employer is entitled to immediately appoint a replacement by giving a Notice to the Contractor with reasons and the name, address and relevant experience of the replacement. This appointment is treated as a temporary appointment until this replacement is accepted by the Contractor, or another replacement is appointed, under this Sub-Clause. This is a new feature within FIDIC 2017.

Agreement or Determination

The process of Agreement or Determination is dealt with in detail within Chapter 10 Employer's and Contractor's Claims.

As stated above, it is critical that the Engineer (Employer's Representative under the Silver Book) acts neutrally between the Parties and not to act for the Employer.

Whenever these Conditions provide that the Engineer is required to agree or determine any matter or Claim, the following procedures apply:

Consultation to reach agreement

The Engineer (Employer's Representative under the Silver Book) is required to consult with both Parties jointly and/or separately and encourage discussion between the Parties in an endeavour to reach agreement on the matter.

The Engineer (Employer's Representative under the Silver Book) provides both Parties with a record of the consultation.

If agreement is achieved, within the time limit for agreement, the Engineer (Employer's Representative under the Silver Book) gives a Notice to both Parties of the agreement, the agreement is signed by both Parties. This Notice is required to state that it is a "Notice of the Parties' Agreement" and must include a copy of the agreement.

If:

a No agreement is achieved within the time limit for agreement, or
b Both Parties advise the Engineer (Employer's Representative under the Silver Book) that no agreement can be achieved within this time limit whichever is the earlier, the Engineer (Employer's Representative under the Silver Book) shall give a Notice to the Parties accordingly proceeds to make a determination.

Engineer's (Employer's Representative under the Silver Book) determination

The Engineer (Employer's Representative under the Silver Book) is required to make a <u>fair determination</u> of a matter or Claim, and give a Notice to both Parties of his determination.

The Notice must state that it is a "Notice of the Engineer's/Employer's Representative's Determination" and describes the determination in detail with reasons and supporting particulars.

Time limits

The Engineer (Employer's Representative under the Silver Book) is required to give a Notice of agreement within 42 days or within such other time limit as proposed by the Engineer and agreed by <u>both</u> Parties.

The Engineer (Employer's Representative under the Silver Book) must give the Notice of his determination within 42 days or within such other time limit as may be proposed by the Engineer and agreed by <u>both</u> Parties.

If the Engineer (Employer's Representative under the Silver Book) does not give the Notice of agreement or determination within the relevant time limit:

i In the case of a Claim, the Engineer (Employer's Representative under the Silver Book) is deemed to have rejected the Claim;
   or
ii The matter is deemed to be a Dispute which may be referred by either Party to the DAAB for its decision.

Effect of the agreement or determination

Each agreement or determination is binding on both Parties and the Engineer (Employer's Representative under the Silver Book) unless and until it is corrected or, in the case of a determination, it is revised under Clause 21 (Disputes and Arbitration).

If, within 14 days after giving or receiving the Engineer's (Employer's Representative under the Silver Book) Notice of agreement or determination, any error of a typographical or clerical or arithmetical nature is found by the Engineer (Employer's Representative under the Silver Book), then he immediately advises the Parties.

If found by a Party, then that Party gives a Notice to the Engineer (Employer's Representative under the Silver Book) identifying the error. If the Engineer (Employer's Representative under the Silver Book) does not agree there was an error, he immediately advises the Parties.

The Engineer (Employer's Representative under the Silver Book) shall, within seven days of finding the error or receiving a Notice under sub-paragraph (b) above (as the case may be), give a Notice to both Parties of the corrected agreement or determination. Thereafter, the corrected agreement or determination shall be treated as the agreement or determination for the purpose of these Conditions.

Dissatisfaction with Engineer's (Employer's Representative under the Silver Book) determination

If either Party is dissatisfied with a determination of the Engineer (Employer's Representative under the Silver Book) the dissatisfied Party may give a Notice of Dissatisfaction (NOD) to the other Party, with a copy to the Engineer (Employer's Representative under the Silver Book) setting out the reason(s) for dissatisfaction, this NOD must be given within 28 days after receiving the Engineer's determination (Employer's Representative under the Silver Book).

If no NOD is given by either Party within the period of 28 days, the determination of the Engineer (Employer's Representative under the Silver Book) is deemed to have been accepted by both Parties and is then final and binding.

If the dissatisfied Party is dissatisfied with only part(s) of the Engineer's (Employer's Representative under the Silver Book) determination the part(s) must be clearly identified in the NOD and this part(s) is deemed to be severable from the remainder of the determination, and the remainder of the determination becomes final and binding on both Parties.

In the event that a Party fails to comply with an agreement of the Parties under this or a final and binding determination of the Engineer (Employer's Representative under the Silver Book), the other Party may refer the failure itself directly to arbitration.

Meetings

The FIDIC 2017 Contracts have new provision for meetings, to try to promote effective communication, co-operation and collaboration between the parties.

The Engineer (Employer under the Silver Book) or the Contractor's Representative may require the other to attend meetings. This is another example of the more collaborative wording of the FIDIC 2017 Contracts.

Others, for example the Employer's other contractors, the personnel of legally constituted public authorities and/or private utility companies and/or Subcontractors, may also attend any such meeting, if requested by the Engineer (Employer under the Silver Book) or the Contractor's Representative.

The Engineer (Employer under the Silver Book) is required to keep a record of each management meeting and supplies copies of the record to those attending, and to the Employer. There is no requirement for the Contractor to sign or accept the record, though he should notify the Engineer (Employer under the Silver Book) of anything that is incorrect or inaccurate.

At any such meeting, and in the record, responsibilities for any actions to be taken must be in accordance with the Contract.

Issue of drawings or instructions

Under the Red Book, the Engineer has a fundamental obligation to provide drawings and/or instructions to the Contractor to allow him to make progress in executing the Works.

If a delay occurs due to the Contractor not receiving any drawings or instruction in time, the Contractor must, under Sub-Clause 1.9, notify the Engineer, including details of what drawing and/or instruction is late, and the delay or disruption caused by that lateness likely to be suffered if it is late.

The Contractor is entitled to an Extension of Time and/or payment of Cost plus Profit; however, the Contractor would not be entitled to any relief if the delay and/or disruption was caused by an error in, or delay in the submission of, any of the Contractor's Documents.

Note that under the Yellow Book and the Silver Book, the Engineer and the Employer, respectively, have different obligations, as they both refer to the Employer's Requirements, so in a sense although there is a Sub-Clause 1.9 it has different obligations.

Under Sub-Clause 1.9 of the Yellow Book, if the Contractor finds an error, fault or defect within the Employer's Requirements as a result of scrutinising them under Sub-Clause 5.1 (i.e. promptly after receiving the Notice from the Engineer stating the Commencement Date (Clause 8.1)), the Contractor gives a Notice to the Engineer, and the Engineer agrees or determines whether or not there is an error and whether an experienced Contractor would have discovered the error, fault or defect before submitting the tender and if an experienced Contractor would not have discovered the error, fault or defect prior to tender, it is judged to be a Variation, and the Contractor, if he suffers delay and/or incurs Cost as a result, is entitled to an Extension of Time and/or payment of Cost plus Profit.

A critical difference is that Sub-Clause 1.9 [Errors in the Employer's Requirements] is included within the Yellow Book, but does not exist within the Silver Book, and this is a major risk for the Contractor.

Other specialists

One role that people often question is that of the "Quantity Surveyor" or "Cost Consultant".

Whilst the FIDIC contracts do not expressly refer to the role of a Quantity Surveyor or a Cost Consultant, it is not unusual, particularly on larger or more complex projects, for the Employer to appoint one pre-contract and/or post-contract to deal with financial matters including the following:

- Pre-contract
- Preparation of Bills of Quantities;
- Pre-contract Cost Planning;
- Preparation of Tender Documents.

Post-contract (on behalf of the Engineer/Employer's Representative)

- Preparing Valuations and dealing with payments;
- Evaluating Variations;
- Evaluating Loss and Expense claims;
- Financial Reporting.

It is important not to expect the Quantity Surveyor or Cost Consultant to deal with all matters of a financial nature.

Employers should also consider employing other specialists on major projects, for examples a forensic delay analyst for complex delays, and a quantum analyst for quantifying additional or changed resources and/or disruption, due to major design changes. These can be appointed as and when required.

The Contractor

The Contractor and the Employer are the contracting parties (referred to in the Contract as a "Party" or the "Parties").

Clause 8 provides the basic statement of what the Contractor agrees to do.

Essentially, this is to complete the Works, and this is an absolute obligation not just that he will do his best. However, the absolute nature of this is substantially reduced by the Works "subject to the provisions of the Contract".

Sub-Clause 4.1 of the Red Book states the Contractor's obligation to execute the Works and requires any Contractor-designed Works to be (when completed) fit for the intended purpose defined in the Contract. Note that most design and build contracts provide for the design to be reasonable skill and care.

This is an important matter to be borne in mind when the tender documents are being written, in particular because of the absence of any reference to purposes which may reasonably be inferred from the Contract. Although the risk is fairly allocated to the Contractor, he may not be able to be indemnified by his insurers or his subcontract designers.

Contractor's design

Under the Red Book, the Contractor is only responsible for the design of the Permanent Works to the extent specified in the Contract.

Note that Contractor's design is discussed in detail in Chapter 5 but is also summarised here.

In the Red Book, Clause 4.1 sub-paragraphs (a) to (f) specify general requirements in respect of such designs, unless they are defined otherwise in the Particular Conditions. These general requirements are also stated in 13.3 (b) as being applicable to any subsequent Contractor-designed "Value Engineering" proposals, unless otherwise agreed by both Parties.

If the Contract specifies that the Contractor is required to design any part of the Permanent Works, then unless otherwise stated in the Particular Conditions:

a The Contractor prepares, and submits to the Engineer for Review, the Contractor's Documents for this part and implements the design during the execution of the Works.

b The Contractor's Documents are required to be in accordance with the Specification and Drawings and to include additional information required by the Engineer to add to the Drawings for the co-ordination of each Party's designs.

  If the Engineer instructs that further Contractor's Documents are reasonably required to demonstrate that the Contractor's design complies with the Contract, the Contractor prepares and submits them promptly to the Engineer at the Contractor's cost.

c Construction of this part cannot commence until a Notice of No-objection is given by the Engineer for all the Contractor's Documents which are relevant to its design and construction of such part.

d If the Contractor has commenced construction of the part of the Works to which such design or Contractor's Documents are relevant, work on this part is suspended, and work shall not resume until a Notice of No-objection is given by the Engineer for the revised documents.

e The Contractor shall be responsible for this part and it shall, when the Works are completed, be fit for such purpose(s) for which the part is intended as are specified in the Contract.

f In addition to the Contractor's undertaking above, the Contractor undertakes that the design and the Contractor's Documents for this part will comply with the technical standards stated in the Specification and Laws (in force when the Works are taken over.

Performance Security

Sub-Clause 4.2 requires the Contractor to provide a Performance Security at his own cost in the amount specified in the Appendix to Tender. Provisions are included for extending the security, and for the Employer's indemnity in respect of any claim which he was not entitled to make.

The contract specifies that *"the Performance Security shall be … in the form annexed to the Particular Conditions or…"*. When preparing the tender documents, the Employer must decide what wording to annex.

The Employer may instruct the Contractor to increase or decrease the Performance Security in the event of Variations that result in a cumulative increase or decrease by more than 20% of the Contract Price.

The Contractor must obtain (at his own cost) a Performance Security to secure his proper performance of the Contract, in the amount and currencies stated in the

Contract Data. If no amount is stated in the Contract Data, the Sub-Clause does not apply.

The Contractor is required to deliver the Performance Security to the Employer, with a copy to the Engineer, within 28 days after receiving the Letter of Acceptance. The Performance Security is to be issued by an entity (e.g. a Bank), and from within a country (or other jurisdiction) to which the Employer gives consent and must be in the form referred to in the Particular Conditions, or in another form agreed by the Employer (but such consent and/or agreement shall not relieve the Contractor from any obligation under this Sub-Clause).

The Contractor must ensure that the Performance Security remains valid and enforceable until the issue of the Performance Certificate and the Contractor has cleared the Site.

If the terms of the Performance Security specify an expiry date, and the Contractor is not entitled to receive the Performance Certificate by the date 28 days before the expiry date, the Contractor extends the validity of the Performance Security until the issue of the Performance Certificate.

Whenever Variations and/or adjustments under Clause 13 [Variations and Adjustments] result in an accumulative increase or decrease of the Contract Price by more than 20% of the Accepted Contract Amount:

a At the Employer's request the Contractor must promptly increase the amount of the Performance Security by a percentage equal to the cumulative increase. If the Contractor incurs Cost as a result of this Employer's request, it is classed as a Variation as if the increase had been instructed by the Engineer.
b In the case of such a decrease, subject to the Employer's prior consent, the Contractor may decrease the amount of the Performance Security in that currency by a percentage equal to the cumulative decrease.

The Employer must not make a claim under the Performance Security, except for amounts to which the Employer is entitled under the Contract in the event of:

a Failure by the Contractor to extend the validity of the Performance Security, as described in this Sub-Clause, in which case the Employer can claim the full amount of the Performance Security;
b Failure by the Contractor to pay the Employer an amount due, as agreed or determined under Sub-Clause 3.7 [Agreement or Determination] or agreed or decided under Clause 21 [Disputes and Arbitration], within 42 days after the date of the agreement or determination or decision or arbitral award (as the case may be);
c Failure by the Contractor to remedy a default stated in a Notice given under Sub-Clause 15.1 [Notice to Correct] within 42 days or other time (if any) stated in the Notice;
d Circumstances which entitle the Employer to terminate the Contract, or if the Contractor removes any defective or damaged Plant from the Site, failure by the Contractor to repair such Plant, return it to the Site, reinstall it and retest it by the date of expiry of the relevant duration stated in the Contractor's Notice. The Employer indemnifies and holds the Contractor harmless against and from all damages, losses and expenses resulting from a claim under the Performance Security to the extent that the Employer was not entitled to make the claim.

Any amount which is received by the Employer under the Performance Security must be taken into account:

i In the Final Payment Certificate, or
ii If the Contract is terminated.

The Employer must return the Performance Security to the Contractor:

a Within 21 days after the issue of the Performance Certificate;
b Promptly after the date of termination if the Contract is terminated.

Subcontractors

Clause 5/ Sub-Clause 4.4 covers Subcontractors, who are subject to the consent. Note that the Contractor does not have to have consent of the Engineer to suppliers or previously named Subcontractors in the Contract.

It is a normal requirement in contracts for subcontractors to be assigned to the Employer in the event of termination by the Employer.

The Contractor is required to execute the Works in accordance with the Contract, as altered or modified by Variations.

Contractor's Representative

The Contractor is required to appoint a qualified, experienced and competent full-time Contractor's Representative, fluent in the language of the Contract, and with full authority to act for the Contractor.

The Contractor's Representative is responsible for issuing and receiving Notices and other communications on behalf of the Contractor.

The Contractor is required to submit the name and particulars of the Contractor's Representative to the Engineer for consent prior to the Commencement Date.

If the Engineer (Employer under the Silver Book) does not respond within 28 days after receiving the submission, by giving a Notice to the Contractor objecting to the proposed person or replacement, the Engineer (Employer under the Silver Book) is deemed to have given his/her consent.

The Contractor must not, without the Engineer's (Employer under the Silver Book) prior consent, appoint a replacement.

The Contractor's Representative may delegate any powers except the authority to issue and receive Notices and instructions.

Any delegation does not take effect until the Engineer (Employer under the Silver Book) has received a Notice from the Contractor's Representative, naming the person; specifying the powers, functions and authority being delegated or revoked; and stating the timing of the delegation or revocation.

Contractor's Documents

The Contractor provides the Plant and Contractor's Documents specified in the Contract, and all Contractor's Personnel, Goods, consumables and other things and services, whether of a temporary or permanent nature, required to fulfil the Contractor's obligations under the Contract.

The Contractor is be responsible for the adequacy, stability and safety of all the Contractor's operations and activities, of all methods of construction and of all the Temporary Works.

Except to the extent specified in the Contract, the Contractor:

- i Is responsible for all Contractor's Documents, Temporary Works and such design of each item of Plant and Materials as is required for the item to be in accordance with the Contract; and
- ii Shall not otherwise be responsible for the design or specification of the Permanent Works.

The Contractor shall, whenever required by the Engineer, submit details of the arrangements and methods which the Contractor proposes to adopt for the execution of the Works. No significant alteration to these arrangements and methods shall be made without this alteration having been submitted to the Engineer.

Preparation and Review

The Contractor's Documents consist of the documents:

- a Stated in the Specification;
- b Required to satisfy all permits, permissions, licences and other regulatory approvals which are the Contractor's responsibility;
- c Described in Sub-Clause 4.4.2 [As-Built Records] and Sub-Clause 4.4.3 [Operation and Maintenance Manuals].

Unless otherwise stated in the Specification, the Contractor's Documents are written in the language for communications defined in the Contract.

The Contractor prepares all the Contractor's Documents, and the Employer's Personnel shall has the right to inspect during the preparation of all these documents, wherever they are being prepared.

If the Specification states that a Contractor's Document is to be submitted to the Engineer (Employer under the Silver Book) for Review, it must be submitted with a Notice from the Contractor stating that the Contractor's Document is ready for Review and that it complies with the Contract.

The Engineer (Employer under the Silver Book) within 21 days after receiving the Contractor's Document and this Notice from the Contractor gives a Notice to the Contractor of No-objection (which may include comments concerning minor matters), or that the Contractor's Document fails to comply with the Contract, with reasons.

If the Engineer gives no Notice within 21 days, the Engineer (Employer under the Silver Book) is deemed to have given a Notice of No-objection to the Contractor's Document.

After receiving a Notice, the Contractor revises the Contractor's Document and resubmits it to the Engineer (Employer under the Silver Book) for Review in accordance with this Sub-Clause and the period of 21 days for Review shall be calculated from the date that the Engineer receives it.

As-built records

The Contractor must prepare a complete set of "as-built" records if required by the Contract. The format, referencing, system of electronic storage and other relevant details of the as-built records must be as stated within the Specification.

The number of copies of as-built records to be submitted by the Contractor under this Sub-Clause is as required under Sub-Clause 1.8 [Care and Supply of Documents].

The as-built records must be kept on the Site and used exclusively for the purposes of the relevant Sub-Clause.

The as-built records are required to be submitted to the Engineer (Employer under the Silver Book) for Review, and the Works are not considered complete, for the purposes of taking-over, until the Engineer has given (or is deemed to have given) a Notice of No-objection to the as-built records.

Operation and maintenance manuals

If no operation and maintenance manuals to be prepared by the Contractor are stated in the Specification, this Sub-Clause shall not apply.

The Contractor is required to prepare the operation and maintenance manuals in the format stated in the Specification.

The operation and maintenance manuals must be submitted to the Engineer (Employer under the Silver Book) for Review, and the Works are not complete for the purposes of taking-over until the Engineer has given (or is deemed to have given) a Notice of No-objection.

Training

The Contractor may be required to carry out training for the Employer's Personnel in the operation and maintenance of the Works if the Specification states.

If training is required, the Contractor is required to provide qualified and experienced training staff, training facilities and all training materials as necessary or as stated in the Specification.

If the Specification requires training to be carried out before taking over, the Works are not considered to be completed for the purposes of taking over until the training has been completed.

Assignment of benefit of subcontract

Should a Subcontractor's obligations extend beyond the expiry of a Defects Notification Period (DNP), and the Engineer (Employer under the Silver Book) instructs the Contractor to assign the benefit of such obligations to the Employer, the Contractor is required to do so.

Co-operation

This is an unusual clause as it requires the Contractor to allow "appropriate opportunities" for carrying out work to various parties employed by the Employer, or of any public authority. This is presumably to encourage the use of local labour. The Contractor is able to recover any associated cost as "Unforeseeable Cost".

The Contractor must, if stated in the Specification or as instructed by the Engineer, co-operate with and allow appropriate opportunities for carrying out work by:

a The Employer's Personnel;
b Any other contractors employed by the Employer; and
c The personnel of any legally constituted public authorities and private utility companies, who may be employed in the carrying out, on or near the Site, of any work not included in the Contract. Such appropriate opportunities may include the use of Contractor's Equipment, Temporary Works, access arrangements which are the responsibility of the Contractor, and/or other Contractor's facilities or services on the Site.

The Contractor is responsible for all construction activities on the Site and is required to use all reasonable endeavours to co-ordinate these activities with those of other contractors as stated in the Specification or as instructed by the Engineer (Employer under the Silver Book).

Note that, if the Contractor suffers any delay and/or incurs Cost as a result of an instruction, to the extent that co-operation, allowance of opportunities and coordination was Unforeseeable, the Contractor is entitled to an Extension of Time and/or payment of such Cost plus Profit.

Setting out

The Contractor is responsible, and liable, for setting out the Works as required.

The Employer is responsible for providing items of reference e.g. levels or datums, but the Contractor must verify their accuracy.

Any delay and/or cost occasioned by the Employer providing incorrect items of reference, e.g. finished floor levels, is recoverable by the Contractor as an Extension of Time and/or payment of Cost plus Profit.

The Contractor is required to set out the Works in relation to the items of reference under Sub-Clause 2.5.

Accuracy

The Contractor is required to:

a Verify the accuracy of all these items of reference before they are used for the Works.
b Promptly deliver the results of the verification to the Engineer.

c Rectify any error in the positions, levels, dimensions or alignment of the Works.
d Be responsible for the correct positioning of all parts of the Works.

So, ultimately, although the Employer provides the items of reference, it is the Contractor's responsibility to ensure that he is working with the correct information.

Errors

Following the Contractor verifying the accuracy of the items of reference, if he finds an error, he is required to give a Notice to the Engineer describing it:

a Within the period stated in the Contract Data, this period is calculated from the Commencement Date;
b As soon as practicable after receiving the items of reference.

Agreement or determination of rectification measures, delay and/or Cost

After receiving a Notice from the Contractor, the Engineer must agree or determine:

a Whether or not there is an error in the items of reference;
b Whether or not an experienced contractor exercising due care would have discovered such an error when examining the Site, the Drawings and the Specification before submitting the Tender, or when examining the items of reference within the period stated in sub-paragraph (a) of Sub-Clause 4.7.2 if the items of reference are specified on the Drawings and/or in the Specification; and

  Note the wording has been changed from the published Contract by the Errata (see Appendix).

  The previous wording within the published Contract was "whether or not (taking account of cost and time) an experienced contractor exercising due care would have discovered such an error when examining the Site, the Drawings and the Specification before submitting the Tender; or if the items of reference are specified on the Drawings and/or in the Specification and the Contractor's Notice is given after the expiry of the period stated in sub-paragraph (a) of Sub-Clause 4.7.2; and"
c What measures (if any) the Contractor is required to take to rectify the error.

The date the Engineer (Employer under the Silver Book) receives the Contractor's Notice under Sub-Clause 4.7.2 [Errors] shall be the date of commencement of the time limit for agreement under Sub-Clause 3.7.3).

If, under sub-paragraph (b) above, an experienced contractor would not have discovered the error, Sub-Clause 13.3.1 [Variation by Instruction] applies to the measures that the Contractor is required to take, and if the Contractor suffers delay and/or incurs Cost as a result of the error, the Contractor is entitled to an Extension of Time and/or payment of such Cost plus Profit.

Under the Silver Book, the Contractor is required to verify the accuracy of the items of reference as with the Red and Yellow Books in accordance with (a), (c) and (d) above i.e.

- Verify the accuracy of all these items of reference before they are used for the Works.
- Rectify any error in the positions, levels, dimensions or alignment of the Works.
- Be responsible for the correct positioning of all parts of the Works.

But there is no requirement to deliver the results to the Employer/Employer's Representative, and there is no relief in terms of an Extension of Time and/or payment of Cost plus Profit if there are any errors by the Employer!

Health and safety obligations

FIDIC 2017 contains a further developed version of the "safety procedures" required within FIDIC 1999.

The Contractor is required to:

a Comply with all applicable health and safety regulations and Laws, so it is not just any regulations contained within the Contract.
b Comply with all health and safety obligations specified in the Contract.
c Comply with all directives issued by the Contractor's health and safety officer.
d Take care of the health and safety of all persons entitled to be on the Site and other places where the Works are being executed.
e Keep the Site, Works and where the Works are being executed, clear of unnecessary obstruction so as to avoid danger to these persons.
f Provide fencing, lighting, safe access, guarding and watching of the Works, until the Works are taken over, and any part of the Works where the Contractor is executing outstanding works or remedying any defects during the DNP.
g Provide any Temporary Works which may be necessary, because of the execution of the Works, for the use and protection of the public and of owners and occupiers of adjacent land and property.

Within 21 days of the Commencement Date and before commencing any construction on the Site, the Contractor is required to submit to the Engineer for information a health and safety manual which has been specifically prepared for the Works, the Site and other places (if any) where the Contractor intends to execute the Works (it is project specific, not a generic document). This is an additional requirement introduced within the FIDIC 2017 Contracts.

This manual is required to be in addition to any other similar document required under applicable health and safety regulations and Laws.

The health and safety manual must set out all the health and safety requirements, stated in the Specification, comply with all the Contractor's health and safety obligations under the Contract and as necessary to effect and maintain a healthy and safe working environment for all persons entitled to be on the Site and other places (if any) where the Works are being executed.

This manual is required to be revised as necessary by the Contractor or the Contractor's health and safety officer, or at the reasonable request of the Engineer. Each revision of the manual shall be submitted promptly to the Engineer.

In addition to the reporting requirement above, the Contractor must submit to the Engineer details of any accident as soon as practicable after its occurrence and, in the case of an accident causing serious injury or death, must inform the Engineer immediately.

The Contractor is required, as stated in the Specification and as the Engineer may reasonably require, maintain records and make reports (in compliance with the applicable health and safety regulations and Laws) concerning the health and safety of persons and any damage to property.

Quality Management and Compliance Verification Systems

This is a new provision within FIDIC 2017, in that the Contractor is required to prepare and operate a Quality Management (QM) System in order to demonstrate that the Works, Materials, Plant and workmanship comply with the requirements of the Contract.

This is a further developed version of the "quality assurance system" required within the FIDIC 1999 Contracts, and mirrors the increasing need for detailed and formal QM and associated reviews to verify compliance.

A QM System is a set of interrelated or interacting elements that companies, and in this case the Contractor, will use to direct, manage and control how quality policies are implemented, and quality objectives are achieved on the project.

In many parts of the world, the QM System will be to ISO 9000 standards, ISO 9000 defining a QM System as "co-ordinated activities to direct and control an organization with regard to the degree to which a set of inherent characteristics fulfills the requirements".

ISO 9001 sets out the basic requirements for a QM System, generally that it enables a company to consistently provide products or services that enhance customer satisfaction whilst meeting applicable statutory and regulatory requirements. It is essentially about providing quality assurance to customers.

It can be used by any organisation, large or small, regardless of its field of activity. In fact, ISO 9001 is implemented by over one million companies and organisations in over 170 countries.

The QM System within FIDIC 2017 must be specifically prepared for the Works and submitted to the Engineer within 28 days of the Commencement Date, also whenever the QM System is updated or revised.

The details of the QM System required are stated in the Specification and shall be in accordance with the details stated in the Specification (if any) and includes the Contractor's procedures to ensure that all Notices and other communications, Contractor's Documents, as-built records, operation and maintenance manuals and contemporary records can be traced to the part of the Works to which they relate.

Also, to ensure the proper co-ordination and management of interfaces between the stages of execution of the Works, and between Subcontractors, and for the submission of Contractor's Documents to the Engineer for Review.

The Engineer carries out a Review of the QM System and may give a Notice to the Contractor stating the extent to which it does not comply with the Contract.

Within 14 days after receiving the Notice, the Contractor must revise the QM System to rectify the non-compliance. If the Engineer does not give such a Notice within 21 days of the date of submission of the QM System, he is deemed to have given a Notice of No-objection.

The Engineer may, at any time, give a Notice to the Contractor stating the extent to which the Contractor is failing to correctly implement the QM System to the Contractor's activities under the Contract. After receiving this Notice, the Contractor immediately remedies such failure.

The Contractor must carry out internal audits of the QM System regularly, and at least once every six months. Checking that the system works is a vital part of ISO 9001. An organisation must perform internal audits to check how its QM System is working.

An organisation may decide to invite an independent certification body to verify that it is in conformity to the standard, but there is no requirement for this. Alternatively, it might invite its Clients to audit the quality system for themselves.

The Contractor must submit to the Engineer a report listing the results of each internal audit within seven days of its completion. Each report includes, where appropriate, the proposed measures to improve and/or rectify the QM System.

If the Contractor is required by the Contractor's quality assurance certification to be subject to external audit, the Contractor immediately gives a Notice to the Engineer describing any failing(s) identified in any external audit.

Compliance Verification System

The Contractor is required to prepare and implement a "Compliance Verification System" to demonstrate that any design, Materials, Employer-Supplied Materials (if any), Plant, work and workmanship comply in all respects with the requirements of the Contract.

The Compliance Verification System must be in accordance with the details stated in the Specification and must include a method for reporting the results of all inspections and tests carried out by the Contractor.

It is not specific as to what the Compliance Verification System should look like, but that is inferred through the wording of the Sub-Clause.

It is essential that the Compliance Verification System is project specific and not generic in nature, so it can be used to manage all aspects of the Works.

The Contractor is required to prepare and submit to the Engineer (Employer under the Silver Book) a complete set of compliance verification documentation for the Works or Section, fully compiled and collated in the manner described in the Specification or, if not so described, in a manner acceptable to the Engineer.

Compliance with the QM System and/or the Compliance Verification System does not relieve the Contractor from any duty, obligation or responsibility under or in connection with the Contract, it is merely a reporting and managing system.

Site Data

Under Sub-Clause 4.10, the Employer is deemed to have made available all relevant data in his possession on sub-surface and hydrological conditions by the Base Date, which is 28 days before the date of submission of the Tender.

The Contractor is responsible for interpreting the data and is be deemed to have inspected and examined the Site. "Deemed to have inspected and examined the Site" does not actually mean that he MUST visit the Site, but that he is "fixed with the knowledge" as if he had.... there is a subtle difference!

As with previous editions of the FIDIC contracts the provisions regarding Site Data and Unforeseeable Physical Conditions need to be read and construed together.

There is an increased obligation upon the Employer to provide information about the effects of climatic conditions at the Site.

"Unforeseeable" is defined as "not reasonably foreseeable by an experienced Contractor by the date for submission of the Tender", allowing nearly 28 days after the Base Date for tenderers to assess the data.

Note that FIDIC 1999 referred to the "date for submission of the Tender" rather than the Base Date.

The Employer is also required to make available the same types of data that comes into his possession thereafter.

The Contractor is responsible for interpreting all data referred to under Sub-Clause 2.5 [Site Data and Items of Reference].

To the extent, which was practicable, the Contractor is deemed to have obtained all necessary information as to risks, contingencies and other circumstances which may influence or affect his Tender or the Works.

To the same extent, the Contractor is deemed to have inspected and examined the Site, access to the Site, its surroundings, the above data and other available information, and to have been satisfied before submitting the Tender as to all matters relevant to the execution of the Works, including:

- a The form and nature of the Site, including sub-surface conditions;
- b The hydrological and climatic conditions, and the effects of climatic conditions at the Site;
- c The extent and nature of the work and Goods necessary for the execution of the Works;
- d The Laws, procedures and labour practices of the Country;
- e The Contractor's requirements for access, accommodation, facilities, personnel, power, transport, water and any other utilities or services.

Sufficiency of the Accepted Contract Amount

The Contractor is deemed to have satisfied himself as to the correctness and sufficiency of the Accepted Contract Amount.

Unforeseeable Physical Conditions

The provisions in respect of Unforeseeable Physical Conditions within the Red and Yellow Books have been restructured, and significant further detail added in a structured format to give clarity to the provisions and their operation.

The principle of Unforeseeable Physical Conditions is a subject which often causes much debate!

Sub-Clause 4.12 defines "physical conditions" and covers those which are "Unforeseeable".

"Physical Conditions" is defined as

> natural physical conditions and physical obstructions (natural or man-made) and pollutants, which the Contractor encounters at the Site during execution of the Works, including sub-surface and hydrological conditions but excluding climatic conditions at the Site and the effects of those climatic conditions.

"Unforeseeable" is defined as "*not reasonably foreseeable by an experienced contractor by the Base Date*".

"Base Date" is defined as "*the date 28 days before the latest date for submission of the Tender*".

If the Contractor encounters physical conditions which it considers to have been Unforeseeable and that the finding will have an adverse effect on the progress and/or increase the Cost of the execution of the Works, the Contractor is required to give a Notice to the Engineer, which shall:

a Be given as soon as practicable to allow the Engineer opportunity to inspect the physical conditions promptly and before they are disturbed;
b Describe the physical conditions, so that they can be inspected by the Engineer;
c Set out the reasons why the Contractor considers the physical conditions to be Unforeseeable as defined under the Contract; and
d Describe the manner in which the physical conditions will have an adverse effect on the progress and/or increase the Cost of the execution of the Works.

The Engineer is then required to inspect and investigate the physical conditions within seven days (or a longer period agreed with the Contractor) after receiving the Contractor's Notice.

The Contractor is then required to comply with any instructions which the Engineer may give for dealing with the physical conditions, which may be a Variation.

If the Contractor suffers delay and/or incurs Cost due to these physical conditions, he is entitled to an Extension of Time and/or payment of such Cost.

The agreement or determination, under Sub-Clause 20.2.5 [Agreement or determination of the Claim], of any Claim under Sub-Clause 4.12.4 [Delay and/or Cost] includes consideration of whether, and to what extent, the physical conditions were Unforeseeable.

The Engineer may review whether other physical conditions in similar parts of the Works were more favourable than could reasonably have been foreseen when the Contractor submitted the Tender.

This provision may be particularly relevant for a tunnel, where there may be adverse conditions, in respect of similar parts of the Works, will not result in a net reduction in the Contract Price (otherwise, an Employer might have been tempted to conceal favourable information from tenderers).

Under the Silver Book, Sub-Clause 4.12 which is titled "Unforeseeable Difficulties" rather than the Red and Yellow Book's Unforeseeable Physical Conditions, the Sub-Clause simply states that except as stated in the Particular Conditions:

> The Contractor is deemed to have obtained all necessary information as to risks, contingencies and circumstances which may influence or affect the Works, the Contractor accepts totally responsibility for having foreseen all difficulties and costs of successfully completing the Works and the Contract Price shall not be adjusted to take account of and Unforeseeable or unforeseen difficulties or costs.

Clearly, the Contractor could not, when compiling and pricing his tender, "foresee all difficulties and costs" so this is a major risk for the Contractor.

The only possibility for the Contractor to perhaps recover Cost (under certain circumstances) or gain an Extension of Time is if the event is of such a magnitude that it can be considered to fall within the definition of an Exceptional Event under Clause 18, for example an earthquake or extreme weather event such as a hurricane, but that is very unlikely, and in an area probe to earthquakes and hurricanes, would not be deemed as "exceptional".

Rights of way and facilities

The Contractor is required to bear all costs and charges in respect of temporary rights of way, including access to Site. He may also obtain at his own risk and cost any other facilities outside the Site.

Avoidance of interference

The Contractor may not interfere "unnecessarily or improperly" with the public, or access to all roads and footpaths.

It does not mean he cannot do it, and there is every likelihood of some disruption to road traffic and pedestrian flows, etc., but he cannot do it "unnecessarily or improperly"!

The Contractor is required to indemnify and hold the Employer harmless against and from all damages, losses and expenses (including legal fees and expenses) resulting from any such "unnecessary or improper" interference.

Access route

The Contractor is deemed to have satisfied himself, at the Base Date (28 days prior to the Tender Date), as to the suitability and availability of the access routes to and from the Site.

The Contractor is now required to take on an additional risk from the previous FIDIC contracts by being required to take all necessary measures to prevent any road or bridge from being damaged by the Contractor's traffic or by the Contractor's Personnel. These measures would include the proper use of appropriate vehicles (conforming to legal load and width limits (if any) and any other restrictions) and routes.

Except as otherwise stated in these Conditions:

a The Contractor is responsible for repair of any damage and any maintenance which may be required for the Contractor's access routes.
b The Contractor provides all necessary signs or directions along access routes, and obtains any permissions or permits which may be required from the relevant authorities, for the Contractor's use of routes, signs and directions.
c The Contractor is responsible for any third-party claims which may arise from its use or otherwise of any access route.
d The Employer does not guarantee the suitability or availability of particular access routes.
e All Costs due to non-suitability or non-availability, for the use required by the Contractor, of access routes shall be borne by the Contractor.

There is now a new provision for a Contractor to claim time and money to the extent that non-suitability or non-availability of an access route arises as a result of changes to that access route by the Employer or by a third party after the Base Date, and if as a result the Contractor suffers delay and/or incurs Cost, the Contractor is entitled to an Extension of Time and/or payment of such Cost.

Transport of goods

The Contractor is required to give the Engineer 21 days' notice prior to the date on which "any Plant or major item of other Goods will be delivered to Site". The

Contractor is responsible for all associated costs in connection with transporting, unloading, distribution and protection. What constitutes a "major item of other Goods" is not defined, and could be the subject of debate.

The Contractor is responsible for packing, loading, transporting, receiving, unloading, storing and protecting all Goods and other things required for the Works, and for customs clearance, permits, fees and charges related to the import, transport and handling of all Goods, including all obligations necessary for their delivery to the Site.

Contractor's Equipment

Equipment is what is normally known in other contracts as "Plant" or "Contractor's Plant", the Contractor being responsible for its provision and care. The Contractor has to have the Engineer's consent to remove major items of Equipment.

The Contractor is required to give a Notice to the Engineer of the date on which any major item of Contractor's Equipment has been delivered to the Site. This Notice must be given within seven days of the delivery date, must identify whether the item of Contractor's Equipment is owned by the Contractor or a Subcontractor or another person and, if rented or leased, identify the rental or leasing entity.

Again, what constitutes a "major item of Equipment" is not defined, and could be the subject of debate.

Clearly there needs to be a "common sense" approach by the Contractor and the Engineer in terms of giving consent, especially on a large Site when there are numerous and frequent equipment movements, deliveries and removals, so there are no delays to progress caused by waiting for consent.

Protection of the environment

The Contractor is required to take all reasonable steps to protect the environment, to comply with the environmental impact statement for the Works (if any) and to limit damage and nuisance to people and property resulting from pollution, noise and other results of the Contractor's operations or activities.

The Contractor is also required to ensure that emissions, surface discharges, effluent and any other pollutants from the Contractor's activities do not exceed either the values indicated in the Specification or those prescribed by applicable Laws.

Note that if the law imposes stricter values than the Specification, the legal requirement will prevail.

Temporary utilities

The Contractor is responsible for the provision of all temporary utilities, including electricity, gas, telecommunications, water and any other services, he may require for the execution of the Works.

If the Employer is to provide utilities for the Contractor's use, the Contractor uses the utilities on the Site for which details and prices are given in the Specification.

The Contractor at its own risk and cost provides any apparatus necessary for the Contractor's use and for measuring the quantities consumed.

The apparatus provided for measuring quantities consumed is subject to the Engineer's consent. The quantities consumed during each period of payment stated in the Contract Data is to be measured by the Contractor, and the amount to be paid by the

Contractor for such quantities (at the prices stated in the Specification) is included in the relevant Statement.

Employer's Equipment and free-issue material

This covers any Equipment to be provided by the Employer. The Contractor is required to pay at the rates stated in the Specification.

Progress reports

The Contractor is required to prepare monthly progress reports, in the format stated in the Specification, and to submit them to the Engineer.

Note that the FIDIC 1999 Contracts required monthly progress reports to be prepared by the Contractor and six copies submitted to the Engineer, so there was also a significant amount of printing/photocopying to be done. The author was never convinced why the Contractor should have to submit six copies of a considerable quantity of paper, in an age where many reports are submitted electronically.

Many Contractors do not have the staff or other resources to comply with the progress reporting obligations within the FIDIC contracts, which will also cascade down the supply chain to Consultants, Subcontractor and Suppliers, who must also provide the numerous and substantial amounts of information required to allow the Contractor to comply.

It is also a substantial task for Engineers to examine and review the progress reports submitted.

In reality, Employers, Engineers and Contractors under the FIDIC 1999 Contracts usually agreed on a "number of copies as required by the parties" rather than "number of copies as required by the contract" though the Contractor would technically be in breach of contract if he did not submit six copies!

Under the FIDIC 2017 Contracts, each progress report is required to be submitted in one paper original, one electronic copy and additional paper copies (if any) as stated in the Contract Data.

The contract prescribes the content of each report which includes progress to date, photographs, critical documents and health and safety records.

The first report shall cover the period up to the end of the first month following the Commencement Date. Reports shall be submitted monthly thereafter, each within seven days after the last day of the month to which it relates.

Reporting shall continue until the Date of Completion of the Works or, if outstanding work is listed in the Taking-Over Certificate, the date on which such outstanding work is completed.

Most of the progress report content will be related directly to the programme submitted under Sub-Clause 8.3, and unless otherwise stated in the Specification, each progress report shall include:

a Charts, diagrams and detailed descriptions of progress, including each stage of design by the Contractor (if any), Contractor's Documents as defined under Sub-Clause 4.4, procurement, manufacture, delivery to Site, construction, erection and testing;

b Photographs and/or video recordings showing the status of manufacture and of progress on and off the Site;

c For the manufacture of each main item of Plant and Materials, the name of the manufacturer, manufacture location, percentage progress and the actual or expected dates of:

i Commencement of manufacture,
ii Contractor's inspections,
iii Tests, and
iv Shipment and arrival at the Site;

d The details described in Sub-Clause 6.10 [Contractor's Records];
e Copies of QM documents, inspection reports, test results and compliance verification documentation;
f A list of Variations, and any Notices given (by either Party);
g Health and safety statistics, including details of any hazardous incidents and activities relating to environmental aspects and public relations;
h Comparisons of actual and planned progress, with details of any events or circumstances which may adversely affect the completion of the Works in accordance with the Programme and the Time for Completion, and the measures being (or to be) adopted to overcome delays.

It is critical that the Engineer/Employer's Agent independently and accurately monitors the Contractor's progress and reports objectively to the Employer.

Whilst the Contractor may issue progress reports they should be fully interrogated and challenged where appropriate, and phrases such as "progress is in line with the programme" may imply that the operation is not in delay, but this may not be the case...... "progress is in line with" what programme?

A Contractor reporting that an operation is say 50% complete can be misleading.

Unless it is a repetitive or linear operation, e.g. 50 similar piles completed out of 100, this form of reporting can be meaningless particularly with complex operations such as Mechanical and Electrical Works.

Also, if a Contractor reports that an operation is 100% complete it should be because it is totally complete....barring any very minor Defects.

Security of the Site

The Contractor is at all times in executing the Works, responsible for the security of the Site, and for keeping unauthorised persons off the Site, and authorised persons being limited to the Contractor's Personnel, the Employer's Personnel and any other personnel identified as authorised personnel by a Notice from the Employer or the Engineer to the Contractor.

Contractor's operations on Site

The Contractor is required to confine his operations to the Site and to keep all Equipment and Personnel within the Site, and off adjacent land.

The Contractor is also required to keep the Site free from all unnecessary obstructions, and to properly store or remove from the Site any Equipment and/or surplus materials.

The Contractor is also required to clear away and remove from the Site any wreckage, rubbish, hazardous waste and Temporary Works which are no longer required.

The term "wreckage" has always been a curious term, but one would assume it would refer to anything damaged beyond repair or any demolished items.

On the issue of a Taking-Over Certificate, the Contractor is required to properly clear away all Contractor's Equipment, surplus material, wreckage, rubbish, hazardous waste and Temporary Works. The Contractor shall leave that part of the Site and the Works in a clean and safe condition. However, the Contractor may retain at locations on the Site agreed with the Engineer, during the DNP, such Goods as are required for the Contractor to fulfil obligations under the Contract.

Archaeological and geological findings

In the event that the Contractor finds fossils, coins, articles of value or antiquity, etc., on the Site they must be placed under the care and authority of the Employer.

There may also be legislative requirements in terms of what has been found.

The Contractor is required to take all reasonable precautions to prevent anyone from removing or damaging any of these findings and is required to give a Notice to the Engineer, allowing the Engineer to promptly inspect the finding as soon as possible.

If the Contractor suffers delay and/or incurs Cost from complying with the Engineer's instructions, he is entitled to an Extension of Time and/or payment of such Cost.

Subcontractors

Clause 5 of the FIDIC 1999 Contracts only dealt with nominated Subcontractors, whereas Clause 5 of FIDIC 2017 now deals with subcontracting and Subcontractor generally, and also has specific limitations as to the value of work that can be subcontracted.

This Clause has been extended from that in the FIDIC 1999 Contracts.

The Contractor must not subcontract:

a Works with a total accumulated value greater than the percentage of the Accepted Contract Amount stated in the Contract Data (if not stated, the whole of the Works); or
b Any part of the Works for which subcontracting is not permitted as stated in the Contract Data.

This is a new provision limiting the value of subcontracted work for a particular by providing a cap within the Contract Data.

The Contractor is fully responsible and liable for the work of all Subcontractors as if they were his own workforce including managing and co-ordinating the acts or defaults of any Subcontractor, any Subcontractor's agents or employees, being as if they were the acts or defaults of the Contractor.

The Contractor is required to obtain the Engineer's prior consent by submitting not less than 28 days before the intended Commencement Date of each Subcontractor's work, the name, address, detailed particulars and relevant experience of each Subcontractor including the work intended to be subcontracted.

The Contractor is not required to get the Engineer's consent to suppliers of Materials, or a subcontract for which the Subcontractor is named in the Contract.

Note that if the Engineer (Employer under the Silver Book) does not respond within 14 days after receiving the Contractor's submission, by giving a Notice objecting to the proposed Subcontractor, the Engineer is deemed to have given his/her consent.

Nominated Subcontractors

A Nominated Subcontractor under the FIDIC contracts is a Subcontractor named in the Specification or whom the Engineer when instructing the expenditure of Provisional Sums instructs the Contractor to employ as a Subcontractor.

The Contractor is not under any obligation to employ a nominated Subcontractor as the Engineer instructs and against whom the Contractor raises "reasonable objection" by giving a Notice to the Engineer, with detailed supporting particulars, no later than 14 days after receiving the Engineer's instruction.

An objection is deemed reasonable if it arises from (among other things) any of the following matters, unless the Employer agrees to indemnify the Contractor against and from the consequences of the matter (this is an important provision as the Employer may take the risk from the Contractor; however, it is difficult to see how the Employer could indemnify the Contractor regarding the proposed Subcontractor, for example, not having sufficient competence, resources, financial strength to carry out the Sub-Contract Works):

- a There are reasons to believe that the nominated Subcontractor does not have sufficient competence, resources or financial strength.
- b The subcontract does not specify that the nominated Subcontractor shall indemnify the Contractor against and from any negligence or misuse of Goods by the nominated Subcontractor, the nominated Subcontractor's agents and employees. Or
- c The subcontract does not specify that, for the subcontracted work (including design, if any), the nominated Subcontractor shall:
    - i Undertake to the Contractor such obligations and liabilities as will enable the Contractor to discharge the Contractor's corresponding obligations and liabilities under the Contract.
    - ii Indemnify the Contractor against and from all obligations and liabilities arising under or in connection with the Contract and from the consequences of any failure by the Subcontractor to perform these obligations or to fulfil these liabilities.

The Contractor is required to pay the nominated Subcontractor the amounts due in accordance with the subcontract, the amounts being included in the Contract Price. Before issuing a Payment Certificate which includes an amount payable to a nominated Subcontractor, the Engineer may request the Contractor to supply "reasonable evidence" that the nominated Subcontractor has received all amounts due in accordance with the previous Payment Certificates, less applicable deductions for retention or otherwise.

Unless the Contractor:

- a Submits this reasonable evidence to the Engineer, or
- b
    - i Satisfies the Engineer in writing that the Contractor is reasonably entitled to withhold or refuse to pay these amounts, and
    - ii Submits to the Engineer reasonable evidence that the nominated Subcontractor has been notified of the Contractor's entitlement, then the Employer may (at the Employer's sole discretion) pay, directly to the nominated Subcontractor, part or all of such amounts previously certified (less applicable deductions)

as are due to the nominated Subcontractor and for which the Contractor has failed to submit the evidence described in sub-paragraphs (a) or (b) above.

Thereafter, the Engineer shall give a Notice to the Contractor stating the amount paid directly to the nominated Subcontractor by the Employer and, in the next IPC after this Notice, shall include this amount as a deduction under sub-paragraph (b) of Sub-Clause 14.6.1 [The IPC].

Staff and labour

Except as otherwise stated in the Specification, the Contractor is required to make arrangements for the engagement of all Contractor's Personnel, and for their payment, accommodation, feeding, transport and welfare.

The Contractor is required to pay rates of wages and observe conditions of labour, which comply with all applicable Laws and are not lower than those established for the trade or industry where the work is carried out.

If no established rates or conditions are applicable, the Contractor shall pay rates of wages and observe conditions which are not lower than the general level of wages and conditions observed locally by employers whose trade or industry is similar to that of the Contractor.

Recruitment of persons

The Contractor must not recruit, or attempt to recruit, staff and labour from amongst the Employer's Personnel.

Neither the Employer nor the Engineer can recruit, or attempt to recruit, staff and labour from amongst the Contractor's Personnel.

The inclusion of the reference to the Engineer is a new provision under the FIDIC 2017 Contracts.

Labour laws

The Contractor must comply with all the relevant labour laws applicable to the Contractor's Personnel, including Laws relating to their employment (including wages and working hours), health, safety, welfare, immigration and emigration, and shall allow them all their legal rights.

The Contractor must require the Contractor's Personnel to obey all applicable Laws, including those concerning health and safety at work.

Working hours

No work shall be carried out on the Site on locally recognised days of rest, or outside the normal working hours stated in the Contract Data, unless:

- a Otherwise stated in the Contract;
- b The Engineer gives consent; or
- c The work is unavoidable or necessary for the protection of life or property or for the safety of the Works, in which case the Contractor shall immediately give a Notice to the Engineer with reasons and describing the work required.

Previously under FIDIC 1999, if the Contractor had to unavoidably work out of hours he had to advise the Engineer, now he has to give a Notice with reasons and describing the work required.

Facilities for staff and labour

Except as otherwise stated in the Specification, the Contractor must provide and maintain all necessary accommodation and welfare facilities for the Contractor's Personnel.

If such accommodation and facilities are to be located on the Site, except where the Employer has given the Contractor prior permission, they are to be located within the areas identified in the Contract.

If any such accommodation or facilities are located elsewhere within the Site, the Contractor must immediately remove them at the Contractor's risk and cost. The Contractor must also provide facilities for the Employer's Personnel as stated in the Specification.

Health and safety of personnel

In addition to the requirements listed under Sub-Clause 4.8, the Contractor is required, at all times, take all necessary precautions to maintain the health and safety of its own personnel.

In collaboration with local health authorities, the Contractor is also required to ensure that:

a Medical staff, first aid facilities and other medical facilities as stated in the Specification are available at all times at the Site and at any accommodation provided for Contractor's and Employer's Personnel; and
b Suitable arrangements are made for all necessary welfare and hygiene requirements, and for the prevention of epidemics.

The Contractor must also appoint a dedicated health and safety officer at the Site, responsible for maintaining health, safety and protection against accidents.

This person is required to be qualified, experienced and competent, and to have the authority to issue directives for the purpose of maintaining health and safety, and the prevention of accidents. Having the authority is very important, as responsibility must come with authority, that person normally reporting direct to the Contractor's Project Director and/or the Contractor's main board.

FIDIC 1999 previously required an accident prevention officer at the Site, and FIDIC 2017 now amplifies the role of the health and safety officer.

Contractor's superintendence

From the Commencement Date until the issue of the Performance Certificate, the Contractor shall provide all necessary superintendence to plan, arrange, direct, manage, inspect, test and monitor the execution of the Works.

Superintendence shall be given by a sufficient number of persons:

a Who are fluent in or have adequate knowledge of the language for communications (defined in Sub-Clause 1.4 [Law and Language]); and b Who have adequate knowledge of the operations to be carried out (including the methods and techniques required, the hazards likely to be encountered and methods of preventing accidents), for the satisfactory and safe execution of the Works.

Contractor's Personnel

The Contractor's Personnel (including Key Personnel, if any) shall be appropriately qualified, skilled, experienced and competent in their respective trades or occupations.

The Engineer may require the Contractor to remove (or cause to be removed) any person employed on the Site or Works, including the Contractor's Representative and Key Personnel (if any), who:

a Persists in any misconduct or lack of care;
b Carries out duties incompetently or negligently;
c Fails to comply with any provision of the Contract;
d Persists in any conduct which is prejudicial to safety, health or the protection of the environment;
e Is found, based on reasonable evidence, to have engaged in corrupt, fraudulent, collusive or coercive practice; or
f Has been recruited from the Employer's Personnel in breach of Sub-Clause 6.3 [Recruitment of Persons].

If appropriate, the Contractor shall then promptly appoint (or cause to be appointed) a suitable replacement. In the case of replacement of the Contractor's Representative, Sub-Clause 4.3 [Contractor's Representative] shall apply. In the case of replacement of Key Personnel (if any), Sub-Clause 6.12 [Key Personnel] shall apply.

Contractor's records

The requirement for the Contractor to keep records and submit each calendar month has been enhanced from FIDIC 1999 and now includes a specific list of records which are to be kept for each work activity shown in the Programme, at each work location and for each day of work.

Unless otherwise proposed by the Contractor and agreed by the Engineer, in each progress report, the Contractor must include records of:

a Occupations and actual working hours of each class of Contractor's Personnel,
b The type and actual working hours of each of the Contractor's Equipment,
c The types of Temporary Works used,
d The types of Plant installed in the Permanent Works, and
e The quantities and types of Materials used.

Clearly, this enhanced requirement is quite an onerous administration requirement for the Contractor and also for the Engineer to check.

Disorderly conduct

The Contractor shall at all times take all necessary precautions to prevent any unlawful, riotous or disorderly conduct by or amongst the Contractor's Personnel, and to preserve peace and protection of persons and property on and near the Site.

Key Personnel

Sub-Clause 6.12 of the FIDIC 2017 Contracts introduces a new requirement for the Employer to identify Key Personnel where stated within the Specification.

If no Key Personnel are stated in the Specification, Sub-Clause 6.12 does not apply.

The Contractor is also required to appoint the persons named in his Tender to the positions of Key Personnel under the Contract.

If none are named, or if an appointed person fails to act in the relevant position of Key Personnel, the Contractor must submit to the Engineer for his consent the name and particulars of another person the Contractor proposes to appoint to such a position. If consent is withheld or revoked by the Engineer, the Contractor must similarly the name and particulars of a suitable replacement for such position.

If the Engineer does not respond within 14 days after receiving the submission, by giving a Notice stating his objection to the appointment of the person with reasons, the Engineer is deemed to have given his consent.

The Contractor must not, without the Engineer's prior consent, revoke the appointment of any of the Key Personnel or appoint a replacement (unless a temporary emergency replacement in the case of death, illness, disability or resignation of the Key Personnel).

All Key Personnel must be based at the Site for the whole time that the Works are being executed.

Finally, all Key Personnel must be fluent in the language for communications defined in Sub-Clause 1.4.

Joint Ventures

On many major international projects where FIDIC contracts are used, the Contractor may be a Joint Venture (JV) consisting of, maybe, a local Contractor joining up with a major overseas Contractor (the author has personal experience of this when working on projects in the Middle East).

Under Sub-Clause 1.14 of the FIDIC 2017 Contract, if the Contractor is a JV:

- a The members of the JV are jointly and severally liable to the Employer for the performance of the Contractor's obligations under the Contract, so each can be held liable for the others performance or non-performance, and any liabilities arising,
- b The JV leader (who has to be named) has the authority to bind the Contractor and each member of the JV,
- c Either the composition of the members or the scope and parts of the Works to be carried out by each member, or the legal status of the JV, can be altered without the <u>prior</u> consent of the Employer.

Limitation of liability

Neither Party shall be liable to the other Party for loss of use of any Works, loss of profit and loss of any contract, nor for any indirect or consequential loss or damage which may be suffered by the other Party in connection with the Contract, other than under:

- a Sub-Clause 8.8 [Delay Damages];
- b Sub-paragraph (c) of Sub-Clause 13.3.1 [Variation by Instruction];
- c Sub-Clause 15.7 [Payment after Termination for Employer's Convenience];

d Sub-Clause 16.4 [Payment after Termination by Contractor];
e Sub-Clause 17.3 [Intellectual and Industrial Property Rights];
f The first paragraph of Sub-Clause 17.4 [Indemnities by Contractor]; and
g Sub-Clause 17.5 [Indemnities by Employer]. The total liability of the Contractor to the Employer under or in connection with the Contract, other than:

    i Under Sub-Clause 2.6 [Employer-Supplied Materials and Employer's Equipment];
    ii Under Sub-Clause 4.19 [Temporary Utilities];
    iii Under Sub-Clause 17.3 [Intellectual and Industrial Property Rights]; and
    iv Under the first paragraph of Sub-Clause 17.4 [Indemnities by Contractor], shall not exceed the sum stated in the Contract Data or (if a sum is not so stated) the Accepted Contract Amount.

This Sub-Clause shall not limit liability in any case of fraud, gross negligence, deliberate default or reckless misconduct by the defaulting Party.

Assignment

Under a FIDIC 2017 Contract, neither Party can assign the whole nor any part of the Contract except with the prior agreement of the other Party. However, a Party may assign its right to a payment due without the prior agreement of the other Party.

Care and supply of documents

The Specification and Drawings remain in the custody and care of the Employer, but two copies of the Contract and of each subsequent Drawing are supplied to the Contractor, who may request further copies at its own cost.

The Contractor's Documents remain in the custody and care of the Contractor, unless and until submitted to the Engineer.

The Contractor is required to supply to the Engineer one paper original, one electronic copy and additional paper copies (if required) as stated in the Contract Data.

The Contractor is required to keep on the Site, a copy of:

a The Contract;
b The records under Sub-Clause 6.10 [Contractor's Records] and Sub-Clause 20.2.3 [Contemporary records];
c The publications (if any) named in the Specification;
d The Contractor's Documents;
e The Drawings; and
f Variations, Notices and other communications given under the Contract. The Employer's Personnel shall have right of access to all these documents during all normal working hours, or as otherwise agreed with the Contractor.

If either Party (or the Engineer) becomes aware of an error or defect in a document which was prepared for use in the execution of the Works, the Party (or the Engineer) promptly gives a Notice of such error or defect to the other Party (or to the Parties if it is the Engineer).

Delayed drawings or instructions

The Contractor is required to give a Notice to the Engineer whenever the Works are likely to be delayed or disrupted due to any necessary drawing or instruction not being issued to the Contractor within a particular time.

The Notice is required to include details of the drawing or instruction, details of why and when it should be issued and details of the nature and amount of the delay or disruption likely to be suffered because it is late.

The Contractor is entitled to an Extension of Time and/or payment of such Cost plus Profit, but only if the delayed issue was not caused by any error or delay by the Contractor.

Notices and other communications

All Notices and other communications required by the Contract are required to be in writing and:

a
   i A paper-original signed by the Contractor's Representative, the Engineer or the authorised representative of the Employer (as the case may be); or
   ii An electronic original generated from any of the systems of electronic transmission stated in the Contract Data (if not stated, system(s) acceptable to the Engineer), where the electronic original is transmitted by the electronic address uniquely assigned to each of such authorised representatives, or both, as stated in these Conditions; and

b If it is a Notice, it shall be identified as a Notice. If it is another form of communication, it shall be identified as such and include reference to the provision(s) of the Contract under which it is issued where appropriate;

c Delivered by hand (against receipt), or sent by mail or courier (against receipt), or transmitted using any of the systems of electronic transmission under sub-paragraph (a)(ii) above; and

d Delivered, sent or transmitted to the address for the recipient's communications as stated in the Contract Data. However, if the recipient gives a Notice of another address, all Notices and other communications shall be delivered accordingly after the sender receives such Notice.

Where these Conditions state that a Notice or NOD or other communication is to be delivered, given, issued, provided, sent, submitted or transmitted, it shall have effect when it is received (or deemed to have been received) at the recipient's current address under sub-paragraph (d) above. An electronically transmitted Notice or other communication is deemed to have been received on the day after transmission, provided no non-delivery notification was received by the sender.

All Notices, and all other types of communication as referred to above, shall not be unreasonably withheld or delayed.

When a Notice or NOD or certificate is issued by a Party or the Engineer, the paper and/or electronic original shall be sent to the intended recipient and a copy shall be sent to the Engineer or the other Party, as the case may be. All other communications shall be copied to the Parties and/or the Engineer as stated under these Conditions or elsewhere in the Contract.

Confidentiality

Contracts by their nature are confidential agreements, and the Parties are required to respect that, but the Contract requires that the Contractor discloses all confidential and other information as the Engineer may reasonably require in order to verify the Contractor's compliance with the Contract. This can be details of staff and other resources being employed, costs and other information which may be required.

Again, due to the confidential nature of contracts, the Contractor must not publish or disclose any particulars of the Contract, except with the Employer's prior consent.

Equally, the Employer and the Engineer must treat all information provided by the Contractor and marked "confidential", as confidential. The Employer must not disclose any such information to third parties, except as may be necessary when exercising the Employer's rights under Sub-Clause 15.2 [Termination for Contractor's Default].

Law and language

The Contract is governed by the law of the country, or other jurisdiction, stated in the Contract Data. If nothing is stated, it is the law of the country where the Site is located.

The ruling language of the Contract is also stated in the Contract Data. The language for communications is that stated in the Contract Data.

The Parties use of each other's documents

The Contractor retains the copyright and other intellectual property rights in the Contractor's Documents and other design documents, made by (or on behalf of) the Contractor.

Similarly, the Employer retains the copyright and other intellectual property rights in the Specification and Drawings and other documents made by (or on behalf of) the Employer.

These documents (in whole or in part) shall not, without the Employer's prior consent, be copied, used or communicated to a third party by the Contractor, except as necessary for the purposes of the Contract.

The Contractor may at its own cost, copy, use and communicate these documents for the purposes of the Contract.

The Contractor is deemed to give to the Employer a non-terminable transferable non-exclusive royalty-free licence to copy, use and communicate the Contractor's Documents (and such other design documents, if any), including making and using modifications of them.

This licence:

a Applies throughout the actual or intended operational life (whichever is longer) of the relevant parts of the Works;
b Entitles any person in proper possession of the relevant part of the Works to copy, use and communicate the Contractor's Documents (and such other design

documents, if any) for the purposes of completing, operating, maintaining, altering, adjusting, repairing and demolishing the Works;

c In the case of Contractor's Documents (and such other design documents, if any) which are in the form of electronic or digital files, computer programs and other software, permits their use on any computer on the Site and/or at the locations of the Employer and the Engineer and/or at other places as envisaged by the Contract; and

d In the event of termination of the Contract:

  i Under Sub-Clause 15.2 [Termination for Contractor's Default], entitles the Employer to copy, use and communicate the Contractor's Documents (and other design documents made by or for the Contractor, if any); or

  ii Under Sub-Clause 15.5 [Termination for Employer's Convenience], Sub-Clause 16.2 [Termination by Contractor] or Sub-Clause 18.5 [Optional Termination], entitles the Employer to copy, use and communicate the Contractor's Documents for which the Contractor has received payment for the purpose of completing the Works and/or arranging for any other entities to do so.

The Contractor's Documents (and other design documents, if any, made by (or on behalf of) the Contractor) shall not, without the Contractor's prior consent, be used, copied or communicated to a third party by (or on behalf of) the Employer for purposes other than those permitted under this Sub-Clause.

4 Quality and Defects Liability

Quality and Defects Liability are covered in the FIDIC 2017 Contracts by:

- *Conditions of Contract for Construction* (The "Red Book")
  - Clause 7 – Plant, Materials and Workmanship
  - Clause 9 – Tests on Completion
  - Clause 10 – Employer's Taking Over
  - Clause 11 – Defects after Taking Over
- *Conditions of Contract for Plant and Design-Build* (The "Yellow Book")
  - Clause 7 – Plant, Materials and Workmanship
  - Clause 9 – Tests on Completion
  - Clause 10 – Employer's Taking Over
  - Clause 11 – Defects after Taking Over
  - Clause 12 – Tests after Completion (Clause 12 is only included within the Yellow and Silver Books)
- *Conditions of Contract for EPC/Turnkey Project* (The "Silver Book")
  - Clause 7 – Plant, Materials and Workmanship
  - Clause 9 – Tests on Completion
  - Clause 10 – Employer's Taking Over
  - Clause 11 – Defects after Taking Over
  - Clause 12 – Tests after Completion (Clause 12 is only included within the Yellow and Silver Books)

Before considering the provisions within the FIDIC 2017 Contracts it is worth reviewing the subject of "Quality" and "Quality Management", particularly in relation to construction.

What is quality?

The term "quality" is often used to describe expensive or prestige products such as opulent jewellery or luxury cars.

However, the term "quality" does not necessarily refer to prestigious products, but merely to the fitness of the product or service to the customer's requirements, whatever those requirements may be.

Quality is therefore best described as "meeting the specific requirements of the customer".

Quality management

Quality management is a major management function within the construction industry.

Unless a Contractor can guarantee the Employer a quality product, whether it be a project, a product or a service, the Contractor cannot compete effectively with others in the modern market.

Quality often stands alongside price as a major factor in Contractor selection by Employers, and comprehensive scoring systems which rate quality alongside price, and other critical matters such as health and safety and financial stability have played a greater part in the past few years and are very much a decider in selecting the right Contractor for the project.

To be competitive and to sustain robust and sustainable business prospects, quality systems need to be designed and maintained which are efficient, and also are able to be proven by evidence, when tendering for work as well as when executing that work.

The role of quality management for a contractor is not an isolated activity, but intertwined with all the operational and managerial processes of that company, effectively this is where the term "total quality management" (TQM) came from.

What is quality control?

The earliest and most basic form of quality management is quality control (QC), the term being defined by an interpretation of its two elements – "quality" and "control".

The concept of being "in control", or having something "under control", is readily understood, we mean we know what we intend to happen, and we have systems in place so that we are confident that we can ensure that it does happen as we intend.

QC introduces inspection and testing to the work being carried out in ensuring that it is undertaken to specified requirements. The quality is often measured by comparing the work actually carried out against the sample provided.

Inspection is the process of checking that what is produced is what is required. It is about the identification and early correction of defects.

The major objectives of QC can be defined as follows:

i To ensure the completed work meets the specification;
ii To reduce Clients' complaints;
iii To improve the reliability of services delivered;
iv To increase Clients' confidence;
v To reduce delivery costs.

QC is primarily concerned with defect detection and correction.

The main QC technique is that of inspection, testing and statistical QC techniques, for example sampling, to ensure that what is delivered and the materials used are within the tolerances specified. However, some of these limits are left to the inspector's judgement and this can be a source of difficulty.

The outcome of an inspection can be viewed in both objective and subjective ways:

Objective inspection

That which is quantifiable and measurable – line, levels, verticality and dimensions.

There are some precise quantified inspections including the testing of plant and machinery, pressure tests in pipework and tests on electrical installations.

Subjective inspection

That which is open to the inspector's interpretation, e.g. cleanliness, fit, tolerances and visual checks.

Traditionally, in construction contracts, there are two sets of documents that are used to determine the required quality of a construction project, these are the specification and the drawings. The Contractor uses these documents during the site operations stage of any project to facilitate "quality" in construction.

The majority of quality checks are undertaken visually. Visual quality checks of each section of construction are undertaken by the Contractors' managers and tradesmen, and then by the Employer's Engineers and Representatives, to ensure it complies with the Drawings and Specification.

Quantifiable and measurable quality checks are also made during the construction stage. These include testing the strength of concrete cubes, checking alignment of brickwork and commissioning of services installations.

The results of these quality checks are recorded and passed to the appropriate authority.

The weakness of QC is the development of the inspection mentality or culture, whereby the construction contractors' operatives and engineers set their standards to that which can "achieve a certain score".

It would be much better if the Contractors' staff and operatives had a clear understanding of the quality required, were able to recognise it themselves and were able to achieve it first time and/or regulate it by self-inspection.

This brings in the concept of quality assurance (QA), which potentially reduces the risks of producing unsatisfactory work and being involved in expensive re-work.

Quality assurance

Whereas QC focuses on defect detection and correction during the execution of the Works, QA focuses on defect prevention.

QA is the process of ensuring that standards are consistently met thereby preventing defects from occurring in the first place.

"Fit for purpose" and "right first time, every time" are the principles of QA, and the reference point for QA is the International Quality Standard ISO 9000 family of standards.

To be certified as operating to the ISO 9000 standard is now virtually seen as essential in today's construction industry. Many clients simply will not do business with companies not certified to ISO 9000.

QA concentrates on the delivery methods and procedural approaches to ensure that quality is built into the system.

Total quality management

This is the third stage and is based on the philosophy of continuously improving goods or services within a company. The key factor is that everyone in the company should be involved and committed from the top to the bottom of the organisation.

The successful total quality managed company ensures that their goods and services can meet the following criteria:

i Be fit for purpose on a consistently reliable basis;
ii Delight the customer with the service which accompanies the supply of goods;
iii Supply a quality service that is so much better than the competition that customers want it regardless of price.

The pursuit of total quality is seen as a never-ending journey of continuous improvement.

Defining quality within Contracts

The Contractor is required to execute the Works in accordance with the Contract, so that the completed Works will be in accordance with the documents forming the Contract.

The Contract must therefore define clearly what is expected of the Contractor in executing the Works. It must cover all aspects of the work, either specifically or by implication, otherwise it may be deemed to be excluded from the contract.

A comprehensive description should therefore be considered, which should be supplemented by specific detailed requirements in the form of Drawings and Specification. If reliance is placed solely on the scope of Works, there is a danger that an item may be missed and could be the subject of later contention.

The Contract will normally be comprised of the following:

Description of the Works

The Contract must include a clear description of what is expected of the Contractor in executing the Works. It must cover all aspects of the project, either specifically or by implication, otherwise it may be deemed to be excluded from the Contract.

A comprehensive description should therefore be considered, which should be supplemented by specific detailed requirements in the form of a specification. If reliance is placed solely on the description of the service, there is a danger that an item may be missed and could be the subject of later contention.

Drawings

The drawings should detail all the work to be carried out by the Contractor and be complementary to the description of the Works and the specification.

Specification

The specification is a written technical description which describes the character and quality of services, materials and workmanship, for work to be executed. It may also lay down the sequence in which various portions of the project are to be executed.

As far as possible, it should describe the outcomes required, rather than how to achieve them, criteria or standards which are required for the work, and complementary to the other constituents of the Contract and any notes or other text on the drawings, and the provisions of the contract.

Responsibilities of the parties

The responsibilities of the contracting parties and their representatives should be clearly defined within the Contract in a manner which will leave no doubt as to the obligations that each is accepting, whilst also considering the written terms of the Conditions of Contract. This will also allow the procedures necessary to enable the contract to progress satisfactorily from inception to completion to be established at the outset of the service.

The contract as a whole will also define the responsibilities for the planning of the Works, timings, resources, the supply of any free issue materials, the issuing of instructions and the form which these instructions are to take, the programming of the Works, the method of measuring and evaluating the Works, the circumstances which will constitute a Variation to the Works and the duties of the parties during the delivery of the project.

Defining testing within Contracts

Testing is a key requirement within a Contract and is usually defined through the specification and often also through applicable law and standards.

Any materials facilities and samples for testing, including who provides them, are also stated within the specification.

Testing can be related to individual Plant and Materials on Site or off Site, including factory testing, testing of individual sections of work, welds and joints once they have been assembled or a multitude of mechanical and electrical tests.

Whilst the specification will set out the procedures and parameters for testing and commissioning, the Contract will set out the Parties' contractual obligations and the consequences of failure to meet them. Many activities, for example electrical testing, will have to comply with current statutory requirements independent of the contract.

Performance testing is usually applicable where the design of Plant and Materials is the responsibility of the Contractor. This can include the operation of the completed plant and the checking of product quality and quantity, fuel consumed, use of energy, waste and by-products, environmental conditions and other aspects such as may be required. In such instances time limits for the rectification of performance defects must be stated together with penalties for failure.

The documentation required in the form of test certificates, warranties and guarantees to enable proper records to be maintained should be stipulated, as will the effect in relation to the guarantee or warranty period and the extent to which it will apply. Many such records are only maintained electronically so access to the relevant asset management systems should be specified for the Client together with the requirements for data transfer at expiry of the service period.

Liability for defects should be defined in the contract together with the period of time for which such liability is to apply and the length of time within which defects are to be remedied.

It is imperative that the specification, produced at tender stage, defines the full extent and timing of tests and inspections required, and the subsequent correction of any defects.

Testing is very often not covered in sufficient detail within the Scope, so it is important for the compiler of the Scope to consider a number of key questions:

i What is the purpose of the test?

What is it intended to prove or disprove? Is the test required before a following activity can be carried out by the Contractor, or another party, or is it required before the Client can use a building or part of the building?

ii What is to be tested?

What elements or components are required to be tested? Is it to be tested in an assembled or unassembled state?

iii How is the test to be carried out, and who is to provide the materials, facilities and equipment to do the test?

Whether the test is a well-known "industry standard" test or not, the exact nature of the test must be clearly defined. Who supplies the equipment, who pays for the power or fuel for the equipment, etc.? It is also important to detail who is to provide the materials and facilities to do the test, and who meets the costs.

iv Who should do the test? Who should be present and who should be invited should they wish to attend when the test is taking place?

There may be a requirement for an independent party or testing house to carry out the test and provide results.

v When and where will the test take place? On Site or off Site?

Clearly state at what stage or on what date the test must take place. Sometimes, tests may have to be carried out on a periodic basis, for example on concrete samples on delivery, and as the concrete cures.

vi What is the expected result or outcome from the test?

What is the test expected to prove? What results are needed to be gathered once the test has been carried out?

vii Who should be advised of the outcome of the test?

The Engineer/Employer's Representative will normally be advised of the outcome of a test, but sometimes a Subcontractor, or a third party outside the contract may need to be informed.

viii What should be done in the event of a successful test e.g. certification.

Does a certificate have to be issued following successful completion of a test? Does that successful outcome mean that the Contractor can proceed to the next stage of the service? Is payment to the Contractor dependent on successful completion of the test?

ix What should be done in the event of an unsuccessful test e.g. rectification or replacement, and further testing.

Does the Contractor have to repeat the test? Is the retest done immediately after the first test, or should there be a period before the next test? Is it possible that even though something has failed a test, it may still be considered acceptable, possibly by the Contractor offering a cost saving?

x What about additional testing i.e. to confirm a suspected defect or the extent of a suspected defect?

There may be a requirement for Plant and Materials to be tested off Site before delivery.

The main documents within the Scope in respect of testing and inspection will normally be the Drawings and Specification.

Limitation periods

Whilst the FIDIC contracts, which we will discuss shortly, have a Time for Completion, then a Defects Notification Period (DNP) following that, during which the Contractor is liable for correcting Defects, most countries also have a statutory limitation period for latent defects which are discovered at a later date, beyond which a claim cannot be made for defective work.

In the UK, for example, statute governs time limits for bringing different types of legal claims, for example, the time limit for actions founded on simple contract is 6 years and for a specialty contract or deed, the time limit is 12 years, in both cases this is measured from the date on which the cause of action accrued.

So the relevant law needs to be considered as well as the Contract, when assessing long-term liability for defects.

Patent and latent defects

Note that within the FIDIC 2017 Contracts a Defect is not defined within Clause 1.

The Engineer's schedule of defects will only list "patent defects" i.e. those that are observable from reasonable inspection at the time, and will not include "latent defects", which may be hidden from reasonable inspection and may come to light at a much later date.

The Contractor's liability for correction of latent defects will depend on the applicable law, and liability is likely to remain despite the issue of the Performance Certificate.

Latent defects are likely to arise after the Engineer/Employer's Representative employment has ended. The Engineer/Employer's Representative may therefore be recalled at a later date to provide information and advice pertaining to any claim against the Contractor relating to a latent defect.

The specific requirements under the FIDIC 2017 Contracts

Manner of Execution

The quality of work including the manufacture, supply, installation, testing and commissioning of Plant and Materials must be as specified in the Contract, executed in a proper and workmanlike manner in accordance with good practice, and with properly equipped facilities and non-hazardous materials.

The author often queries what would be termed a "*proper and workmanlike manner*", which is actually a quite common terms in construction contracts.

This term would appear to suggest that the Contractor uses a degree of effort, efficiency, knowledge and experience in executing the Works as required by the Contract, probably akin to a designer using reasonable skill and care in carrying out his design.

You cannot strictly define it, but practitioners will have an opinion as to whether the Contractor is, or is not doing it!

Note that the FIDIC 1999 Contracts only applied to the manufacture of Plant, production and manufacture of materials and generally to the execution of the Works.

Samples

Sub-Clause 7.2 has not changed from the FIDIC 1999 Contracts in requiring the Contractor to submit samples to the Engineer (Employer under the Silver Book) for consent, properly labelled.

Inspection

The Employer's Personnel (note the Employer's Personnel rather than the Engineer) have the right to full access to all parts of the Site during normal working hours as stated in the Contract Data, and at all other times.

The Contractor is required to give a Notice to the Engineer (Employer under the Silver Book) whenever Materials, Plant or work are ready for inspection and before they are covered up, the Employer's Personnel then respond promptly. If they do not, the Engineer (Employer under the Silver Book) may give Notice to the Contractor that they do not require to do so.

Note that, although the Notice goes to the Engineer (Employer under the Silver Book), it is the Employer's Personnel who carry out the inspection, not the Engineer (or the Employer under the Silver Book).

If the Engineer (Employer under the Silver Book) does not issue the Notice, the Contractor may proceed with covering up that part of the Works.

Note that FIDIC 2017 now requires the Contractor to make available records including photographs and and/or video recordings.

If the Contractor fails to give the Notice to the Engineer (Employer under the Silver Book), then it is required to uncover, reinstate and make good at its own expense.

Finally, there is now an express right for the Contractor to cover up work or material which would otherwise need to be inspected if the Engineer (Employer under the Silver Book) does not respond to the Contractor's Notice advising that an inspection may be carried out, or the Employer's Personnel do not attend to make the inspection at the notified time.

Testing by the Contractor

The Contractor must provide all apparatus, assistance, documents, temporary supplies of electricity and water, experienced and competent staff, etc., to carry out the specified tests. The requirement for temporary supplies of electricity and water, and that the Contractor's staff must be competent enough to ensure the tests are carried out properly, is additional to that within the FIDIC 1999 Contracts.

The Contractor is required to give a Notice to the Engineer (Employer under the Silver Book) within a reasonable time before carrying out such tests, previously under the FIDIC 1999 Contracts he just had to carry out the tests.

Note that, although the Notice is issued to the Engineer (Employer under the Silver Book), it is actually the Employer's Personnel who may attend.

The Engineer (Employer under the Silver Book) may instruct as a Variation, a changed location and/or timing of tests, or additional testing, unless what is tested is shown not to be in accordance with the Contract. Obviously, a Variation may give rise to a payment to the Contractor unless it shows that the work was not in accordance with the Contract.

It would seem that instructing a changed location/timing of a test is somewhat bureaucratic, unless the tests are critically changed,

The Engineer (Employer under the Silver Book) is required to give not less than 72 hours' notice (this was formerly 24 hours under the FIDIC 1999 Contracts) if he wishes to attend the tests. If the Contractor is delayed or incurs costs in respect of this, he can claim an Extension of Time and/or Cost plus Profit under Clause 20.2.

If the Contractor causes any delay to tests and the delay causes the Employer to incur costs, the Employer is entitled payment of these costs from the Contractor under Sub-Clause 20.2. This is a new provision.

The Contractor is required to forward certified reports of the tests to the Engineer (Employer under the Silver Book). When the tests have been passed, the Engineer (Employer under the Silver Book) endorses the Contractor's test certificate, or issues a test certificate to the Contractor, to that effect. If the Engineer (Employer under the Silver Book) has not attended the tests, he is deemed to have accepted the test results as accurate.

Defects and rejection

If any work fails a test, the Engineer (Employer under the Silver Book) gives a Notice to the Contractor describing the failure. The Contractor is then required to make and submit a proposal for remedial work, which the Engineer (Employer under the Silver Book) may review and state if the proposal does <u>not</u> comply with the Contract. The Contractor then submits a further proposal for the Engineer's Review if required.

If the Engineer (Employer under the Silver Book) does not give a Notice within 14 days after receiving the Contractor's proposal (or revised proposal), the Engineer (Employer under the Silver Book) is deemed to have given a Notice of No-objection.

If the Contractor fails to submit a proposal, or fails to carry out the proposed remedial work to which the Engineer (Employer under the Silver Book) has given (or is deemed to have given) a Notice of No-objection, the Engineer may:

- a Instruct the Contractor to carry out remedial work.
- b Reject the Plant, Materials, Contractor's design (if any) or workmanship by giving a Notice to the Contractor, with reasons, in which case Sub-Clause 11.4 [Failure to Remedy Defects] applies.

After remedying defects, if the Engineer (Employer under the Silver Book) requires any items to be retested, the tests are repeated in accordance with requirements, and at the Contractor's risk and cost.

If the rejection and retesting causes the Employer to incur additional costs, the Employer is entitled to payment of these costs from the Contractor.

Remedial work

The Engineer (Employer under the Silver Book) may instruct the Contractor to:

- a Repair or remedy (which may require its removal and return to Site), or permanent removal from the Site;
- b Replace any Plant or Materials which are not in accordance with the Contract;
- c Repair or remedy, or remove and re-execute, any other work which is not in accordance with the Contract;
- d Carry out any remedial work which is <u>urgently</u> required for the safety of the Works, whether because of an accident, unforeseeable event or otherwise.

The Contractor is required to comply with the instruction at its own cost, except any work under sub-paragraph (c) above which is attributable to:

- i The Employer or the Employer's Personnel. If the Contractor suffers delay and/or incurs Cost in carrying out such work, the Contractor is entitled to an Extension of Time and/or payment of Cost plus Profit.

ii An Exceptional Event, in which case Sub-Clause 18.4 [Consequences of an Exceptional Event] applies.

This is a "softening" of the FIDIC 1999 Approach which was that the Contractor was required to carry out remedial work at its own cost, now he can recover the Cost plus Profit.

If the Contractor fails to comply with the Engineer's (Employer under the Silver Book) instruction, the Employer may employ and pay others to carry out the work.

The Employer is entitled to payment by the Contractor of all costs arising from this failure.

Ownership of Plant and Materials

Sub-Clause 7.7 defines the date when each item of Plant and Materials becomes the property of the Employer i.e. at the earlier of:

a When the Plant or Materials are delivered to the Site;
b When the Contractor is paid the value of the Plant and Materials after Employer's Suspension;
c When the Contractor is paid the amount determined for unfixed Plant and Materials.

Note the reference to the Laws of the Country, which may determine at what point the Plant and Materials become the Employer's property. The issue of "title" can be significantly different under differing legislations.

Also, note that 80% of the amount determined is included in an Interim Payment Certificate (IPC) (the balance will only be paid when the Plant or Materials are incorporated into the Works).

Royalties

Unless the Specification (Employer's Requirements in the Yellow and Silver Books) states otherwise, the Contractor must pay all royalties, rents and other payments in connection with the Works for:

a Natural Materials obtained from outside the Site;
b The disposal of material from demolitions and excavations and of other surplus material (whether natural or man-made), except to the extent where disposal areas within the Site are stated in the Specification (Employer's Requirements in the Yellow and Silver Books).

Tests on Completion

Contractor's obligations

The Contractor is required to carry out all Tests on Completion after submitting the As-Built Records (if applicable) and Operation and Maintenance Manuals (also if applicable).

There is a new provision under FIDIC 2017 in that the Contractor must submit, not less than 42 days before the tests commence, a detailed test programme giving details of when each test is carried out, its duration and the resources that it intends to use to complete the tests. This is a useful addition for clarity purposes.

The Engineer (Employer under the Silver Book) may review the proposed test programme and may give a Notice to the Contractor stating the extent to which it does not comply with the Contract.

The Contractor must then respond to that Notice within 14 days revising the test programme to rectify the non-compliance.

If the Engineer (Employer under the Silver Book) does not give a Notice within 14 days after receiving the test programme, he is deemed to have given a Notice of No-objection. The Contractor must not commence the tests until a Notice of No-objection is given (or is deemed to have been given) by the Engineer (Employer under the Silver Book).

The Contractor should also give a Notice to the Engineer (Employer under the Silver Book), of not less than 21 days, of the date after which the Contractor will be ready to carry out each of the Tests on Completion.

The Contractor then commences the tests within 14 days after this date, or whatever the Engineer (Employer under the Silver Book) instructs to proceed in accordance with his test programme.

Unless otherwise stated within the Employer's Requirements, the Yellow and Silver Books prescribe three testing stages which must be carried out in sequence:

i Pre-commissioning tests including appropriate inspections to verify that each item of work can proceed to the next stage;
ii Commissioning tests including operational tests stated within the Employer's Requirements to demonstrate that the works can be used safely;
iii Trial operation to demonstrate that the Works perform reliably.

Again, a new provision under the FIDIC 2017 Contracts, as soon as the Works or any specified Section have passed the Tests on Completion, the Contractor submits a certified report of the results to the Engineer (Employer under the Silver Book).

The Engineer (Employer under the Silver Book) carries out a Review of the report and may give a Notice to the Contractor stating the extent to which the results of the tests do not comply with the Contract. If the Engineer (Employer under the Silver Book) does not give a Notice within 14 days after receiving the test results, the Engineer (Employer under the Silver Book) is deemed to have given a Notice of No-objection.

In considering the results of the Tests on Completion, the Engineer (Employer under the Silver Book) is required to make allowances for the effect of any use of (any part of) the Works by the Employer on the performance or other characteristics of the Works.

Delayed tests

Sub-Clause 9.2 deals with delayed tests, whether the delay is caused by the Engineer (Red Book), the Employer's Personnel (Yellow and Silver Books) or the Contractor.

If the tests are "delayed" by the Engineer or by the Employer's Personnel, or by a cause for which the Employer is responsible, and the Contractor has given a Notice to that effect, Sub-Clause 10.3 [Interference with Tests on Completion] applies (see below).

If the Tests on Completion are delayed by the Contractor, the Engineer (Employer under the Silver Book) may give a Notice to the Contractor requiring him to carry out the tests within 21 days after receiving the Notice.

The Contractor must carry out the tests within this period of 21 days, and must give not less than 7 days prior Notice of dates to the Engineer (Employer under the Silver Book).

If the Contractor fails to carry out the Tests on Completion within the period of 21 days:

- a After a second Notice is given by the Engineer (Employer under the Silver Book) to the Contractor, the Employer's Personnel may proceed with the tests.
- b The Contractor may attend and witness the tests.
- c Within 28 days of these tests being completed, the Engineer (Employer under the Silver Book) sends a copy of the test results to the Contractor.
- d If the Employer incurs additional costs as a result of the testing, he is entitled to payment by the Contractor of the costs reasonably incurred.

Whether or not the Contractor attends, the Tests on Completion are deemed to have been carried out as if they were in the presence of the Contractor and the results of the tests are accepted as accurate.

Retesting

If any work fails a test, the defect must be repaired and the test repeated under the same conditions.

Failure to pass Tests on Completion

If the Works, or any Section fails the Tests on Completion, the Engineer (Employer under the Silver Book) is entitled to order further repetition of tests, to reject the Works, to reject the relevant Section or to issue a Taking-Over Certificate if the Employer requests.

If the Engineer (Employer under the Silver Book) issues a Taking-Over Certificate the Contractor is required to proceed with his obligations under the Contract, and the Employer may claim payment from the Contractor, or a reduction to the Contract Price.

If the Works, or a Section, are rejected, the Employer can recover costs as if the rejected part of the Works or Section had been omitted, as a Variation.

Employer's Taking Over

Clause 10 specifies the procedures for certification of completion and for the Employer's Taking Over.

Taking Over of the Works and Sections

The Works are taken over by the Employer when:

- a The Works have been completed in accordance with the Contract, including the passing of the Tests on Completion.
- b If applicable, the Engineer (Employer under the Silver Book) has given (or is deemed to have given) a Notice of No-objection to the As-Built Records submitted.

c If applicable, the Engineer (Employer under the Silver Book) has given (or is deemed to have given) a Notice of No-objection to the operation and maintenance manuals.

d If applicable, the Contractor has carried out the training as described under Sub-Clause 4.5 [Training]. Obviously, this may not be applicable, but again another important provision.

(b), (c) and (d) are critical pre-requisites; it is not just a requirement to have physically constructed the Works.

e A Taking-Over Certificate for the Works has been issued, or is deemed to have been issued in accordance with this Sub-Clause.

The Contractor may apply for a Taking-Over Certificate by giving a Notice to the Engineer (Employer under the Silver Book) not more than 14 days before the Works will, in the Contractor's opinion, be complete and ready for taking over.

If the Works have been divided into Sections, the Contractor may similarly apply for a Taking-Over Certificate for each Section.

The reference to the supply of As-Built Records, Operation and Maintenance Manuals and associated No-objection by the Engineer (Employer under the Silver Book), and also training is a new provision under FIDIC 2017.

If any Part of the Works is taken over, the remaining Works or Section is not be taken over until all the conditions described in sub-paragraphs (a) to (e) above have been fulfilled.

Note that "Part" is under FIDIC 2017 a defined term under Clause 1 meaning "*a part of the Works or Section (as the case may be) which is used by the Employer and deemed to be taken over under Sub-Clause 10.2*".

The Engineer (Employer under the Silver Book) is required, within 28 days after receiving the Contractor's Notice, either to:

i Issue the Taking-Over Certificate, stating the date on which the Works or Section was completed in accordance with the Contract, except for any minor outstanding work and defects (which should be listed in the Taking-Over Certificate) and which will not substantially affect the safe use of the Works or Section for their intended purpose (either until or whilst this work is completed and these defects are remedied), or

ii Reject the application by giving a Notice to the Contractor, with reasons.

This Notice should specify in detail the work, which is required to be done, the defects required to be remedied and/or the documents required to be submitted by the Contractor (whichever are applicable) to enable the Taking-Over Certificate to be issued.

The Contractor then completes this work, remedies the defects and/or submits such documents (whichever are applicable) before giving a further Notice under this Sub-Clause.

If the Engineer (Employer under the Silver Book) does not issue the Taking-Over Certificate or reject the Contractor's application within this period of 28 days, and if the conditions described in sub-paragraphs (a) to (d) above (where applicable) have been fulfilled, the Works or Section is deemed to have been completed in accordance with the Contract on the 14th day after the Engineer (Employer under the Silver Book) receives the Contractor's Notice of application and the Taking-Over Certificate is deemed to have been issued.

Taking-Over Parts

This is the equivalent of "partial possession" in other Contracts, allowing the Employer to take over part of the Works before the Taking-Over Certificate has been issued for the whole of the Works or for a Section.

If the Employer does so, the Part taken over is deemed to be taken over, the Contractor is no longer responsible, but he is liable (see Chapter 9) for the care of that Part, and the Engineer (Employer under the Silver Book) issues a Taking-Over Certificate.

The Contractor is able to recover any additional Cost plus Profit associated with this action.

The Engineer (Employer under the Silver Book) may, at the sole discretion of the Employer, issue a Taking-Over Certificate for any part of the Permanent Works. Note the term "sole discretion", the Employer does not have to ask anyone.

Note that the Employer must not use any Part of the Works until the Engineer (Employer under the Silver Book) has issued a Taking-Over Certificate for this Part.

However, if the Employer does use any Part of the Works before the Taking-Over Certificate is issued the Contractor gives a Notice to the Engineer (Employer under the Silver Book) to that effect, identifying the Part and describing its use, and:

a That Part is deemed to have been taken over by the Employer as from the date on which it is used.
b The Contractor ceases to be responsible for the care of that Part from this date, when responsibility passes to the Employer.
c The Engineer (Employer under the Silver Book) immediately issues a Taking-Over Certificate for the Part, and any outstanding work to be completed (including Tests on Completion) and/or defects to be remedied is listed in the Taking-Over Certificate.

After the Engineer (Employer under the Silver Book) has issued a Taking-Over Certificate for a Part, the Contractor must be given the earliest opportunity to take such steps as necessary to carry out the outstanding work and/or remedial work for any defects listed in the Taking-Over Certificate.

The Contractor must carry out these Works as soon as practicable and, in any case, before the expiry date of the relevant DNP.

If the Contractor incurs Cost as a result of the Employer taking over and/ or using a Part, the Contractor can, as stated above, claim payment of Cost plus Profit.

If the Engineer (Employer under the Silver Book) issues a Taking-Over Certificate for any Part of the Works, or if the Employer is deemed to have taken over a Part the Delay Damages for completion of the remainder of the Works are reduced. Similarly, the Delay Damages for the remainder of the Section.

This reduction is calculated as the proportion which the value of the Part (except the value of any outstanding works and/or defects to be remedied) bears to the value of the Works or Section as a whole. This is a lot easier to calculate than some contracts which reefer to valuing the "usefulness" of the Part which has been taken over in comparison to the whole of the Works or a Section, dependent on how the Delay Damages have been calculated.

The Engineer (Employer under the Silver Book) agrees or determines this reduction in the Delay Damages.

The date the Engineer (Employer under the Silver Book) receives the Contractor's Notice is the date of commencement of the time limit for agreement. The provisions of this paragraph shall only apply to the daily rate of Delay Damages, and shall not affect the maximum amount of these damages.

Interference with Tests on Completion

In the event that the Contractor is prevented, for more than 14 days, from carrying out the Tests on Completion by the Employer's Personnel or by any cause for which the Employer is responsible:

a The Contractor is required to give a Notice to the Engineer (Employer under the Silver Book) describing the prevention.
b The Employer is deemed to have taken over the Works or Section on the date when the Tests on Completion would "otherwise" have been completed.
c The Engineer (Employer under the Silver Book) is required to immediately issue a Taking-Over Certificate for the Works or Section.

  What about the As-Built Records, Drawings, etc.?

  Also, note there is "deemed" take-over, but not a deemed Taking-Over Certificate, it still relies on the Engineer (Employer under the Silver Book) issuing the Taking-Over Certificate, though it should be done "immediately", but what happened if the Engineer (Employer under the Silver Book) does not issue it immediately, or at all?

Note that the Sub-Clause refers to the failure to carry out any "performance test that is not possible due to available operating conditions during trial operation".

It is not clear what this refers to, but subject to the Contractor issuing a Notice, there will be a deemed Taking Over of the Works or Section.

NB: The Contract refers to the 14 Days either being a continuous period or multiple periods which total more than 14 Days, so this Clause could be difficult to administer.

After the Engineer (Employer under the Silver Book) has issued the Taking-Over Certificate, the Contractor is required to carry out the Tests on Completion as soon as practicable and, in any case, before the expiry date of the DNP.

The Engineer (Employer under the Silver Book) must give a Notice to the Contractor, of not less than 14 days, of the date after which the Contractor may carry out each of the Tests on Completion.

If the Contractor suffers delay and/or incurs Cost as a result of being prevented from carrying out the Tests on Completion, the Contractor is entitled to an Extension of Time and/or payment of such Cost plus Profit.

Surfaces requiring reinstatement (Red and Yellow Books only)

A Taking-Over Certificate (unless stated otherwise) does not include for completion of surfaces requiring reinstatement.

Sub-Clause 11.11 deals with the Contractor being required to clear the Site after the issue of the Performance Certificate.

Note that this Sub-Clause does <u>not</u> exist under the Silver Book.

Defects after Taking Over

Completion of outstanding work and remedying defects

The Contractor is required to:

- a Complete any work which is outstanding on the relevant Date of Completion, within the time(s) stated in the Taking-Over Certificate, or within any other reasonable time instructed by the Engineer (Employer under the Silver Book).
- b Execute all work required to remedy defects or damage, of which a Notice is given to the Contractor by (or on behalf of) the Employer on or before the expiry date of the DNP for the Works or Section or Part (as the case may be).

Note that the provision applies not just to the DNP for the Works or Section, but also to a Part (as the case may be).

If a defect appears, or damage occurs during the relevant DNP, a Notice is given to the Contractor by (or on behalf of) the Employer.

Promptly:

- i The Contractor and the Employer's Personnel are required to jointly inspect the defect or damage.
- ii The Contractor must then prepare and submit a proposal for the necessary remedial work.
- iii The second, third and fourth paragraphs of Sub-Clause 7.5 [Defects and Rejection] apply i.e. the Engineer (Employer under the Silver Book) reviews the Contractor's proposal stating whether the proposal complies with the Contract, and the Contractor may be required to submit a revised proposal.

  The author is not positive towards cross-referencing Clauses and Sub-Clauses within Contracts as this could lead to confusion between the parties, and particularly when Clauses/Sub-Clauses are amended (which is not encouraged!) the links can be lost.

  This is an important matter, so for clarity, why not just repeat what Sub-Clause 7.5 states, within this Sub-Clause?

Cost of remedying defects

All defects are to be rectified at the Contractor's cost. If the defect has arisen which is not the Contractor's fault, or for which he is not responsible, it is treated as a Variation and the Contractor is paid according to the rules for valuing Variations.

Note that the Contractor is also responsible for remedying defects that arise from improper operation and maintenance which was attributable to matters for which the Contractor is responsible (i.e. As-Built Records, Operation and Maintenance Manuals and Training).

If the Contractor considers that the defect is attributable to any other cause, the Contractor is required to promptly give a Notice to the Engineer (Employer under the Silver Book) and the Engineer (Employer under the Silver Book) then agrees or determines the cause, the date of the Notice is the date of commencement of the time limit for agreement under Sub-Clause 3.7.3.

If it is agreed or determined that the work is attributable to a cause other than those listed above, Sub-Clause 13.3.1 [Variation by Instruction] applies as if the work had been instructed by the Engineer.

Extension of DNP

Under Sub-Clause 1.1.27 (Sub-Clause 1.1.24 under the Silver Book) "Defects Notification Period" or "DNP" is defined as

> the period for notifying defects and/or damage in the Works or a Section or a Part (as the case may be) under Sub-Clause 11.1 [Completion of Outstanding Work and Remedying Defects], as stated in the Contract Data (if not stated, one year), and as may be extended under Sub-Clause 11.3 [Extension of Defects Notification Period]. This period is calculated from the Date of Completion of the Works or Section or Part.

So, the default period is "one year", and applies to the Works, a Section or a Part.

Employers should consider carefully what they require for, DNP for the Works, each Section or each Part, rather than relying on this "deemed provision".

The Employer is entitled to an extension of the DNP for the Works, or a Section or a Part, if the Works, Section, Part or a major item of Plant cannot be used for the intended purpose(s) by reason of a defect or damage which is attributable to any of the matters under sub-paragraphs (a) to (d) of Sub-Clause 11.2 [Cost of Remedying Defects]. This would be subject to Sub-Clause 20.2 [Claims For Payment and/or EOT].

This is more restrictive than the FIDIC 1999 Contracts which allowed the Employer to have an extension to the DNP wherever the defects or damage affected the Works; however, this has now been changed within the FIDIC 2017 Contracts in that the Employer is only entitled to extension of the DNP if the Works, Section, Part or a major item of Plant cannot be used for the intended purposes, and it is attributable to any of the matters under Sub-Clause 11.2 as stated above and also it is subject to Clause 20.2.

Also, a DNP cannot be extended by more than a period of <u>two years</u> after the expiry of the DNP stated in the Contract Data.

If delivery and/or erection of Plant and/or Materials was suspended under Sub-Clause 8.9 [Employer's Suspension] or Sub-Clause 16.1 [Suspension by Contractor], the Contractor's obligations under this Clause shall not apply to any defects or damage occurring <u>more than two years</u> after the DNP for the Works, of which the Plant and/or Materials form part, would otherwise have expired. This is an important point.

Failure to remedy defects

Under a new provision within FIDIC 2017, if the Contractor delays remedying any defect or damage, a date may be fixed by (or on behalf of) the Employer, on or by which the defect or damage is to be remedied.

A Notice of this date must be given to the Contractor by the Employer allowing the Contractor a reasonable time to remedy the defect or damage.

If the Contractor fails to remedy the defect or damage by the date stated in this Notice, the Employer may (and each of these could generate some debate!):

a Carry out the work or have the work carried out by others at the Contractor's cost, but the Contractor then has no responsibility for this work. The Employer is entitled to payment by the Contractor of the costs reasonably incurred by the Employer in remedying the defect or damage (costs reasonably incurred always being subjective).
b Accept the damaged or defective work, in which case the Employer is entitled to a reduction in the Contract Price.

What that reduction in the Contract Price amounts to could be the subject of much debate!

c Require the Engineer to treat any part of the Works which cannot be used for its intended purpose(s) under the Contract by reason of this failure as an omission.
"Cannot be used for its intended purpose(s)", how would this be defined and calculated?
d Terminate the Contract as a whole with immediate effect if the defect or damage deprives the Employer of substantially the whole benefit of the Works. The Employer is then entitled to recover from the Contractor all sums paid for the Works, plus financing charges and any costs incurred in dismantling the same, clearing the Site and returning Plant and Materials to the Contractor.
Presumably, the Contractor then owns the Plant and Materials, through it is unlikely that he will be able to do anything with them!
If the Employer terminates, presumably it is not subject to the rules under Clause 15 [Termination by Employer].

As stated these provisions are quite subjective in terms of calculating the outcome, so it is important that the Contractor prevents such debates by executing the Works properly and if necessary remedying defects properly as required by the Contract.

Remedying of defective work off Site

If, during the DNP, the Contractor considers that any defect or damage in any Plant cannot be remedied on the Site and it has to be dismantled and taken off Site, the Contractor may give a Notice to the Employer requesting consent to remove the defective or damaged Plant off the Site for the purposes of repair.

The change under FIDIC 2017 is in the Contractor considering that the defect or damage cannot be remedied on Site and also having to give details as listed below.

Within the Notice, the Contractor is required to clearly identify each item of defective or damaged Plant, and also to give full details of:

a The defect or damage to be repaired;
b The location where the defective or damaged Plant is to be taken for repair;
c The transportation which is to be used (and the insurance cover for such transportation);
d The proposed inspections and testing that will be carried out off Site;
e The planned duration required before the repaired Plant can be returned to the Site;
f The planned duration for reinstallation and retesting of the repaired Plant.

When the Employer gives consent, the Contractor may remove from the Site such items of Plant as are defective or damaged.

Note that the Employer may require the Contractor to increase the amount of the Performance Security by the full replacement cost of the defective or damaged Plant.

Further tests after remedying defects

Within seven days of completing the remediation of any defect or damage, the Contractor must give a Notice to the Engineer (Employer under the Silver Book) describing the remedied Works, Section, Part and/or Plant and the proposed repeated tests.

Within seven days after receiving this Notice, the Engineer (Employer under the Silver Book) gives a Notice to the Contractor either:

- a Agreeing with the proposed testing; or
- b Instructing that repeated tests are necessary to demonstrate that the remedied Works, Section, Part and/or Plant comply with the Contract.

If the Contractor fails to a Notice within the seven days, there is a default provision in that the Engineer (Employer under the Silver Book) may give a Notice to the Contractor within 14 days after the defect or damage is remedied, instructing the repeated tests that are necessary to demonstrate that the remedied Works, Section, Part and/or Plant comply with the Contract.

All repeated tests under this Sub-Clause, if applicable, are to be carried out in accordance with the terms applicable to the previous tests i.e. by the Contractor giving Notice and the Engineer (Employer under the Silver Book) responding, except that they are carried out at the risk and cost of the Party liable for the cost of the remedial work.

Right of access after taking over

Until 28 days after issue of the Performance Certificate, the Contractor has the right of reasonable access to the Works as reasonably required in order to comply with this Clause, except as may be inconsistent with the Employer's reasonable security restrictions.

Under new provisions within the FIDIC 2017 Contracts, whenever the Contractor intends to access any part of the Works during the relevant DNP:

- a The Contractor is required to request access from the Employer by giving a Notice, describing the parts of the Works to be accessed, the reasons for such access and the Contractor's preferred date for access.

  This Notice is required to be given in reasonable time in advance of the preferred date for access, taking due regard of all relevant circumstances including the Employer's security restrictions; and
- b Within seven days after receiving the Contractor's Notice, the Employer is required to respond, giving a Notice to the Contractor either:
  - i Stating the Employer's consent to the Contractor's request; or
  - ii Proposing reasonable alternative date(s), with reasons.

If the Employer fails to give this Notice within seven days, the Employer is deemed to have given consent to the Contractor's access on the preferred date stated in the Contractor's Notice, so it is critical that this is dealt with by the Employer within the required time scale.

Note that, if the Contractor incurs additional Cost as a result of any unreasonable delay (how long it has to be to be delayed to be an "unreasonable delay" is tested on its merits for each request) is by the Employer in permitting access to the Works by the Contractor, the Contractor is entitled to payment of Cost plus Profit.

Contractor to search

Within the FIDIC 1999 Contracts, the Engineer (Employer under the Silver Book) may require the Contractor to search for the cause of a Defect, under the FIDIC 2017 Contracts it becomes a specific instruction by the Engineer (Employer under the Silver Book). This is common in other construction contracts, and is often referred to as "opening up" or "uncovering".

If instructed, the Contractor must search for the cause of any defect, under the direction of the Engineer (Employer under the Silver Book).

The Contractor is required to carry out the search on the date(s) stated in the Engineer's (Employer under the Silver Book) instruction or other date(s) agreed with the Engineer (Employer under the Silver Book).

Unless the defect is to be remedied at the cost of the Contractor, the Contractor is entitled to payment of the Cost plus Profit of carrying out the search.

If the Contractor fails to carry out the search in accordance with the instruction, the search may be carried out by the Employer's Personnel, and may be at the Contractor's Cost.

The Contractor is given a Notice of the date when such a search will be carried out, and the Contractor may attend at his own cost.

Under a new FIDIC 2017 Contract provision, if the defect is to be remedied at the cost of the Contractor, the Employer is entitled to payment by the Contractor of the costs of the search reasonably incurred by the Employer.

Performance Certificate

The issue of the Performance Certificate confirms that the Contractor has fulfilled all his obligations under the Contract (FIDIC 1999 referred to the Contractor "completing his obligations").

The Certificate is issued within 28 days after the latest of the expiry dates of the DNPs.

The Engineer (Employer under the Silver Book) issues the Performance Certificate to the Contractor (with a copy to the Employer and to the DAAB) within 28 days after the latest of the expiry dates of the DNPs, or as soon as the Contractor has:

a Supplied all the Contractor's Documents.
b Completed and tested all the Works (including remedying any defects) in accordance with the Contract.

It is curious that the Performance Certificate now has to be issued not only to the Parties (Employer and Contractor) but also to the Dispute Avoidance/Adjudication Board (DAAB), presumably in case there is a dispute about it, or a related matter?

Presumably, the Contractor's Documents must have not only been supplied, but have received "No-Objection" to its contents.

Under a new provision within the FIDIC 2017 Contracts, if the Engineer (Employer under the Silver Book) fails to issue the Performance Certificate within the period of 28 days, the Performance Certificate is deemed to have been issued on the date 28 days after the date on which it should have been issued i.e. a total of 56 days.

Unfulfilled obligations

Although this seems a contradiction to mention "unfulfilled obligations" as the Contractor has surely fulfilled his obligations, but there may be outstanding items such as clearance of the Site, Final Payment Certificate and the possibility of latent defects.

The FIDIC 1999 Contracts were silent on liability for latent defects, so it was left to whatever the relevant legislation stated.

Under the FIDIC 2017 Contracts, the Contractor is not liable for any defects or damage occurring more than two years after expiry of the DNP for the Plant, except if prohibited by law or in any case of fraud, gross negligence, deliberate default or reckless misconduct.

As stated above, the Parties should be particularly aware of the relevant legislation, particularly in respect of future long-term liability for latent defects, and take legal advice accordingly. In most cases statutory liability will prevail over contractual liability.

Clearance of Site

After the issue of the Performance Certificate, the Contractor must:

- a Remove any Contractor's Equipment, surplus material, wreckage, rubbish and Temporary Works from the Site. (The term "wreckage" has often caused some discussion amongst practitioners, but it clearly something that is damaged and/or surplus to requirements!)
- b Reinstate all parts of the Site which were affected by the Contractor's activities during the execution of the Works.
- c Leave the Site and the Works in the condition stated in the Specification (Employer's Requirements in the Yellow and Silver Books) (and if not stated, in a clean and safe condition).

If the Contractor fails to comply with the above within 28 days after the issue of the Performance Certificate, the Employer may (where permitted by the Law, and this must be checked as it could vary under differing legislations) sell or otherwise dispose of any remaining items and/or may reinstate and clean the Site (as necessary), at the Contractor's cost.

The Employer is entitled to payment by the Contractor of the costs reasonably incurred in connection with, such sale or disposal and reinstating and/or cleaning the Site, less an amount equal to any proceeds from the sale.

Tests after Completion (Clause 12) (Yellow and Silver Books only)

Note that Clause 12 "Tests after Completion" is only included within the Yellow and Silver Books, Clause 12 under the Red Book being entitled "Measurement and Valuation".

For building and civil engineering works "Tests on Completion" tend to be more critical to show that the project is complete in terms of construction <u>and is able to be used as intended</u>.

"Tests after Completion" are normally more specific to process contracts where there is a requirement for the Contractor to prove "tests in use" in that the facility continues to be working and performing reliably as it should for some time after it is all completed. These tests could be carried out on different occasions under different conditions.

Clause 12 under the Yellow and Silver Books is, like many of the FIDIC 2017 Contract Clauses, significantly more detailed than FIDIC 1999 in terms of what is required of the Parties.

The Clause now refers specifically to the "Employer's Requirements" rather than, more generically to "the Contract", so the Parties know where to look within the Contracts, and this also highlights the criticality of drafting the Employer's Requirements to cover all of what is required with the Contract.

For clarity, the Engineer (Employer under the Silver Book) also now has to provide a test programme to the Contractor showing the estimated timing of the tests, in order that the Contractor can plan to attend as required.

Also, for clarity, the Contractor has to provide and submit a detailed test programme to the Employer, who then has the opportunity to notify the Contractor of any non-compliance with the requirements of the Contract.

Procedure for Tests after Completion

If Tests after Completion are specified in the Employer's Requirements:

Under the Yellow Book

The Employer must:

a Provide all electricity, water, sewage (if applicable), equipment, fuel, consumables, instruments, labour, materials, and suitably qualified, experienced and competent staff, as are necessary to carry out the Tests after Completion efficiently and properly.

b Carry out the Tests after Completion in accordance with:

   i The Employer's Requirements;
   ii The O&M Manuals to which the Engineer (Employer under the Silver Book) has given (or is deemed to have given) a Notice of No-objection;
   iii Such guidance as the Contractor may be required to give during the course of these tests, and in the presence of such Contractor's Personnel as either Party may reasonably request.

So, the responsibility for hosting and carrying out these tests is placed on the Employer.

The timing of the Tests after Completion is as stated in the Employer's Requirements (if not stated, as soon as is reasonably practicable) after the Works or Section (as the case may be) has been taken over under Clause 10 [Employer's Taking Over].

The Engineer must give a Notice to the Contractor, of not less than 21 days, of the date on which, and place at which, the Tests after Completion will be carried out.

This Notice must also include a test programme showing the estimated timing for each of the tests. Unless otherwise agreed with the Contractor, these tests are required to be carried out on this date.

If the Contractor does not attend at the time and place stated in the Engineer's (Employer under the Silver Book) Notice (or otherwise agreed with the Contractor), the Employer may proceed with the Tests after Completion, and they are deemed to have been made in the Contractor's presence, and the Contractor is deemed to have accepted the results as accurate.

The results of the Tests after Completion are compiled and evaluated by both Parties, with appropriate account taken of the effect of the Employer's prior use of the Works.

Under the Silver Book

The Employer must provide all electricity, water, sewage (if applicable), equipment, fuel, consumables, instruments, labour, materials, and suitably qualified, experienced and competent staff, as are necessary to carry out the Tests after Completion efficiently and properly.

The Contractor must:

- a Provide all apparatus, assistance, documents, etc., to carry out the test.
- b Submit to the Employer not less than 42 days before the testing is intended to start, and detailed test programme showing the timing and resources for the tests.
- c Give a Notice to the Employer of not less than 21 days of the date when the Contractor is ready to carry out the tests.
- d Not commence the tests until after a Notice of No-objection is given to the Contractor's test programme.
- e Commence the tests within 14 days after the date stated in the Notice (see (c)).
- f Carry out the tests in accordance with the test programme, the Employer's Requirements and then if applicable the O&M manuals for which the Employer has given his No-objection.

The testing must all be done in the presence of the Employer's Personnel.

The results of the Tests after Completion are compiled and evaluated by both Parties, with appropriate account taken of the effect of the Employer's prior use of the Works.

Delayed tests

If the Contractor incurs Cost as a result of any unreasonable delay by the Employer in carrying out the Tests after Completion, the Contractor is entitled to payment of Cost plus Profit.

If, for reasons not attributable to the Contractor, a Test after Completion on the Works or any Section cannot be completed during the DNP (or any other period agreed by both Parties), then the Works or Section is deemed to have passed this Test after Completion.

Retesting

If the Works, or a Section, fail to pass the Tests after Completion:

a Sub-paragraph (b) of Sub-Clause 11.1 [Completion of Outstanding Work and Remedying Defects] applies; and
b After remedying any defect or damage, Sub-Clause 11.6 [Further Tests after Remedying Defects] applies.

If this failure and retesting are attributable to the Contractor and cause the Employer to incur additional costs, the Employer is entitled to payment of these costs by the Contractor.

Failure to pass Tests after Completion

If:

a The Works, or a Section, fail to pass any or all of the Tests after Completion; and
b If applicable Performance Damages are set out in the Schedule of Performance Guarantees, the Employer is entitled subject to payment of these Performance Damages by the Contractor in full satisfaction of this failure.

   Note that the Employer has to recover the Performance Damages by way of a Claim, with the Notice requirements that entail, including the 28 day time bar (when does the 28 days start?). If the Contractor pays these Performance Damages to the Employer during the DNP, then the Works or Section is deemed to have passed these Tests after Completion.

If the Works, or a Section, fail to pass a Test after Completion and, by giving a Notice to the Employer, the Contractor proposes to make adjustments or modifications to the Works or such Section:

i The Contractor may be instructed by a Notice given by the Employer that right of access to the Works or Section cannot be given until a time that is convenient to the Employer.
ii The Contractor remains liable to carry out the adjustments or modifications and to satisfy this Test, within a reasonable period of receiving the Notice.
iii If the Contractor does not receive a Notice during the relevant DNP, the Contractor is relieved of the obligation to make such adjustments or modifications and the Works or Section is deemed to have passed this Test after Completion.

If the Contractor incurs additional Cost as a result of any unreasonable delay by the Employer in permitting access to the Works or Section by the Contractor, either to investigate the causes of a failure to pass a Test after Completion or to carry out any adjustments or modifications, the Contractor is entitled to payment of any such Cost Plus Profit.

5 Design responsibility and liability

Contractor's design is covered in the FIDIC 2017 Contracts by:

- *Conditions of Contract for Construction* (The "Red Book")
  - Sub-Clause 4.1
- *Conditions of Contract for Plant and Design-Build* (The "Yellow Book")
  - Clause 5
- *Conditions of Contract for EPC/Turnkey Project* (The "Silver Book")
  - Clause 5

Introduction

The general principle with the FIDIC 2017 Contracts is that when using the Red Book, the Contractor undertakes the construction work in accordance with the Employer's design. However, the Works may include some Contractor-designed civil, mechanical, electrical and/or construction works.

When using the Yellow Book and the Silver Book, the Contractor designs and provides the Works in accordance with the Employer's Requirements.

Before considering the specific design provisions under the FIDIC 2017 Contracts, let us first consider the principles of Design and Build.

The concept of Design and Build

Whilst many believe "Design and Build" where the Contractor is wholly responsible for the design as well as the building of the project, to be a recent innovation, it is a procurement method which has been in use for a very long time.

Design and Build can be used for any construction project, whether it be buildings, infrastructure, process plants, etc.

In the building industry, for example, before the emergence of architecture as a profession, Employers used to procure buildings by a process of Design and Build, the Architect appointing the various artisans, labour and materials to construct the projects, so the Architect was effectively acting as the Contractor. In the nineteenth century there was a separation of responsibility for design from responsibility for constructing the project that led to the emergence of so called "traditional contracting".

The separation of "Design" from "Build" generally within the construction industry has been the source of many problems however, and has for a long time been the source of many issues and disputes, particularly in managing and co-ordinating the two doilies, and remains so, despite the more recent efforts of many to combine the two disciplines.

As recently as the 1970s it became increasingly apparent to Employers that the "traditional contracting" process, where the design team was managed by the Architect or the Engineer, who remained in the Employer's team for the full duration of the project, did not always fulfil their needs.

In response to this, many Employers who undertook construction work regularly began to look to contracts that reduced their risk, and more specifically in this case that transfer design responsibility to the Contractor, hence the growth of the design and build concept, with its inherent advantages at operational level:

- The design and construction periods can overlap, leading to faster delivery of the project.
- The Contractor can utilise its experience and preferred methods of construction to build rationally designed projects which minimise costs and programme durations.
- The Temporary Works/Permanent Works interface and influence of design is rationalised as both remain with the Contractor, and therefore, the Permanent Works should be more "buildable".
- The traditional design/construction interface and the risks associated with it are transferred to the Contractor.
- The management of the design risk by the Contractor can result in greater certainty of the time, cost and performance and project objectives being met.

When Design and Build first came into use, Employers provided Employer's Requirements, which specified what was required, so that the Contractor carried out the design from the start and then continued into the construction phase.

More recently and now common practice, the Employer's design team who had taken the design to a certain stage is novated to the Contractor, who is then responsible for completing the design and construction of the project using that design team.

From the Employer's perspective, there are several advantages and disadvantages in carrying out work using Design and Build:

Advantages of Design and Build

Price certainty

Design and Build Contracts are lump sum, where the Contractor commits himself to a lump sum price, which includes the cost of Consultants to deliver the design element for the project. Although it is a lump sum price the Contract Price can be adjusted to incorporate Variations, Claims and any other express provisions within the Contracts such as price escalation, changes in Laws, etc.

Single point responsibility

The primary interface is between the Employer (and/or the Engineer or Employer's Representative) and the Contractor. There is no need (or right) for the Employer (and/or the Engineer or Employer's Representative) to appoint, liaise with, manage or pay for separate design consultants and external third parties such as regulatory authorities. If there is an issue that issue is primarily between the Employer and the Contractor.

Good For "Non Expert" clients

Appointing, liaising with and managing a wide range of Consultants requires a high degree of knowledge and experience by Employers, who then find themselves having to appoint expert advisors to enable them to make these appointments.

Most Employers do not wish to "become actively involved in the construction industry", they just want the Contractor to design and build the project.

Contractor's Expertise

Many Contractors have a wealth of knowledge and experience, not only of construction, but also of design and the co-ordination with construction, statutory requirements, value engineering, etc. The Design and Build process capitalises on that knowledge and experience. The emergence of Design and Build and more recently Early Contractor Involvement (ECI) has brought to the fore that knowledge and experience.

The Design and Build process increases the opportunities for harnessing the benefit of the Contractor's experience during the design stages of the project.

Many of the developments in procurement processes, and much of the work in fields of study on value management, value engineering and buildability have had this purpose in mind. The benefits of the integration of designers and builders are more economic buildings as well as a more economic and effective production process.

Quicker Build Time

The traditional contracting process requires the design to be complete, or at least at an advanced stage, then Bills of Quantities and other tender documents have to be prepared before going out to tender.

With Design and Build, the Employer's Requirements are drafted, there is no requirement to provide Bills of Quantities, tender documents are prepared, then the project goes out to tender. Design has either already commenced prior to tender and the designers are novated to the successful Contractor, or the Contractor commences the design process from the point he is appointed and manages the design process as construction progresses, thereby leading to time savings in the tender process and also during the construction phase.

Disadvantages of Design and Build

The author is of the opinion that these disadvantages below are "perceived" ather than reality, and can be largely resolved:

Contractor designs and builds to a price

One reported disadvantage of Design and Build is that where there is a conflict between aesthetic quality and ease of fabrication, the latter will take precedence.

A further criticism has been that a Design and Build Contractor will put in the minimum design effort required to win the contract.

These two criticisms suggest strongly that quality, particularly design quality, will suffer under this procurement process. However, this is not a valid criticism of the process itself, but rather the quality of Employer's Requirements and of some of the people who may exercising it. As such, it is a criticism which can be made about any process.

Clear, concise and complete Employer's Requirements will enable the brief to be met at the right price.

As stated above, Design and Build Contracts are lump sum where the Contractor commits himself to a lump sum price, usually in competition with others with the Contract Price only changed under various provisions under the Contract.

The Contractor will use the Employer's Requirements as a basis for his pricing. He is not going to include additional features which could increase his tender price.

It is important that when assessing tenders, Employers should consider what the Contractor is offering for the price, and any additional value within the offer, for example durable materials, low energy costs, low maintenance cost, rather than just the price itself.

Employer's reduced control of design and quality

Many Employers perceive a loss of control of design and quality as a disadvantage.

Again, it starts with a robust set of Employer's Requirements that are sufficiently robust, and also if the right Contractor is appointed, this should not be a disadvantage.

Also, if the Employer is novating the designer(s), the design needs to have been developed sufficiently before the novation.

Also, on larger projects, the Employer may retain its own shadow design team to check on design and quality of construction.

If we consider the appointment of an Employer's shadow design team, whilst the Contractor carries full responsibility for completing the design and carrying out and completing the construction, it is not unusual, particularly on larger and/or more complex projects, for the Employer to retain his original design team or to appoint a new team, where the original team has been novated, to advise him on design, quality and other compliance issues.

If something goes wrong in the future, and there have been cases of structural collapse or major fires which have involved deaths and major injuries, it is not sufficient to merely state that "the design was the Contractor's responsibility" as a problem may arise which may lead to legal action where the Employer and other parties may also be implicated, sometimes many years later in the case of incorrect design or latent defects.

This would be further complicated in cases where the Contractor has ceased trading since carrying out the design and construction of the building or facility.

Conversely, some Employers perceive reduced control of design and quality as reduced involvement in design and quality and therefore an advantage because, as stated previously, they do not wish to become involved in the construction process.

Employer's reduced control of change costs

Again, because Design and Build is a lump sum procurement method, there is no Bill of Quantities to check on the price of changes. Essentially, the best advice to

Employers in this respect is not to change the Employer's Requirements once they have appointed the Contractor.

There will never be the amount of detail that you would have within a bill of quantities, but a well-structured pricing schedule will assist the Employer to obtain some certainty of the value of a change.

Not good for complex projects

There is a perception amongst some Employers that Design and Build is good for more straightforward projects, but for larger, more complex projects they would rather have the design team "on their side".

There is a wealth of large, complex projects around the world that have been carried out successfully using Design and Build as the procurement method.

The key is again clear, concise and complete Employer's Requirements.

Longer tender period

Employers must allow a longer period for Contractors to prepare their tender as they are not pricing Bills of Quantities as with traditional contracts, but considering design issues, planning, other statutory requirements and the ensuing costs.

Expensive for contractors to prepare tenders

As stated above, Employers must allow a longer period for Contractors to prepare their tender as they are carrying out significantly more work in preparing their tenders, and often liaising with design and consultants who offer their services on a "no win – no fee" basis; however, Consultants can only survive on this basis if the Contractor has a successful strike rate in winning tenders.

Dealing with planning permission and other statutory consents

One issue that needs to be considered with Design and Build is the matter of planning permission and various consents required prior to building a project and which Party should be responsible for obtaining them.

There are various consents that may be required for a project, for example:

Planning permission/planning consent

Outline planning permission (permission to build something in a certain location) is usually the responsibility of the Employer, but on a Design and Build contract, detailed planning permission (permission to build something for which details are given) may be the responsibility of the Employer, or may be assigned to the Contractor.

In some cases, the Contractor may also be responsible for gaining outline planning permission for the project; however, this should be a "two stage" process so that if outline planning permission is not obtained, the second stage would not proceed.

The Employer's Requirements must clearly state the status of any planning permission, and who is responsible for obtaining any further consents and for discharging any planning conditions.

Some Planning Conditions, such as colour or type of external materials, must be discharged by the Contractor in designing and building the project....but the Contractor must be made aware of them.

However, other Planning Conditions, such as future use of the development and maintaining access for others, can only be discharged by the Employer.

Building Control/Building Regulation Approval

Again, on a Design and Build contract it is usually the Contractor's responsibility to ensure that the design, and the construction of the project ultimately complies with Building Control requirements, including the various statutory regulations.

Depending on what stage the design is at Contractor appointment stage, and any novation provisions, the responsibility for building control can change from initially the Employer to the Contractor.

Rights of Light/Oversailing Rights

A right to light is separate from planning permission. It is the Contractor's responsibility to ensure that the design and construction comply with these requirements.

Environmental ratings

If the Employer requires the design and construction of the building to achieve certain environmental ratings, then this must be clearly stated within the Employer's Requirements. The Contractor's design and construction, then, has to meet the required rating.

Reasonable Skill and Care v Fitness for Purpose

Another area to consider with Design and Build is the terms "Reasonable Skill and Care" and "Fitness for Purpose".

"Reasonable skill and care" means designing to the level of an ordinary competent person exercising a particular skill.

This is normally achieved by a Consultant designer following accepted practice and complying with British Standards, Codes of Practice, etc.

In the absence of any express terms to the contrary, a designer will normally be required to design using "reasonable skill and care".

"Fitness for purpose" means producing a building, part of a project or a product fit, for its intended purpose. This is an absolute duty independent of negligence. In the absence of any express terms to the contrary, a Contractor or Subcontractor will normally be required to design "fit for purpose".

Defining the purpose of a project or parts of it is extremely difficult, it is not sufficient to state that the project is a highway for vehicles, a hospital for sick people or a drainage scheme to drain water and/or reprocess water. It has to be defined in an objective way, so it can be measured as to whether the Contractor's design and construction was truly fit for the intended purpose. The Contractor's design then has to satisfy the requirements, including statutory compliance.

Contrast this, as stated above, with the level of liability of a consultant (whether acting for an Client or a Contractor) in providing a design service, where there is an implied term that the consultants will carry out the service, in this case design, with reasonable care and skill, which means designing to the level of an ordinary, but competent person exercising a particular skill.

Some contracts will limit the Contractor's Liability to that of a consultant i.e. reasonable skill and care.

So, in the absence of any express terms to the contrary, a designer will normally be required to design using "reasonable skill and care". This is normally achieved by the designer following accepted practice and complying with national standards, codes of practice, etc.

In a Design and Build contract clearly, despite the Contractor believing that it has offset its design obligations and liability to its designing consultant, it has to be aware that it and its consultant have differing levels of care and liability.

Within the FIDIC contracts, note the differing Sub-Clause 4.1.

The differing wording of Clause 4.1:

- FIDIC 1999 Clause 4.1 (Contractor's General Obligations) states that

  > the Contractor shall design, execute and complete the Works in accordance with the Contract, and shall remedy any defects in the Works. When completed, the Works shall be fit for the purposes for which the Works are intended as defined in the Contract.

- FIDIC 2017 Clause 4.1 (Contractor's General Obligations) states that

  > the Contractor shall execute the Works in accordance with the Contract. When completed, the Works (or Section or Part or major item of Plant, if any) shall be fit for the purpose(s) for which they are intended, as defined and described in the Employer's Requirements (or, where no purpose(s) are so defined and described, fit for their ordinary purpose(s)).

So FIDIC 1999 provided a far wider obligation on the Contractor by stating "as defined in the Contract" rather than the FIDIC 2017 principle of "as defined and described in the Employer's Requirements", which is more specific, though if no purpose is defined in the Employer's Requirements, or a purpose is stated outside the Employer's Requirements it rests on a potentially far wider "fit for their ordinary purposes".

It is vital that Employers fully describe the purpose of the Works within the Employer's Requirements objectively rather than subjectively so the Contractor knows what his obligations are, rather than have any reliability on the fall back "fit for their ordinary purposes".

Note also that under Sub-Clause 17.4 (Indemnities by Contractor) the Contractor is required to indemnify the Employer against all acts, errors or omissions by the Contractor that result in the Works (or Section or Part or major item of Plant) not being fit for the purpose(s) under Sub-Clause 4.1 [Contractor's General Obligations].

Note that the burden of proof is on the Employer to identify a relevant act, error or omission as opposed to simply demonstrating that the works are not fit for purpose.

Novation of designers

It is common, as stated above, for Employers to appoint Consultants to carry out early design, prior to inviting tenders.

These designers may be involved in obtaining planning consents, preparing feasibility proposals, and also alternative design proposals.

The tenderers then prepare their bids on the basis that the Consultant's team will be novated to the successful Contractor, who will then be responsible for appointing the team under a new agreement.

It is critical that the Employer, and his advisors, ensure that the wording of all novation agreements is established and agreed with Consultants before going out to tender, so that the Contractor, the Employer and the Designers who are to be novated are fully aware of the requirements, and in particular the Contractor is able to fully allow in their price and programme for that novation.

Employer's Requirements and Contractor's Proposals

Employer's Requirements

The Employer sets out his or her needs in whatever degree of detail is deemed appropriate within the "Employer's Requirements". This can be anything from a performance-related statement of objectives to a detailed and prescriptive statement of what the Contractor is to provide.

The Employer's Requirements should include all relevant criteria, including quality, performance criteria and testing, but need not specify any matters which would be imposed on the Works by the applicable law. Quality should be specified in terms that are not so detailed as to reduce the Contractor's design responsibilities, not so imprecise as to be difficult to enforce, and not subject to anybody's future opinions which tenderers may consider impossible to forecast.

The Contractor responds with the "Contractor's Proposals", which set out the way he or she proposes to design and construct the Works.

When preparing the Employer's Requirements, it is important that the correct balance of information should be considered. If it is too prescriptive it will lead to all the tendering Contractors submitting similar designs and prices. If not detailed enough, then none of the tendering Contractors will produce a design which is as the Employer intended.

The scale of design work required from the Contractor depends on the extent to which the Employer has already commissioned such work from others. The Contractor's design input varies from one contract to another, ranging from the development of a schematic design provided by others into a workable final design for construction, to a full design process including proposals, sketch schemes and production information based on a schedule of requirements and/or performance criteria. There will usually be some negotiation between the Employer and the Contractor, with the aim being to settle on an agreed set of Contractor's Proposals.

These proposals will include the Tender Price, as well as the way it has been calculated.

The relationships between the consultants can be of several types, for example, an Architect may be employed (or other designer) to put together some initial proposals.

These would form the basis of the Employer's Requirements, and the Employer could subsequently assign or novate the Architect to the Contractor.

Alternatively, the Employer may approach directly a Design and Build Contractor who employs in-house consultants.

It is this inherent flexibility that is one of the strengths of Design and Build as a method of procurement. This flexibility happens not just between projects, but also within projects. For example, an Employer with a prestigious office building may have some detailed requirements, comprehensively designed by an architectural team, concerning the facade and the entrance lobby to the building. Once the Employer's Requirements and the Contractor's Proposals are consistent, the Contract can be executed, and the Contractor can implement the work.

The Contractor will be responsible for undertaking the design work in the Contractor's Proposals, including the appointment of consultants if the Contractor does not have the necessary skills in-house. As stated already, the Employer may also choose to appoint consultants to monitor the various aspects of the work, essentially as a "shadow" design team, but this is not always the case.

Some practitioners state that the disadvantage of Design and Build is where there is a conflict between aesthetic quality and ease of fabrication, the latter will take precedence. A further criticism has been that a Design and Build Contractor will put in the minimum design effort required to win the contract.

These two criticisms suggest strongly that quality, particularly architectural quality, will suffer under this procurement process. However, this is not a valid criticism of the process itself, but rather the quality of Employer's Requirements and of some of the people who may exercise it.

As the Employer's Requirements will be the primary document of reference it is important that this document is as complete as possible at the point of tender.

It is the responsibility of the Engineer (or Employer's Representative under the Silver Book) to check that the Employer's Requirements typically cover the following:

- Details of the Site, its boundaries and any site surveys or reports;
- Project description including any drawings/specifications already prepared;
- Purpose for which the building or facility is to be used;
- Statements of functional requirements, e.g. type and number of buildings, density and mix of accommodation, schematic layouts and/or drawings;
- Specific requirements as to finishes and any other elements;
- The pricing schedule;
- Requirements regarding submission of the Contractor's Documents;
- Requirements regarding records the Contractor's Representative is to keep;
- Requirements for "as-built" drawings and other post-completion documents;
- Functions to be carried out by the Engineer (Employer's Representative under the Silver Book) and any other members of the Employer team;
- Details of any Provisional Sums;
- Statement of any planning restraints, e.g. restrictive covenants, and any permissions already granted;
- Access restrictions;
- Availability of public utilities;
- Form of Contract and any amendments and any legal documents;
- Method of presentation of the Contractor's Proposals within the Tender and post-award;
- Various other information to be included in the contract, e.g. Project control requirements and templates;
- Any other matters likely to affect the preparation of the Contractor's Proposals.

As an example, the Employer wishes to invite tenders for the Design and Build of a major government office building.

The Employer has certain essential requirements, but wishes to attract some innovation in that Contractors will interpret the shape, size and appearance of the office building in different ways subject to the approval of external bodies such as the local planners and other legislative bodies.

The Employer's Requirements should therefore comprise the following groups:

Essentials – Requirements that the Employer must have, for example, any specific requirements including the purpose of the building; the number of people using the office and the mix of needs such as open plan/cellular offices; and the need for other facilities within the office, for example restaurant/canteen facilities, meeting rooms, fitness rooms, car parking, etc.

Any other essentials should also be clearly stated, otherwise a tenderer may not include them.

Preferences – Requirements that the Employer would prefer to have. For example, the Employer may have a preference for a certain type of heating or cooling system, but the Contractor may offer an alternative which is more efficient, cheaper to run, etc.

The Employer may also state within his preferences that "equal and approved" may be acceptable subject to the Contractor proving that it truly is "equal" and also that he is prepared to approve it. As part of this proof the Contractor should consider issues such as life cycle costing. For example, if the Contractor can prove that the alternative heating or cooling system is equally efficient and easy to operate, but in the longer term the alternative requires a higher level of maintenance then the Employer will need to be made aware of this requirement. It is suggested that the preferences should not apply to critical aspects of the Project.

Innovations – Something the Employer may not have thought of that the Contractor believes could win the tender for him!

This is open to the Contractor to provide as added value his particular bid. For instance, one Contractor may offer additional facilities, or particularly "green" environmental design/construction over and above the requirements of the Employer's Requirements and/or the relevant legislation.

If a Contractor feels that he can exceed the Employer's Requirements it is essential that the Contractor's Proposals comply with the document (a compliant bid), but list separately how he could exceed it, and any time and/or price implications should the Employer wish to take up the options.

With regard to the Contractor's Proposals, he will wish to demonstrate that his bid is compliant with the Employer's Requirements, though it is important that he is not prescriptive about exactly what he will provide, as his design may evolve and develop as the construction progresses. He may find new equipment and materials, or he may wish to use some monies to offset any problems that may arise.

The author has been recently involved with drafting the Employer's Requirements for a major office project.

The Employer was interviewed and a questionnaire drafted to consider the following as a basis for the Employer's Requirements:

Quality

What are the overall requirements?

1. The size and use of the building. If no specific size of building is known, details of exactly what the Employer intends to use the proposed building for and expected number of occupants.
2. Open plan or cellular offices;
3. Any specific requirements, for example canteen facilities, gym, etc. Number of floors, clear span or sub-divisions;
4. Need for future flexibility and expansion;
5. Specific heating requirements, lighting levels for various parts of the building, air conditioning requirements, running costs and energy conservation requirements;
6. Car parking facilities;
7. Maintenance, running costs and life cycle cost factors. How long is the building expected to last?

Time

1. When will the Employer require the building?
2. Is the starting date and/or completion date critical to the Employer?
3. Consider pre-tender activities such as planning consents, preparation of tender documents and also other operations which need to be carried out before the Contractor can be appointed and start on Site.

Cost

1. What is the budget for the building? How restricted is that budget? Is there a cost plan and expected cash flow projections for the building? How long will it take to prepare one?
2. Is the project being wholly funded by the Employer or is partial funding coming from another source, for example grant aid? How much influence do the funders have in terms of the design and construction of the building?

From these initial considerations the documentation to be prepared for tender can be established:

1. What drawings are to be included within the Employer's Requirements? Broad sketch plans, showing size and general layout only? Or detailed design drawings?
2. Information about the Site. How detailed will it be? Who is responsible for accuracy?
3. Who will apply for outline and later detailed planning consents? Possible delays with consents? Revised applications due to changes in design suggested by Contractor?
4. Performance Specifications? Or specification of work and materials? Outline specifications or detailed at this stage?
5. Any major products, specialist or supply items over which the Employer wishes to exercise some choice or control. Nomination is difficult with design build and creates complications.
6. Employer to provide an outline time programme, or left to the Contractor to submit with his tender?

7 Sectional Completions? Split Responsibilities and Insurances.
8 Constraints. Access, space, working hours, sequence of working.
9 Which main and secondary options are to be used?
10 Design and construction liability. Any special insurance requirements or performance bonds? Design warranties?

From the above it can be readily seen that it is vital that the Employer defines how much information and detail is required in the returned tender and it is crystallised within the Employer's Requirements.

The Contractor's Proposals (where relevant)

In simple terms, the Contractor's Proposals should reflect and directly respond to the Employer's Requirements.

The representative of the Employer must check that the Employer's Requirements clearly, concisely and correctly interpret the Employer's brief.

If the Employer's Requirements are detailed, the Contractor's Proposals will be detailed, if they are vague then the Contractor's Proposals will also be vague, so the Employer may not get what he or she expected.

As stated above, it is important to note that if the Contractor's Proposals differ, or fall short of the Employer's Requirements.

The Contractor's Proposals within his tender should incorporate the following:

- Plans, elevations, sections or typical details;
- Information about the structural design;
- Outline drawings of incorporated services;
- Specifications for materials and workmanship;
- Any alternative solutions departing from the brief;
- A method statement and construction programme;
- The price;
- A staged payment schedule (if appropriate);
- Proposed main Subcontractors and Suppliers.

It is important to note that in most contracts, if the Contractor's Proposals subsequently fall short of the Employer's Requirements then the Employer's Requirements will prevail.

The FIDIC 2017 requirements in respect of design

Red Book – Clause 4.1

Within the Red Book, specifically in respect of any design to be carried out by the Contractor, Sub-Clause 4.1 states that except to the extent specified within the Contract, the Contractor:

i Is responsible for the Contractor's Documents, Temporary Works (e.g. temporary support works, formwork, etc.) and such design as is required for the item to be in accordance with the Contract.
ii Is not otherwise responsible for the design or specification of the Permanent Works.

So, the default is that the Contractor does not have responsibility for the design of the Permanent Works.

Sub-paragraphs (a) to (d) (below) specify the general requirements in respect of any design to be carried out by the Contractor, unless otherwise defined in the Particular Conditions.

These general requirements are also stated in 13.3 (b) as being applicable to any subsequent Contractor-designed "value engineering" proposals, unless otherwise agreed by both Parties.

Under the Red Book, if the Contractor is required to design any part of the Permanent Works then unless the Particular Conditions state otherwise:

a The Contractor is required to prepare and submit to the Engineer for Review the Contractor's Documents (see below) for the part that the Contractor is responsible for designing, and any other documents necessary for him to complete and use the design during the execution of the Works and to instruct the Contractor's Personnel.

b The Contractor's Documents are required to be in accordance with the Specification and Drawings (so the Employer should make sure he has clearly stated the requirements within the Specification and Drawings) and also to include additional information which may be required by the Engineer to add to the Drawings for co-ordination of each Party's designs.

  Very often design problems arise, not within each Party's designs, but at the interface between them and the co-ordination between different designers.

c Construction must not commence until a Notice of No-objection is given.

  Note that the Engineer does not approve the Contractor's design, but he may issue a Notice stating that he does not object to the design.

  As noted in Chapter 2, the FIDIC 2017 Contracts have a new term "No-objection", which is defined in Sub-Clause 1.1.55 (Silver Book Sub-Clause 1.1.48) as meaning that "*the Engineer/Employer has no objection to the Contractor's Documents, or other documents submitted by the Contractor under these Conditions, and such Contractor's Documents or other documents may be used for the Works*".

  The Engineer, having reviewed the Contractor's Documents, does not actually "approve" or "accept" them, but does "not object" to them.

  In doing so the Engineer is not trying to absolve himself of liability, by only giving "No-objection", but what he is doing is not objecting to the design, but at the same time not absolving the Contractor of his liability in this case for his design responsibility and the associated Contractor's Documents, and this is how the FIDIC 2017 No-objection notice should be viewed and applied.

d The Contractor may modify any design or Contractor's Documents which have previously been submitted for review by giving a notice to the Engineer, with reasons for the modification.

e The Contractor must ensure that when the Works are completed they are fit for the intended purpose, or where no purpose is intended or prescribed, fit for their ordinary purpose (see also above)

  This is an important matter to be borne in mind when the tender documents are being written, in particular because of the absence of any reference to purposes which may reasonably be inferred from the Contract. Although the risk is fairly allocated to the Contractor, he may not be able to be indemnified by his insurers or his subcontract designers.

- f The Contractor undertakes that the design and the Contractor's Documents comply with the technical standards stated within the Specification and Laws.
  Again, an important point, the Contractor's design does not just have to comply with the Contract, but also with the relevant Laws.
- g If the Contractor has to submit As-Built Records and/or Operation and Maintenance Manuals, then he must do so in sufficient detail for the Employer to operate, maintain, dismantle, reassemble, adjust and repair the relevant part.
- h If the Contractor is required to provide Training, the Contractor carries out training of the Employer's Personnel in the operation and maintenance of this part.

Yellow and Silver Books – Clause 5

General Design Obligations

Within the Yellow and Silver Books, Clause 5 is entitled "Design" and covers all Contractor's design-related provisions.

As stated above, when using the Yellow Book and the Silver Book, the Contractor designs and provides the Works in accordance with the Employer's Requirements.

The requirement under FIDIC 2017 for the Contractor to provide details of his designers is significantly more than in FIDIC 1999.

Under FIDIC 2017, the Design must be prepared by designers who:

- a Are qualified, experienced professionals and competent in the disciplines of the design for which they are responsible;
- b Comply with the criteria stated in the Employer's Requirements (if any are stated);
- c Are qualified and entitled under applicable Laws to design the Works. This obviously covers the basic qualifications, but there may also be a requirement for specific permits or licences or grades to be able to carry out design in a specific country.

The Contractor must also submit to the Engineer (Employer under the Silver Book) for his consent, the name, address, particulars and experience of each proposed designer, unless otherwise stated in the Employer's Requirements.

Finally, the Contractor warrants that the Contractor, the Contractor's designers and design Subcontractors all have the experience, capability and competence necessary to carry out the design as required.

The Contractor is also required to undertake that the designers (and any design Subcontractors where relevant) will be available to attend discussions with the Engineer and/or the Employer at all reasonable times until the issue of the Performance Certificate.

So, in all there are quite prescriptive and significant requirements within the FIDIC 2017 in respect of the designers themselves.

After receiving a Notice to commence the Works, the Contractor is required to scrutinise the Employer's Requirements, and if he discovers any error, fault or other defect Sub-Clause 1.9 [Errors in the Employer's Requirements] shall apply.

Under Sub-Clause 1.9 (see also Chapter 2) if the Contractor finds an error, fault or defect in the Employer's Requirements he gives a Notice to the Engineer within the period stated in the Contract Data (if not stated, 42 days) calculated from the Commencement Date. The Contractor may notify later than the period stated see (b) below.

The Engineer then determines whether or not there is an error, fault or defect in the Employer's Requirements, and also whether or not an experienced contractor exercising due care would have discovered the error, fault or other defect:

a When examining the Site and the Employer's Requirements before submitting the Tender; or
b If the Contractor's Notice is given after the period stated in the Contract Data, when scrutinising the Employer's Requirements under Sub-Clause 5.1 [General Design Obligations]; and
c What measures (if any) the Contractor is required to take to rectify the error, fault or defect.

If, under sub-paragraph (b) above, an experienced contractor would not have discovered the error, fault or other defect:

i Sub-Clause 13.3.1 [Variation by Instruction] applies to the measures that the Contractor is required to take (if any), and
ii If the Contractor suffers delay and/or incurs Cost as a result of the error, fault or defect, the Contractor is entitled to an Extension of Time and/or payment of Cost plus Profit.

Note that Sub-Clause 1.9 is not included within the Silver Book.

The Silver Book states that the Employer is not responsible for any error, inaccuracy or omission with the Employer's Requirements, and any data or information received by the Contractor from the Employer does not relieve the Contractor of any obligation to provide the Works.

The only exclusions are:

a Data and information which is specifically stated within the Contract as being the responsibility of the Employer;
b Definitions of intended purpose of the Works;
c Criteria for testing and performance of the completed Works;
d Data and information which cannot be verified by the Contractor.

Sub-Clause 5.2 within the FIDIC 1999 Contracts was one long clause, but has now been subdivided within the FIDIC 2017 Contracts into three parts:

5.2.1 – Preparation by Contractor
5.2.2 – Review by Engineer (Employer under the Silver Book)
5.2.3 – Construction

Contractor's Documents

The provisions related to Contractor's design and submission and review of the Contractor's Documents have been restructured from the FIDIC 1999 Contracts, and further detail added in a structured format, so that the Contractor submitting the documents, and the Engineer (under the Yellow Book 2017) or the Employer (under the Silver Book 2017) reviewing the documents know exactly what to do, and by when.

The Contractor's Documents consist of the documents:

a Specified in the Employer's Requirements (this needs to be carefully drafted with the Employer's Requirements);
b Required for the Contractor to satisfy all permits, permissions, licences and other statutory/regulatory approvals which are his responsibility under the Contract;
c Described in respect of all "As-Built Records" and "Operation and Maintenance Manuals".

Unless otherwise stated in the Employer's Requirements, the Contractor's Documents must be written in the language for communications defined in the Contract.

The Contractor must prepare all the Contractor's Documents, and any other documents necessary to complete and implement the design during execution of the Works and to instruct the Contractor's Personnel. The Employer's Personnel have the right to inspect the preparation of all these documents.

Review by Engineer (Employer under the Silver Book)

The Engineer (Employer under the Silver Book) has a Review Period of not exceeding 21 days, which is calculated from the date on which the Engineer (Employer under the Silver Book) receives a Contractor's Document and a Contractor's Notice.

"Contractor's Document" excludes any of the Contractor's Documents which are not specified in the Employer's Requirements as being required to be submitted for Review, but includes all documents on which a specified Contractor's Document relies for completeness.

"Contractor's Notice" means the Notice which states that the relevant Contractor's Document is considered by the Contractor to be ready for Review, and that it complies with the Employer's Requirements and the Conditions, or the extent to which it does not do so.

The Engineer (Employer under the Silver Book), within the 21 day Review Period, must give a Notice to the Contractor:

a Of No-objection (which may include comments); or
b That the Contractor's Document fails to comply with the Employer's Requirements and/or some other part of the Contract, with reasons.

Note, as previously stated, that the Engineer no longer approves the Contractor's Documents, he gives a Notice of no-Objection.

If the Engineer (Engineer under the Silver Book) gives no Notice within the 21-day Review Period, the Engineer is deemed to have given a Notice of No-objection to the Contractor's Document (provided that all other Contractor's Documents on which that Contractor's Document relies (if any) have been given, or are deemed to have been given, a Notice of No-objection).

If the Engineer (Employer under the Silver Book) instructs that further Contractor's Documents are reasonably required to demonstrate that the Contractor's design complies with the Contract, the Contractor must prepare them and submit them promptly to the Engineer at the Contractor's cost.

If the Engineer (Employer under the Silver Bok) gives a Notice that the Contractor's Documents fail to comply with the Employer's Requirements and/or the Contract, the Contractor must revise the Contractor's Document and resubmit it to the Engineer for Review, and the Review Period is calculated from the date that the Engineer receives it.

The Contractor is not entitled to an Extension of Time for any delay caused by this resubmission and subsequent Review by the Engineer.

If the Employer incurs additional costs as a result of the resubmission and subsequent Review, the Employer shall be entitled subject to Sub-Clause 20.2 [Claims for Payment and/or EOT] to payment by the Contractor of the costs reasonably incurred.

Construction

Except for Contractor's Documents under Sub-Clause 5.6 [As-Built Records] and Sub-Clause 5.7 [Operation and Maintenance Manuals], for each part of the Works requiring Contractor's Documents to be submitted for Review:

a Construction of such a part cannot commence until a Notice of No-objection is given (or is deemed to have been given) by the Engineer for all the Contractor's Documents which are relevant.

   This is a very critical provision.

b Construction of such a part shall be in accordance with these Contractor's Documents.

c The Contractor may modify any design or Contractor's Documents which have previously been submitted for Review, by giving a Notice to the Engineer with reasons. If the Contractor has commenced construction of the part of the Works to which such design or Contractor's Documents are relevant:

   i Work on this part is suspended.

   ii The provisions of Sub-Clause 5.2.2 [Review by Engineer/Employer] apply as if the Engineer (Employer under the Silver Book) had given a Notice in respect of the Contractor's Documents under sub-paragraph (b) of Sub-Clause 5.2.2.

   iii Work on this part shall not resume until a Notice of No-objection is given (or is deemed to have been given) by the Engineer (Employer under the Silver Book) for the revised documents.

Contractor's undertaking

The Contractor is required to undertake that the design is in accordance with the Laws of the Country, in addition to the documents forming the Contract. This is critical.

Technical standards and regulations

The Contractor's Documents and the execution of the work are to be in accordance with the Contract, and the applicable Laws in force when the Works or Section or Part is taken over.

References in the Contract to published standards are to the edition applicable on the Base Date i.e. 28 days prior to the tender submission date, unless stated otherwise.

If changed standards come into force in the Country after the Base Date, the Contractor promptly gives a Notice to the Engineer and submits proposals for compliance with those changed standards. Note that changed standards do not always apply to existing contracts.

To the extent that:

a The Engineer (Employer under the Silver Book) considers that compliance is required, and compliance requires change(s) to the execution of the Works; and
b The Contractor's proposals for compliance constitute a Variation, then the Engineer (Employer under the Silver Book) initiates a Variation.

Training

The Contractor is required to provide training of the Employer's Personnel in the operation and maintenance of the Works if and as required by the Employer's Requirements.

The Contractor is required to provide qualified and experienced staff to carry out the training, also adequate training facilities and all training materials as necessary and/or as stated in the Employer's Requirements.

As-Built documents

The Contractor's obligations to provide "As-Built Records" and "Operation and Maintenance Manuals" have been expanded and are now set out separately in Sub-Clauses 4.4.2 and 4.4.3.

The Contractor is required to prepare and update a complete set of "As-Built" Records and to issue them to the Engineer (Employer under the Silver Book). This would be in the form of drawings, specifications, contact details, etc.

The format of the documents, any referencing system, electronic storage and other relevant details of the As-Built Records must be strictly as stated in the Employer's Requirements.

The Contractor is required to submit to the Engineer (Employer under the Silver Book):

a The As-Built Records for the Works or Section before the commencement of the Tests on Completion; and
b Updated As-Built Records to the extent that any work is executed by the Contractor:
    i During and/or after the Tests on Completion, but before the issue of any Taking-Over Certificate; and
    ii After taking over, before the issue of the Performance Certificate.

Operation and Maintenance Manuals

The Contractor is to supply, prior to commencement of Tests on Completion, Operation and Maintenance Manuals. The Works are not deemed to be complete for taking over until the Engineer (Employer under the Silver Book) has received them.

The format of the Operation and Maintenance Manuals is stated in the Employer's Requirements and must be in sufficient detail for the Employer to:

i Operate, maintain, dismantle, reassemble, adjust and repair the Plant; and
ii Operate and maintain the Works to ensure that the performance of the Works, Section and/or Plant continues to comply with the performance criteria specified in the Employer's Requirements and the Schedule of Performance Guarantees.

It must also include an inventory of spare parts required for the Employer's future operation and maintenance of the Plant.

Before commencement of the Tests on Completion, the Contractor must submit provisional operation and maintenance manuals for the Works or Section (as the case may be) to the Engineer.

If, during the Tests on Completion, any error or defect is found in the provisional Operation and Maintenance Manuals, the Contractor must promptly rectify the error or defect at the Contractor's risk and cost.

Before the issue of any Taking-Over Certificate, the final operation and maintenance manuals must be submitted to the Engineer (Employer under the Silver Book).

Design error

Any design errors, ambiguities, inconsistencies, etc., in the Contractor's Design or Contractor's Documents are to be corrected by the Contractor at his own cost.

6 Variations, Measurement and Payment Procedures

Variations, Measurement and Payment Procedures are covered in the FIDIC 2017 Contracts by:

- *Conditions of Contract for Construction* (The "Red Book")
  - Clause 12 – Measurement and Valuation
  - Clause 13 – Variations and Adjustments
  - Clause 14 – Contract Price and Payment
- *Conditions of Contract for Plant and Design-Build* (The "Yellow Book")
  - Clause 13 – Variations and Adjustments
  - Clause 14 – Contract Price and Payment
- *Conditions of Contract for EPC/Turnkey Project* (The "Silver Book")
  - Clause 13 – Variations and Adjustments
  - Clause 14 – Contract Price and Payment

Clause 12 is only included within the Red Book, is entitled "Measurement and Valuation" and covers all payment-related subjects:

Measurement and Valuation

The Works are to be measured and valued for payment in accordance with the Contract.

The Engineer decides what is to be measured and when, and issues a Notice to the Contractor not less than seven days before it is to be measured, stating what is to be measured and when.

The Contractor's Representative is then required to attend, and to supply any particulars required by the Engineer.

If, having been given notice, the Contractor fails to attend, the Engineer's measurement is deemed accepted as accurate.

If the Contractor attends, or examines the measurement records and the Engineer and Contractor are unable to agree the measurement, then the Contractor gives a Notice to the Engineer setting out the differences.

If the Contractor fails to give Notice within 14 days after attending the measurement or receiving the measurement records the Contractor is deemed to have accepted.

If the Contractor issues a Notice, the Engineer proceeds to agree or determine the measurement under Sub-Clause 3.7 and within the time limit for agreement.

Until such time as the measurement is agreed, the Engineer issues a provisional measurement for the purpose of payment.

In reality, it is usually the Contractor who physically carries out the measurement of the Works and submits it to the Engineer, rather than the Engineer actually doing the measurement.

The method of measurement is stated within the Contract Data, or in accordance with the Bills of Quantities or other Schedule.

Valuation of the Works

A rate in the Bills of Quantities or price for an item of work may be amended if:

a The item is not included within the Bills of Quantities or other Schedule (so there was never a rate or price for it);

b

- i The measured quantity of the item is changed by more than 10% from the quantity of this item in the Bill of Quantities or other Schedule;
- ii This change in quantity multiplied by such specified rate for this item exceeds 0.01% of the Accepted Contract Amount (the original amount in the Letter of Acceptance);
- iii This change in quantity directly changes the Cost per unit quantity of this item by more than 1%; and
- iv This item is not specified in the Contract as a "fixed rate item".

Note that all FOUR conditions have to apply:

c The work is instructed as a Variation.

The amended rate may be from relevant similar work in the Bill of Quantities or other pricing schedule (with adjustments where necessary), for example dependent on the quantity or the conditions under which the work is carried out, or by assessing reasonable Cost plus Profit.

The cost which has been incurred and is not covered within the omission of any work is determined by the Engineer.

Variations and Adjustments

Right to vary

The Engineer (Employer under the Silver Book) may vary the Works any time before the issue of the Taking-Over Certificate for the Works.

The Contractor is required to comply with the Variation, subject to the following exceptions:

a The varied work was Unforeseeable (note that "Unforeseeable" has a strict definition i.e. "*not reasonably foreseeable by an experienced contractor by the Base Date*", which is 28 days before the tender submission date).

b The Contractor cannot obtain the required Goods for the Variation.

c It will adversely affect the Contractor in terms of Health and Safety Obligations or Protection of the Environment.

There are two additional exceptions under the Yellow and Silver Books:

- i It will have an adverse impact on the Schedule of Performance Guarantees.
- ii It may adversely affect the Contractor's obligation to complete the Works fit for the purpose(s) stated in the Contract.

This express right of the Contractor, to object via a Notice to the Engineer (Employer under the Silver Book), is new to FIDIC 2017, previously the clause was open ended.

Once the Contractor has given a Notice to the Engineer (Employer under the Silver Book), the Engineer (Employer under the Silver Book) may then cancel, confirm or vary the instruction.

Note that the FIDIC 1999 Contracts (Sub-Clause 13.1) only allowed the Contractor to object to a Variation if it could not readily obtain the Goods required for the Variation.

The FIDIC 2017 Yellow and Silver Books also include that the Variation may adversely affect the Contractor's fitness for purpose obligation in respect of the completed Works.

Note that Sub-Clause 13.1 now expressly provides for the Parties to agree to the omission of any work to be carried out by the Employers or others.

Under Clause 13.3, if the Parties agree to an omission of any work which is to be carried out by others, the Contractor is entitled to loss of profit and any other losses as a result of the omission. Note that this is specifically excluded from Sub-Clause 1.15 which states that neither Party is liable to the other for loss of profit.

Value engineering

The principle of value engineering is about eliminating or reducing unnecessary costs, while also maintaining or improving function and quality of the finished product, essentially increasing the value of that product.

But value engineering is not just about reducing costs, for example, there may be time savings during construction, or capital cost may be increased initially to give cost savings, efficiencies or other enhancements over the design life of a product i.e. life cycle costs.

Sub-Clause 13.2 allows the Contractor to submit a proposal to the Engineer (Employer under the Silver Book) which could

- a Accelerate completion of the Works.

  Acceleration can either mean increasing resources, working faster or working longer hours so that Completion can be achieved by the Completion Date, in effect the Contractor is catching up to mitigate or recover a delay.

  However, acceleration in this case is where the Contractor includes within his proposal for increasing resources, working faster or working longer hours so that Completion can be achieved before the Completion Date. It should be established that as the Employer is paying or part paying for the acceleration measures, the Contractor is not seeking to cover his own delays "under cover" of a proposal to accelerate completion of the Works.
- b Reduce the cost to the Employer of executing, maintaining or operating the Works.

  This could be a short-term saving, or a long-term saving over the design life of a component, for example, energy savings on electrical components.

c Improve the efficiency or value of the completed Works.
d Or provide some other benefit.

The Contractor is required to prepare and submit a proposal at his own expense, and the Engineer (Employer under the Silver Book) responds as soon as practicable with his consent or otherwise.

If the Engineer (Employer under the Silver Book) gives his consent, it is treated as a Variation by Instruction, and will include consideration by the Engineer (Employer under the Silver Book) of any sharing of benefit, costs or otherwise as stated in the Particular Conditions.

Note that as any benefit to the Contractor is established through the Engineer's (Employer under the Silver Book) consideration, this may conclude with zero benefit to the Contractor!

Note that the FIDIC 1999 Red Book provided for the Contractor to be paid a fee equalling half (50%) of the difference between:

i The reduction of the contract value, and
ii Any reduction in value to the Employer, for example reduction in quality, anticipated life or operation efficiencies.

If (ii) exceeds (i) no fee is applicable.

It may be appropriate for a fixed percentage of any cost savings to be included within the Contract as previously included within the FIDIC 1999 Contracts, so that the Contractor may be motivated to submit value engineering proposals knowing the financial benefit, not only to the Employer, but also to himself, as proposals may have a significant costs in their drafting and submission, particularly where it may involve a re-design by the Contractor.

Unless otherwise agreed, the Contractor is responsible for any re-design at his own cost and the rules regarding Contractor's design.

Variation procedure

Note first that the valuation of a Variation does not require the Contractor to give a Notice under Clause 20.2 as if it was a claim, or to deal with it in that manner.

However, any claim for a breach of contract does require a Notice. So, if a Contractor has given Notice and the Engineer (Employer under the Silver Book) has confirmed the instruction, the Contractor should immediately give notice under Clause 20.2 to entitle a claim for additional payment and/or an extension of time if the Engineer (Employer under the Silver Book) later declines or fails to include the costs in its valuation of the Variation.

Significantly, if the Engineer (Employer under the Silver Book) issues the Contractor with a Notice but it is not referred to as a Variation, but the Contractor considers that it is a Variation, the Contractor can immediately notify the Engineer that it considers that a Variation has been instructed. This prevents any misunderstanding between the Parties.

If the Engineer (Employer under the Silver Book) does not respond either to confirm or revoke the instruction within seven days, the Engineer (Employer under the Silver Book) will be deemed to have revoked the instruction.

The Engineer (Employer under the Silver Book) may initiate Variations of two types:

i "Variation by Instruction" – which are to be carried out as instructed;
ii "Variation by Request for Proposal" – for which the Contractor is required to provide a price before receiving an instruction to carry out the work.

NB: the FIDIC 1999 Contracts provided for a Variation to be instructed, or a proposal requested, the FIDIC 2017 Contracts deal with the matter in two parts: in greater detail and in a more defined way.

Variation by Instruction

The Engineer (Employer under the Silver Book) may instruct a Variation by giving a Notice to the Contractor in accordance with the Contract.

The Contractor is required to proceed with the work as instructed subject to the exceptions stated above, and within 28 days to submit to the Engineer (Employer under the Silver Book) detailed particulars including:

i A description of the varied work which is to be carried out, including details of the resources and methods which the Contractor proposes to adopt;
ii A programme for the work instructed;
iii The Contractor's proposal for adjustment to the Contract Price, with supporting particulars, including any prolongation costs to be incurred as a result of carrying out the work.

As stated above, if the instruction is an omission of work, the Contractor may include within his proposal an amount for loss of profit or any other losses as a result of the omission. Note that the Contract generally excludes liability for loss of profit, consequential loss, etc., but this is one of the few exceptions.

The Engineer (Employer under the Silver Book) then, under Sub-Clause 3.7, agrees or determines:

i Any Extension of Time;
ii Any adjustment to the Contract Price.

Variation by Request for Proposal

The Engineer (Employer under the Silver Book) may request a proposal, before instructing a Variation, by giving a Notice to the Contractor.

This facility was available to the Engineer (Employer under the Silver Book) under the FIDIC 1999 Contracts, but is now more defined under the FIDIC 2017 Contracts.

The Contractor is required, similar to a Variation by Instruction, to respond to the Notice as soon as practicable, by either:

a Submitting a proposal, which include the matters previously stated i.e.

  i A description of the varied work which is to be carried out, including details of the resources and methods which the Contractor proposes to adopt;

- ii A programme for the work instructed;
- iii The Contractor's proposal for adjustment to the Contract Price, with supporting particulars, including any prolongation costs to be incurred as a result of carrying out the work.

b Giving reasons why the Contractor cannot comply.

If the Contractor submits a proposal, the Engineer (Employer under the Silver Book) must, as soon as practicable, respond by giving a Notice to the Contractor giving his consent to the proposal, or otherwise.

If the Engineer (Employer under the Silver Book) accepts the proposal, which can be with or without comments, the Engineer (Employer under the Silver Book) then instructs the Variation based on that proposal.

Note that, if the Engineer (Employer under the Silver Book) does not accept the proposal, again with or without comments, and if the Contractor has incurred Cost as a result of submitting the proposal, the Contractor is entitled to payment of such Cost.

Provisional Sums

Provisional Sums are used in contracts where there are elements of work which are not designed or cannot be sufficiently defined at the time of tender but may be required to be done by the Contractor, and therefore, a sum of money is included by the Employer within the Bill of Quantities or other pricing document to cover the item.

When the item is defined or able to be properly defined, the Engineer instructs regarding the Provisional Sum, the Contractor is given the information which allows him to price it, the Provisional Sum is omitted and the price included in its stead (see also Chapter 12).

The Engineer (Employer under the Silver Book) may give instructions regarding Provisional Sums with the Contract Price being adjusted accordingly.

For each Provisional Sum, the Engineer (Employer under the Silver Book) may instruct:

- a Work to be executed by the Contractor, and for which adjustments to the Contract Price are agreed or determined as a Variation by Instruction (see above); and/or
- b Plant, Materials, works or services to be purchased from a nominated Subcontractor (see Clause 5) or otherwise, and which there is included in the Contract Price.

Note that in the case of (b) the amount due to the Contractor includes:

- i The actual amounts paid (or due to be paid) by the Contractor; and
- ii A sum for overhead charges and profit, calculated by applying the relevant percentage rate stated in a Schedule, or within the Contract Data.

The Engineer's (Employer under the Silver Book) instruction may include a requirement for the Contractor to submit quotations from the Contractor's suppliers and/or subcontractors for all, or some, of the work. This is a new provision within the FIDIC 2017 Contracts.

The Engineer (Employer under the Silver Book) may then give a Notice either instructing the Contractor to accept one of these quotations or revoking the instruction.

If the Engineer (Employer under the Silver Book) does not respond within seven days of receiving the quotations, the Contractor is entitled to accept any of these quotations at his discretion.

Daywork

For work of a minor or incidental nature, particularly where nothing exists within the Contract to compare a price with (perhaps some minor demolition which is covered by a Variation), the Engineer (Employer under the Silver Book) may instruct that a Variation be executed on a daywork basis and the work is valued according to the Daywork Schedule. Note that such work has to be specifically instructed to be carried out on a Daywork basis.

Before ordering Goods for such work (other than any Goods already priced in the Daywork Schedule), the Contractor submits one or more quotations from the Contractor's suppliers and/or subcontractors to the Engineer (Employer under the Silver Book).

The Engineer (Employer under the Silver Book) may instruct the Contractor to accept one of these quotations. It is not common for quotations to be required for work carried out on daywork basis, but this provision is within the FIDIC 1999 and now 2017 Contracts.

If the Engineer (Employer under the Silver Book) does not instruct the Contractor differently within seven days of receiving the quotations, the Contractor is entitled to accept any of these quotations at his discretion.

The Contractor is required to deliver each day to the Engineer (Employer under the Silver Book) accurate statements in duplicate (and one electronic copy), including records of the resources used in executing the previous day's work.

One copy of each statement, if correct and agreed, is be signed by the Engineer (Employer under the Silver Book) and promptly returned to the Contractor.

If not correct or agreed, the Engineer (Employer under the Silver Book) agrees or determines the resources, and the date the Works are completed by the Contractor is the date of commencement of the time limit for agreement under Sub-Clause 3.7.3.

In the next Statement submitted for payment, the Contractor then submits priced statements of the agreed or determined resources to the Engineer (Employer under the Silver Book), together with all applicable invoices, vouchers and accounts or receipts in substantiation of any Goods used in the daywork (other than Goods priced in the Daywork Schedule).

Unless otherwise stated in the Daywork Schedule, the rates and prices in the Daywork Schedule are deemed to include taxes, overheads and profit.

Adjustments for Changes in Laws

This Sub-Clause was formerly titled in FIDIC 1999 "Adjustments for Changes in Legislation".

Under Sub-Clause 1.4 (Law and Language), the Contract is governed by the law of the country stated in the Contract Data. The ruling language of the Contract and the language for communications are also stated in the Contract Data.

Sub-Clause 1.13 (Compliance with Laws) requires that the Parties comply with all applicable laws (which is obvious), but specifically that (unless stated within the Specification) the Employer is responsible for obtaining planning, zoning, building permit, licences and/or approvals for the Permanent Works, and any other permits, permissions, licenses and/or approvals described in the Specification as being obtained by the Employer, and indemnifies the Contractor against the failure to do so.

The Contractor is required to give all notices, pay taxes, duties and fees, etc., and to and obtain all other permits, permissions, licences and/or approvals in relation to the execution of the Works and indemnifies the Employer against the failure to do so.

The Contractor is required to provide any assistance and all documentation, as stated in the Specification, or reasonably required by the Employer, to allow the Employer to obtain any permit, permission, licence, approval, etc.

The Contractor is also required to comply with all permits, permissions, licences and/or approvals obtained by the Employer.

If the Contractor suffers delay and/or incurs Cost as a result of the Employer's delay or failure to obtain any permit, permission, licence, approval, etc., the Contractor is entitled to an Extension of Time and/or payment of Cost plus Profit.

The Changes in Laws provisions have been extended to apply to changes in permits to be obtained by the Employer or a change in the requirements for any permit to be obtained by the Contractor.

Similarly, if the Employer incurs additional costs as a result of the Contractor's failure to comply, the Employer is entitled to payment of these costs by the Contractor.

Note that the Contract Price that is increased or decreased due to changes in Laws of the Country after the Base Date is paid to the Contractor.

The Sub-Clause defines what constitutes changes in laws, including the introduction of new Laws, judicial or governmental interpretation of Laws, any permit, permission of approval required to be obtained by the Employer or the Contractor. This is an expansion of what was previously provided under FIDIC Sub-Clause 13.7 and could now include changes in laws outside the country where the project is located so could include for materials and/or plant purchased from abroad.

If the Contractor suffers delay, he is entitled to an Extension of Time and/or payment of Cost, subject to giving the appropriate Notice under Sub-Clause 20.2.

There is a change under the FIDIC 2017 Contracts in that, if there is a decrease in Cost, the Employer is entitled to a reduction in the Price, so it provides a remedy either way.

Note that if there is a Change in the Laws which requires an adjustment to the execution of the Works, the Contractor can promptly give a notice to the Engineer (Employer under the Silver Book) or the Engineer (Employer under the Silver Book) can give a Notice to the Contractor. There is no time scale for issuing such Notice, other than promptly.

This may then be dealt with as a Variation.

Adjustments for changes in cost

The FIDIC 1999 Contracts (Sub-Clause 13.8) contained a complex formula for calculating the adjustments for changes in cost, based upon the assumed availability of suitable cost indices. The formula defined the adjustment multiplier (Pn) to be applied to the contract value of each month's work.

This has now been replaced within the FIDIC 2017 Contracts with a reference to "Schedule(s) of cost indexation" which may (or may not if the Sub-Clause is not applicable) be included in the Contract, the Contractor being paid within Payment Certificates for increases/decreases in cost by reference to the Schedule(s) calculated in the currencies in which the Contract Price is payable.

The new provision provides a degree of flexibility for the Parties to agree on payment for fluctuations.

The Contractor is responsible, and is deemed to have allowed, for any increases/decreases in cost not covered by the Schedule(s).

The Engineer (Employer under the Silver Book) uses a provisional index when issuing Payment Certificates until the current cost index is available.

If the Contractor fails to complete the Works within the Time for Completion, the adjustment of prices beyond the Time for Completion is calculated using either:

a Each index or price applicable on the date 49 days before the expiry of the Time for Completion of the Works; or
b The current index or price, whichever is more favourable to the Employer.

The Contract Price

Sub-Clause 14.1 sets out the financial principles of the Contract:

i Red Book

  a The Contract Price is set by the value of the Works in accordance with Sub-Clause 12.3 (Valuation of the Works) (see above), and subject to adjustments, additions and/or deductions, for example, for Variations, Claims, adjustments for Changes in Laws or Costs, etc., in accordance with the Contract.
  b The Contractor is required to pay all taxes, duties and fees.
  c Any quantities in the Bills of Quantities are estimated, and subject to remeasurement. Although they are estimated they must reasonably reflect the type of work to be done, and the relevant quantity is expected, though it may not be possible to be totally accurate.
  d The Contractor is required to submit to the Engineer within 28 days of the Commencement Date a breakdown of each lump sum in the Schedules that the Engineer (Employer under the Silver Book) can take account of for the purpose of preparing Payment Certificates.

ii Yellow Book

  a The Contract Price is the lump sum Accepted Contract Amount, and subject to adjustments, additions and/or deductions again, for example, for Variations, Claims, adjustments for Changes in Laws or Costs, etc., in accordance with the Contract.
  b The Contractor is required to pay all taxes, duties and fees.
  c Any quantities which may be set out in a Schedule are estimated quantities and not to be taken as actual and correct.
  d Any quantities or price data set out in a Schedule shall only be used for the purposes stated within the Schedule.

iii Silver Book

- a Payment for the Works is based on the lump sum Contract Price stated in the Contract Agreement, and subject to adjustments, additions and/or deductions, for example, for Variations, Claims, adjustments for Changes in Laws or Costs, etc., in accordance with the Contract.
- b The Contractor is required to pay all taxes, duties and fees.
- c Any quantities which may be set out in a Schedule are estimated quantities and not to be taken as actual and correct, and are only to be used for the purposes stated within the Schedule.

Whilst the Contract Price for the Red, Yellow and Silver Books is different, the payment process for the Red and Yellow Books is the same, but for the Silver Book is different.

Advance payment

Sub-Clause 14.2 is only applicable if an advance payment is stated in the Contract Data.

The advance payment is an interest-free loan normally to cover mobilisation, enabling work, placing orders for materials and subcontractors, and also for early design.

If an advance payment is to be made, the Contractor is required to obtain an advance payment Guarantee at its own cost, valid and enforceable until the advance payment has been repaid and submits it to the Employer with a copy to the Engineer.

Note that the Advance Payment Guarantee is "*in the form annexed to the Particular Conditions*".

This was originally stated in the published Contracts as "*based on the sample form within the tender documents*" but has been amended through the Errata (see Appendix).

If the advance payment specifies its expiry date, and the Contractor has not repaid it within 28 days prior to its expiry date, the Contractor may extend the validity of the Advance Payment Guarantee, and provide documentary evidence to the Employer of that extension. If the Contractor does not issue this documentary evidence seven days prior to the expiry date, the Employer is entitled to claim the full amount of the advance payment that has not been repaid at the time.

The Engineer, under the Red and Yellow Books, issues an Advance Payment Certificate (note that under the Silver Book, the Employer makes the payment) within 14 days after:

- a The Employer has received the Performance Security and the Advance Payment Guarantee, and
- b The Engineer (Employer under the Silver Book) has received a copy of the Contractor's application for advance payment.

The advance payment is repaid through pre-agreed percentage deductions in Payment Certificates:

- a Deductions commence within the Interim Payment Certificate (IPC) in which the total of all certified interim payments the same as the currency of the advance payment exceeds 10% of the portion of the Accepted Contract Amount payable in that currency less Provisional Sums.

b Deductions are made at the amortisation rate of one quarter of the amount of each IPC in the currencies and proportions of the advance payment until the advance payment has been repaid.

If the advance payment has not been repaid before the issue of the Taking-Over Certificate for the Works, or before termination under Clause 15 or Clause 16 or Clause 18 (Exceptional Events), the whole of the balance then becomes due and payable by the Contractor to the Employer.

Application for interim payment

The Contractor is required to submit a Statement (one paper copy, one electronic copy and any additional copies stated in the Contract Data) to the Engineer (in a form acceptable to the Engineer) (under the Silver Book it is submitted to the Employer in a form acceptable to the Employer) at the end each period of payment stated in the Contract Data, showing in detail the amounts to which he considers himself entitled.

Note the FIDIC 1999 Contracts required that the Contractor submit a Statement in six copies to the Engineer (Employer under the Silver Book).

As has been stated previously in Chapter 2 within the section on Progress Reports, the author has never understood why six copies were required, and who needed them.

These Statements require a considerable amount of effort to compile them, then there was also a significant amount of printing/photocopying to be done using a considerable quantity of paper, in an age where many reports are submitted electronically.

The Statement must include:

i The estimated contract value of Works executed, including Variations;
ii Any amounts due to changes in Laws and for changes in Costs;
iii Any amount to be deducted for retention;
iv Any amounts to be added/deducted for advance payments;
v Any amounts due for Plant and Materials intended for the Works;
vi Any other amounts to added/deducted;
vii Any amounts to be added for Provisional Sums;
viii Any amounts to be added for release of Retention Money;
ix Any amounts to be deducted for the Contractor's use of utilities provided by the Employer;
x Deduction of any amounts certified in previous payment Certificates.

Note that under the FIDIC 1999 Contracts, any monies awarded to the Contractor by the Dispute Adjudication Board (DAB) had to be included within the Statement, whereas under the FIDIC 2017 Contracts if the Contractor is awarded any money by a Dispute Avoidance/Adjudication Board (DAAB) (Clause 21) no Statement is required to be issued by the Contractor.

Schedule of Payments

Payment due to the Contractor may be pre-defined within the Contract as a Schedule of Payments which become an estimate of contract values for the purposes of payment. There is then no provision for payment for unfixed materials.

If the Engineer (Employer under the Silver Book) does not agree that progress of the Works is in accordance with the Schedule of Payments, he can agree or determine revised instalments.

Within the FIDIC 1999 Contracts, the Engineer (Employer under the Silver Book) could revise a Schedule of Payments only if progress was slower than expected, but the Engineer (Employer under the Silver Book) may now amend the Schedule of Payments if it "differs", which could include the Contractor making faster progress; however, it is the Engineer (Employer under the Silver Book) who "finds" the difference rather than the Contractor informing him, although there is of course no problem with the Contractor informing the Engineer (Employer under the Silver Book) of his faster progress.

If there is no Schedule of Payments, the Contractor submits estimates of the payments which it expects to become due during each period of three months.

The first estimate is submitted within 42 days after the Commencement Date, with revised estimates submitted at intervals of three months, until the issue of the Taking-Over Certificate for the Works.

Plant and Materials intended for the Works

This Sub-Clause provides for Plant and Materials where listed in the Contract Data when shipped and/or payment when delivered to Site for incorporation into the Works to be paid.

The Contractor is required to include:

- An amount to be added for Plant and Materials which have been shipped or delivered to the Site for incorporation in the Permanent Works;
- An amount to be deducted when the contract value of such Plant and Materials is included as part of the Permanent Works.

These provisions were previously included within the FIDIC 1999 Contracts, provided the Contractor could provide satisfactory evidence to the Engineer (Employer under the Silver Book).

The Engineer (Employer under the Silver Book) agrees or determines each amount to be added for Plant and Materials if the following conditions are fulfilled:

a The Contractor has:

   i Kept satisfactory records which are available for inspection by the Engineer (Employer under the Silver Book);
   ii Submitted written evidence demonstrating that the Plant and Materials comply with the Contract and/or the compliance verification documentation under Sub-Clause 4.9.2;
   iii Submitted a statement of the Cost of acquiring and shipping or delivering (as the case may be) the Plant and Materials to the Site, supported by satisfactory evidence, and either;

b The Plant and Materials:

   i Are those listed in the Contract Data for payment when shipped. This is a critical pre-condition;

ii Have been shipped to the Country, en route to the Site, in accordance with the Contract; and
iii Are described in a clean shipped "bill of lading" or other evidence of shipment, which has been submitted to the Engineer together with:

- Evidence of payment of freight and insurance;
- any other documents reasonably required by the Engineer (Employer under the Silver Book);
- a written undertaking by the Contractor that the Contractor will deliver to the Employer (prior to submitting the next Statement) a bank guarantee in a form and issued by an entity to which the Employer gives consent, in amounts and currencies equal to the amount due under this Sub-Clause.
- This guarantee is required to be in a similar form to the form described in Sub-Clause 14.2.1 [Advance Payment Guarantee] and valid until the Plant and Materials are properly stored on Site and protected against loss, damage or deterioration.

a The Plant and Materials:

i Are those listed in the Contract Data for payment when delivered to the Site. This is a critical pre-condition;
ii Have been delivered to and are properly stored on the Site; are protected against loss, damage or deterioration; and appear to be in accordance with the Contract.

The amount agreed takes account of the evidence and documents required, and of the contract value of the Plant and Materials.

Note that the Engineer (Employer under the Silver Book) has no obligation to certify any payment under this Sub-Clause until the Employer has received the bank guarantee.

The sum to be certified by the Engineer (Employer under the Silver Book) in an IPC is to be the equivalent of 80% of this agreed or determined amount.

Note that property in Materials and Plant does not pass to the Employer until they are fully paid for, so this 80% provision means that the Contractor retains ownership far longer than one might expect or wish for.

The currencies for this certified sum must be the same as those in which payment will become due when the contract value is included.

At that time, the Payment Certificate includes the applicable amount to be deducted which must be equivalent to, and in the same currencies and proportions as, this additional amount for the relevant Plant and Materials.

Issue of the IPC (Red and Yellow Book)/Interim Payments (Silver Book)

Sub-Clause 14.6 has been restructured and expanded within the FIDIC 2017 Contracts, which introduce the phrase "IPC" (Interim Payment Certificate).

Note that no IPC is issued under the Silver book, the Employer issues a Notice.

Provided the Employer has received and approved the Performance Security, and the Contractor has appointed the Contractor's Representative (additional requirement from FIDIC 1999), the Engineer (Employer under the Silver Book) issues an IPC to the Employer (copied to the Contractor) (under the Silver Book, a Notice from the

Employer) within 28 days of receiving each Contractor's Statement. The IPC must clearly show any difference between the Contractor's Statement and the amount certified for payment.

Note that under FIDIC 1999 the Certificate was only issued to the Employer, and the Engineer was not expressly required to explain the difference to the Contractor between what the Contractor had applied for, and what it was to be paid.

There is provision in the Contract for a minimum amount of Payment Certificate to be stated in the Appendix to Tender to prevent having to issue Certificates for minor amounts.

An IPC (interim payment under the Silver Book) must not be withheld for any other reason, although:

a If anything supplied or work done by the Contractor is not in accordance with the Contract, the estimated cost of rectification or replacement (whichever is relevant) may be withheld until it has been completed (a quite subjective calculation).
b If the Contractor fails to perform anything in accordance with the Contract, the value of this failure may be withheld until the work or obligation has been performed (another subjective calculation).
c Note that, if the Engineer (Employer under the Silver Book) finds any significant error or discrepancy in the Statement or supporting documents, the amount of the IPC may take account of the extent to which this error or discrepancy has prevented or prejudiced proper investigation of the amounts in the Statement until such error or discrepancy is corrected in a subsequent Statement. This can be quite another quite subjective provision in terms of what the Engineer (Employer under the Silver Book) can do, rather than a simple adjustment of the error.

For each amount withheld, in the supporting particulars for the IPC the Engineer (Employee under the Silver Book) is required to detail his calculation of the amount and state the reasons for it being withheld.

The Engineer (Employer under the Silver Book) may, within any Payment Certificate, make any correction or modification that should properly be made to any previous Payment Certificate.

Note that a Payment Certificate is not deemed to indicate the Engineer's (Employer under the Silver Book) acceptance, approval, consent or Notice of No-objection to any Contractor's Document or to (any part of) the Works (this is a common provision). The Contractor remains responsible (and liable) for all of his obligations under the Contract

There is a new provision in that if the Contractor considers that an IPC does not include any amounts to which it is entitled, these amounts are to be identified in the next Statement. The Engineer (Employer under the Silver Book) then makes any correction or modification that should properly be made in the next Payment Certificate.

If:

a The Contractor is not satisfied that this next Payment Certificate includes the identified amounts, and
b The identified amounts do not concern a matter for which the Engineer (Employer under the Silver Book) is already carrying out his/her duties under Sub-Clause 3.7 [Agreement or Determination],

For example, there may be a claim by the Employer against the Contractor which is in the process of being determined by the Engineer (Employer under the Silver Book). The Employer or Engineer cannot deduct an amount for the claim until it has been properly determined.

the Contractor may, by giving a Notice, refer the matter to the Engineer for Agreement or Determination.

Payment

- The Employer is required to pay the advanced payment (in accordance with the Advance Payment Certificate) and within the period stated in the Contract Data, and if none is stated, within 21 days after the Employer receives the Advance Payment Certificate.
- Payments certified within each IPC are made within the period stated in the Contract Data, if no period is stated, within 56 days after the Engineer (Employer under the Silver Book) receives the Statement with supporting documents from the Contractor, for an IPC and 28 days after the Employer receives the IPC for a Partially Agreed Final Statement (PAFS, which is at best confusing!).
- Payments certified within the Final Payment Certificate (FPC) (note there is no FPC under the Silver Book) within the period stated in the Contract Data if no period is stated, within 56 days after the Engineer (Employer under the Silver Book) receives the Statement with supporting documents from the Contractor.

Payment of the amount due in each currency is to be made into the bank account, nominated by the Contractor, and in the payment country for this particular currency as specified in the Contract.

Delayed payment

If the Contractor does not receive payment from the Employer as required by the Contract (above), then he is entitled to be paid financing charges on the amount due until that amount due has been paid in full.

The amount of the financing charges is calculated from the date the payment period expired, rather than the date of the IPC and is calculated at the annual rate of 3%, unless a different rate is stated in the Contract Data.

Note that the financing charges are due to the Contractor who is entitled to payment from the Employer, without the need for the Contractor to submit a Statement or any formal Notice.

Note also that under some legislations the Contractor may suspend the Works or terminate the Contract for late payment, subject to certain conditions, so this should be considered by both Parties.

Release of retention money

Normal rules apply here i.e. first half of Retention Money repaid on completion of the Works; second half repaid at the end of the Defects Notification Period.

The first half of Retention Money is certified by the Engineer (Employer under the Silver Book) upon the issue of the Taking-Over Certificate, the outstanding balance being certified upon the expiry of the Defects Notification Periods.

After the issue of the Taking-Over Certificate for:

- a The Works, the Contractor includes for release of the first half of the Retention Money in a Statement;
- b A Section, the Contractor includes for release of the relevant percentage of the first half of the Retention Money in a Statement.

After the latest of the expiry dates of the Defects Notification Periods, the Contractor includes the second half of the Retention Money in a Statement promptly after such latest date.

If a Taking-Over Certificate was (or was deemed to have been) issued for a Section, the Contractor includes the relevant percentage of the second half of the Retention Money in a Statement promptly after the expiry date of the Defects Notification Period for that Section.

In the next IPC after the Engineer (Employer under the Silver Book) receives such a Statement, the Engineer (Employer under the Silver Book) certifies the release of the corresponding amount of Retention Money.

However, when certifying any release of Retention Money, if any work remains to be executed under Clause 11 [Defects after Taking Over], the Engineer (Employer under the Silver Book) is entitled to withhold certification of the estimated cost of this work until it has been executed.

Note, the relevant percentage for each Section is the percentage value of the Section as stated in the Contract Data.

If the percentage value of a Section is not stated in the Contract Data, no percentage of either half of the Retention Money is to be released under this Sub-Clause in respect of such Section.

Note that under the FIDIC 1999 Contracts, the Engineer (Employer under the Silver Book) certified the payment of the first portion of Retention Money when the Taking-Over Certificate was issued then the final portion after the latest of the expiry dates of the Defects Notification Periods, but under the FIDIC 2017 Contracts the Contractor has to issue a Statement first, so payment will be later under the FIDIC 2017 Contracts. This new provision marks a considerable negative change as far as Contractors are concerned.

Statement at Completion

Within 84 days after the Date of Completion of the Works, the Contractor submits to the Engineer (Employer under the Silver Book) a Statement, with supporting documents, showing:

- a The value of all work done in accordance with the Contract up to the Date of Completion of the Works;
- b Any further sums which the Contractor considers to be due at the Date of Completion of the Works;

c An estimate of any other amounts which the Contractor considers will have become due after the Date of Completion of the Works under the Contract or otherwise.

These estimated amounts are shown separately (to those of sub-paragraphs (a) and (b) above) and include estimated amounts for:

i Claims for which the Contractor has submitted a Notice under Sub-Clause 20.2;
ii Any matter referred to the DAAB under Sub-Clause 21.4;
iii Any matter for which a Notice of Dissatisfaction has been given under Sub-Clause 21.4.

This Sub-Clause has always been required to include any amounts the Contractor considers to be due, but within the FIDIC 2017 Contracts it has been expanded to include estimated amounts for which the Engineer (Employer under the Silver Book) is considering, or which has been, or is about to be referred to the DAAB.

The Engineer (Employer under the Silver Book) then issues an IPC in accordance with Sub-Clause 14.6 [Issue of IPC].

Final Statement

The provision for a Final Statement has been restructured from the FIDIC 1999 Contracts.

Submission by the Contractor of any Statement is not to be delayed by reason of any referral under Sub-Clause 21.4 for a DAAB's Decision, or any arbitration under Sub-Clause 21.6, it must be issued anyway.

(i) Draft Final Statement

Within 56 days after the issue of the Performance Certificate, the Contractor must submit to the Engineer (Employer under the Silver Book), a draft Final Statement.

This Statement must:

a Be in the same form as Statements previously submitted by the Contractor under the Contract;
b Be submitted in one paper-original, one electronic copy and additional paper copies (if any) as stated in the Contract Data;
c Show in detail, with supporting documents:

  i The value of all work done in accordance with the Contract,
  ii Any further sums which the Contractor considers to be due at the date of the issue of the Performance Certificate,
  iii An estimate of any other amounts which the Contractor considers have or will become due after the issue of the Performance Certificate, under the Contract or otherwise, including estimated amounts, by reference to the matters described in sub-paragraphs (c) (i) to (iii) of Sub-Clause 14.10 [Statement at Completion]. These estimated amounts are to be shown separately (to those of sub-paragraphs (i) and (ii) above).

Except for any amount under sub-paragraph (iii) above, if the Engineer (Employer under the Silver Book) disagrees with or cannot verify any part of the draft Final Statement, the Engineer (Employer under the Silver Book) promptly gives a Notice to the Contractor.

The Contractor is then required to submit further information as the Engineer (Employer under the Silver Book) may "reasonably" require within the time stated in the Notice, and must make such changes in the draft as may be agreed between them.

(ii) Agreed Final Statement

If there are no amounts under Sub-Clause 14.11.1 [Draft Final Statement], the Contractor prepares and submit to the Engineer the Final Statement as agreed.

However if:

a There are amounts, and/or
b Following discussions between the Engineer (Employer under the Silver Book) and the Contractor, it becomes evident that they cannot agree any amount(s) in the draft Final Statement, the Contractor prepares and submits to the Engineer a Statement, identifying separately: the agreed amounts, the estimated amounts and the disagreed amount(s), referred to as the "Partially Agreed Final Statement".

Whilst the notion of a "Partially Agreed Final Statement" seems to introduce a level of uncertainty it does formalise what is agreed and what is not agreed, so it does provide a workable part-time solution.

Note the difference between payment for an FPC which is 56 days after receipt by the Employer, whereas a PAFS which then falls under the IPC process is paid 28 days after receipt by the Employer.

Discharge

The FIDIC 1999 Contracts required that the Contractor, when submitting the Final Statement, submitted a written discharge confirming that the Final Statement represented full and final settlement of all monies due under the Contract. And became effective when the Contractor had received the Performance Security and the outstanding balance of monies due.

Under the FIDIC 2017 Contracts, the discharge can exclude amounts in respect of any Dispute for which a DAAB is proceeding or any arbitration which is in progress, but not amounts still being dealt with by the Engineer (Employer under the Silver Book) and have not yet become Disputes, which would be a risk for Contractors, but is required before a FPC is issued!

When submitting the Final Statement, or the PAFS, the Contractor is required to submit a discharge which confirms that the total of such Statement represents full and final settlement of all monies due to the Contractor under or in connection with the Contract.

As stated, this discharge may state that the total of the Statement is subject to any payment that may become due in respect of any Dispute for which a DAAB proceeding or arbitration is in progress under Sub-Clause 21.6 [Arbitration] and/or that it becomes effective after the Contractor has received:

a Full payment of the amount certified in the FPC; and
b The Performance Security.

If the Contractor fails to submit this discharge, it is deemed to have been submitted and to have become effective when the conditions of sub-paragraphs (a) and (b) have been fulfilled.

Note that if the Contractor does not submit a discharge, there is a deeming provision that if the Contractor has received full payment of the amount certified in the FPC and the Performance Security then the discharge is deemed to have been fulfilled. Note, however, that the FPC cannot be issued until the discharge is provided!

A discharge under this Sub-Clause does not affect either Party's liability or entitlement in respect of any Dispute for which a DAAB proceeding or arbitration is in progress under Clause 21.

Issue of the FPC (Final Payment under the Silver Book)

The FPC is issued by the Engineer to the Employer, with a copy to the Contractor within 28 days after receiving the Final Statement or the PAFS, and the discharge under Sub-Clause 14.12 (under the Silver Book the Employer gives a Notice to the Contractor).

The FPC (Notice under the Silver Book) states:

a The amount which the Engineer (Employer under the Silver Book) considers is finally due, including any additions and/or deductions which have become due under Sub-Clause 3.7 [Agreement or Determination].
b After giving credit to the Employer for all amounts previously paid by the Employer and for all sums to which the Employer is entitled, and after giving credit to the Contractor for all amounts (if any) previously paid by the Contractor and/or received by the Employer under the Performance Security, the balance (if any) is due from the Employer to the Contractor or from the Contractor to the Employer. (This is a new provision not previously provided for within the FIDIC 1999 Contracts.)

If the Contractor has <u>not</u> submitted a draft Final Statement within the time specified, the Engineer requests the Contractor to do so.

If the Contractor fails to do so within a period of 28 days, the Engineer (Employer under the Silver Book) issues the FPC (Employer under the Silver Book gives a Notice), for such an amount as the Engineer (Employer under the Silver Book) considers to be due.

If:

i The Contractor has submitted a PAFS, or
ii No PAFS has been submitted by the Contractor but a draft Final Statement submitted by the Contractor is deemed by the Engineer to be a PAFS,

the Engineer then issues an IPC (the Employer under the Silver book makes a payment).

Cessation of Employer's liability

The Employer is not liable to the Contractor for any matter or thing under or in connection with the Contract or execution of the Works, except to the extent that the Contractor shall have included an amount expressly for it in:

a The Final Statement or PAFS; and
b (except for matters or things arising after the issue of the Taking-Over Certificate for the Works) the Statement under Sub-Clause 14.10 [Statement at Completion].

Note that under FIDIC 2017, unless the Contractor makes or has made a Claim under Sub-Clause 20.2 [Claims for Payment and/or EOT] in respect of an amount or amounts under the FPC within 56 days of receiving a copy of the FPC the Contractor is deemed to have accepted the amounts so certified. The Contractor is again strongly minded to submit his claims in a timely manner to prevent this occurring.

The Employer then has no further liability to the Contractor, other than to pay the amount due under the FPC and to return the Performance Security to the Contractor.

However, this Sub-Clause does not limit the Employer's liability under the Employer's indemnification obligations, or the Employer's liability in any case of fraud, gross negligence, deliberate default or reckless misconduct by the Employer.

Note that "gross negligence" can be interpreted in differing ways under differing legislations.

Currencies of payment

As FIDIC is an international contract it provides for stating, within the Contract Data, the currency that the Contract Price is to be paid in.

Sometimes the Contract Price may be paid in more than one currency, for example, in the Middle East it may be paid in U.S. Dollars ($), which reflects the oil price, and also in the country's local currency.

If the Contract Price is paid in more than one currency, the proportions or amounts of the Local and Foreign Currencies, and the fixed rates of exchange to be used for calculating the payments, are as stated in the Contract Data. If there are no rates of exchange stated within the Contract Data, the rates are prevailing on the Base Date and published by the central bank of the Country.

If an amount is agreed in respect of a Variation, value engineering, Delay Damages and any other payments to or from the Parties, the payment is in the applicable currencies also specified.

7 Progress, Delays, Extensions of Time and Completion

Progress, Delays, Extensions of Time and Completion are covered in the FIDIC 2017 Contracts by:

- *Conditions of Contract for Construction* (The "Red Book")
  - Clause 8 – Commencement, Delays and Suspension
- *Conditions of Contract for Plant and Design-Build* (The "Yellow Book")
  - Clause 8 – Commencement, Delays and Suspension
- *Conditions of Contract for EPC/Turnkey Project* (The "Silver Book")
  - Clause 8 – Commencement, Delays and Suspension

Introduction

Commencement of Works

The Engineer (Employer under the Silver Book) is required to give the Contractor not less than 14 days' notice before the Commencement Date, this normally being within 42 days after the Contractor receives the Letter of Acceptance.

Previously, within the 1999 FIDIC contracts, the notice period required was seven days.

Note that the Contractor is not specifically required to start the Works on the Commencement Date, but as soon as practicable after it.

The Contract requires that the Contractor proceeds with the Works with "due expedition" which is an old term which means that the Contractor has to apply himself to continuing with the work throughout the project through to completion. Other contracts use the phrase "regularly and diligently", which essentially means the same thing.

Time for Completion

The Contractor is required to complete the whole of the Works within the Time for Completion, which is stated in the Contract Data as a specific period of days.

Any Sections must also be described within the Contract Data, which defines what the Section is, the percentage of the Accepted Contract Amount that the Section represents (for the purpose of calculating release of Retention), the time scale (in days) for completing each Section and any Delay Damages applicable to that Section.

The issue of whether a construction project has achieved physical Completion has long been the source of many disputes within the construction industry, and can often be quite subjective.

This is a critical event as most of the Standard Forms of Contract rely upon the issue of a Certificate issued by the Employer's Contract Administrator (in the case of FIDIC contracts the Engineer or the Employer/Employer's Representative) to trigger such matters as the release of the first half of retention, and stop liability for liquidated damages/delay damages, and care and insurance of the Works.

Given the importance of this matter it is somewhat surprising that contracts used throughout the construction industry do not define "completion" and also, they use terms such as "Practical Completion" or "Substantial Completion", which then leaves the matter to the discretion of the Employer's Contract Administrator.

Various attempts have been made by practitioners, the courts and others over the years to form a definitive opinion for Completion, conclusions including that the obligation on the Contractor to complete the Works by the Date fixed for Completion must be an obligation to complete the Works in the sense in which the words are "practically complete" i.e. it can be used as intended. If completion meant "down to the last detail", however trivial and unimportant, it would constitute a penalty clause and would be unenforceable.

Because a project can seldom if ever be built as precisely as required by the Contract, and if completion is something which occurs only after all defects, shrinkages and other faults have been remedied and a Certificate to that effect has been given, it would make the liquidated damages/delay damages provision unworkable".

Under the FIDIC contracts, the Contractor not only has to physically complete the Works, but the Works have to pass various tests, and also, amongst other provisions, there may be a requirement for As-Built documentation before a Taking-Over Certificate can be issued.

The Contractor's programme generally

The programme often referred to in the geographical areas FIDIC covers as "the schedule" is an often understated, and to be honest, undervalued element of a construction contract, some published contracts saying very little about what the Contractor should produce and submit, and the status of that programme within the Contract.

Having said that, what the Contractor should produce and submit should be at least what the Contract requires, in order to properly describe in advance what he is going to do, what he requires others to do in order that he can do it, when and how he is planning to do it and with what resources.

This then allows for effective progress forecasting and reporting, and is the Contractor's best ally should he submit a claim, not only for delays, but also for additional cost recovery.

In fact, a properly constructed and maintained programme will make life very difficult for the Contractor who wishes to submit a claim, and for the Engineer/Employer's Representative who has to determine any entitlement.

The FIDIC 2017 contracts, which we will be discussing shortly in terms of programme requirements, have been substantially enlarged in terms of programme requirements, which is a welcome improvement, and the author believes that the FIDIC 2017 contracts are starting to appear as more of a "best practice project management" tool in terms of programme, risk management and claims, rather than just a series of obligations upon the Parties.

Whilst programme should be updated on a regular basis, it is essential that the Parties never lose sight of the baseline programme, the one that was submitted at the commencement of the work, before all the changes, delays, risks, etc., came into being, so that if necessary it can be compared with a current programme when submitting claims, i.e. what was intended at the commencement of the Works versus what actually occurred.

The type of programme to be submitted by the Contractor to fulfil his obligations under Clause 8.3 of the FIDIC 2017 contracts is not prescribed within the contract, other than that the Sub-Clause refers to one paper copy, one electronic copy and additional copies stipulated in the Contract Data.

Also, that the programme is required to be prepared and submitted using the software stated in the Specification, or the software acceptable to the Engineer.

A variety of techniques are available for programming a project, and these need to be considered in the light of the type of project in hand, the Contractor's preferences and any stipulations in the contract documentation.

Briefly, the types of programme in common use include:

- i Bar Charts,
- ii Elemental Trend Analysis ("Line of Balance"),
- iii Project Network Techniques Precedence Diagrams, Network Analysis, Critical Path Analysis.

Bar charts

The bar chart (Gantt chart) is the most commonly used programme within the construction industry, introduced by Henry Laurence Gantt in the early 1900s, and they have now become the basis of the modern bar chart, developed as a graphical representation of construction sequences.

Bar charts are the most commonly used method of depicting relationships between construction activities. They are a simple format readily understood at all levels of management and can be used to show all stages of the planning process, from pre-tender/pre-contract and on into contract planning. They also assist in showing the relationship between the pre-tender programme, master programme and short-term programmes.

Bar charts are well suited to depicting construction sequences and can be used to link the programme prepared at the tender stage with the master programme and likewise with the short-term planning throughout the contract period. Bar charts are easily and readily updated at weekly and monthly time intervals. A colour coding system may be introduced for progress recording which provides an accurate record of the contract's progress for future reference purposes.

They can also be easily updated, at weekly/monthly time intervals for review purposes and progress reports.

Resources may be readily related to the rate of working – labour histograms, value forecasts (value/time), and cumulative labour and plant forecasts in the form of project budgets.

The programme may also form the basis of financial/cash flow forecasting for both the Employer and the Contractor.

The use of computer software enables bar chart relationships to be readily developed and amended on the computer screen. Major projects within the larger construction organisations are currently being managed with the aid of such software. The software allows the master programme, stage programmes and short-term planning procedures to be linked together to form a comprehensive planning and reporting tool.

Developments in project management computer software have led to resurgence in the use of bar charts – especially linked bar chart relationships.

The software enables the user to present bar charts in a professional manner for the pre-tender and master programmes which can be readily updated on site with the aid of a notebook computer.

Relationships used on linked bar charts include:

- Finish to start relationships;
- Start to start relationships;
- Finish to finish relationships.

Elemental Trend Analysis ("Line of Balance")

A line of balance diagram comprises a series of inclined lines which represent the rate of working between repetitive activities in a construction sequence.

The application of line of balance to construction was pioneered in the 1960s, and became recognised as being the best planning method for repetitive work.

It is not unusual to see bar chart displays incorporating a line of balance diagram to illustrate the programming of any repetitive sections of the Works.

Line of balance is a visual display of the rate of working from one activity to another.

The ideal line of balance display shows all balance lines running parallel to each other; in practice this is often difficult to achieve.

Project Network Analysis – The Critical Path Method

In early 1957, the basic principles of CPM (Critical Path Method) were developed.

A test group was set up to apply the new technique to a chemical plant project in Kentucky.

The development of PERT (Program Evaluation and Review Technique) was originated by the Special Projects Office for the Polaris missile programme.

In the UK, development work on CPM was undertaken by the Building Research Establishment and a number of papers were published by Nuttall and Jeans in 1960.

The initial applications developed were only suitable for the large complex projects, and required the assistance of a mainframe computer to analyse the data.

Through the discipline of CPM, the user can achieve better planning due to the logical approach undertaken during the development of the construction sequence.

Identification of critical activities on which effort and resources can be applied aids the Contractor's management. This enables "management by exception" to be applied to critical activities.

CPM allows the planner to express his ideas in graphical form. Recent developments in project management software allow networking on the screen to be carried out.

The planner has the facility to assign priorities to labour, plant, material and Subcontractor resources to each activity on the network.

Bar chart analysis aids understanding at site management level for both the Contractor and Client.

The effect of changes and variations can be evaluated, and time cost optimisation analysis undertaken.

Cash flow assessments and valuation forecasting information may be output in graphical format and readily updated during the project.

Using notebook and site-based computers, the progress situation may be speedily analysed, and a variety of management reports made available.

Some terms used in programming

Float

Float is spare time in a programme, normally shown between activities (free float) and at the end of the project (terminal float). There is no legal or contractual definition of float, but the Contractor will normally build some float in the programme to allow some flexibility within that programme in case of variations to the work, risks which may materialise or his own incorrect forecasts or inefficiencies.

Float is affected by many things including re-sequencing, fluctuation in outputs and resources, varied and additional works or relaxed contractual constraints which could also benefit the project programme as a whole.

Most people interpret that "float" belongs to the Contractor as it is his programme, and he has the right to finish in accordance with his programme. In effect, it depends on what the float is for and where it appears in the programme that will determine any ownership of that float. Many practitioners are of the opinion that free float is available to whoever needs it first, whereas terminal float belongs to the Contractor, which the author is in agreement with.

Critical path

In a programme there will be a series of activities for which there is no float, these activities are "critical" to the work being completed on time.

The critical path is simply the longest path of work activities through the programme from start to end. It is the shortest duration of the contract programme. Critical path activities have no float. If critical path activities are delayed, they will lead to a delay to the overall completion of the project; this is a factor we will consider in the next section when we look at extension of time claims.

Time and any delays are very important aspects of any construction contract.

The FIDIC 2017 programme requirements

The FIDIC 1999 Contracts had fairly detailed provisions regarding time and the submission of the Contractor's Programme, those provisions have now been significantly increased within the FIDIC 2017 Contracts.

The Contractor is required to submit an initial programme to the Engineer (Employer under the Silver Book) within 28 days of receiving the Notice stating the Commencement Date under Clause 8.1.

The new Clause 8.3 now refers to an "initial programme" and "each revised programme", whereas the FIDIC 1999 Contracts only referred to a "detailed time programme", which was revised whenever the previous programme was inconsistent with actual progress.

The initial programme is to be submitted by the Contractor in one paper copy, one electronic copy and additional copies where stipulated in the Contract Data.

The Contract stipulates that the Contractor prepares the programme using the software stated in the Specification (Employer's Requirements under the Silver Book), or the software acceptable to the Engineer (Employer under the Silver Book).

The Contractor is also required to submit revised programmes showing the actual progress of the works. The programme will change as Variations, delays, risks, etc., come on stream so the previous programme becomes, in effect, out of date and does not reflect the manner in which the Contractor will achieve its obligations.

Note that whenever the programme is revised it is resubmitted, and must show the elements required in the initial programme submission (stated below).

The requirements within the programme submission have been expanded considerably, and are much more detailed than within FIDIC 1999, which only required details of the order in which the Contractor intended to carry out the Works including any design, stages of work for Nominated Subcontractors, the sequence and timing of inspections and tests and a supporting report which included a general description of the methods which the Contractor intended to adopt, and details of the Contractor's estimate of Personnel and Contractor's Equipment.

Under FIDIC 2017 Clause 8.3, the Contractor is required to include within his initial and revised programmes (changes from the FIDIC 1999 Contracts shown):

a The Commencement Date (as referred to in the Engineer's Notice), and the Time for Completion of the Works and each Section (where applicable) as stated within the Contract Data. These dates were not expressly required in FIDIC 1999, but they are critical and Contractors included them.
b The date access and possession of each part of the Site is to be given to the Contractor in accordance with the Contract Data. Note that if these dates are not included within the Contract Data, the date(s) the Contractor requires the Employer to provide them. Again, these were not expressly required in FIDIC 1999, but Contractors included them.
c The order or sequence in which the Contractor intends to carry out the Works including anticipated timing of design stages (under the Red Book the Contractor may have design responsibility, under the Yellow and Silver Books the Contractor has full design responsibility), preparation and submission of Contractor's Documents, procurement, manufacture, inspection, delivery to Site, construction, erection, installation, any work to be undertaken by any Nominated Subcontractor and testing. These requirements were previously included in FIDIC 1999.
d The Review periods for any submissions stated in the Specification (Employer's Requirements under the Silver Book) or required by the Conditions. It is critical that the Contractor allows sufficient time for submission of documents for Review by the Engineer, and for any comments or rejections to be dealt with. These were not expressly required in FIDIC 1999, but Contractors tended to include them.
e The sequence and timing of inspections or tests specified in the Contract. Again, it is critical that the Contractor allows sufficient time for inspection and tests by the

Engineer, and for any comments or rejections to be dealt with. This was required in FIDIC 1999. These were not expressly required in FIDIC 1999.

f For a revised programme, the sequence and timing of any remedial work to which the Engineer has given a Notice of No-objection. These were not expressly required in FIDIC 1999.

g All activities (to the level of detail stated in the Specification) logically linked and showing the earliest and latest start and finish dates for each activity the float (if any) and the critical path(s). These were not expressly required in FIDIC 1999, and are a welcome addition.

h The dates of all locally recognised days of rest and holiday periods (if any). These were not expressly required in FIDIC 1999, but many Contractors included them, for example Christmas, and Eid holidays together with Ramadan working restrictions and shortened working hours.

i All key delivery dates of Plant and Materials. These were not expressly required in FIDIC 1999.

j For a revised programme and for each activity, the actual progress to date and the effects of delay on other activities. These were not expressly required in FIDIC 1999.

k A supporting report which includes:

  i A description of all the major stages of the execution of the Works (what is a major stage is not defined, but common sense will hopefully prevail);
  ii The methods that the Contractor proposes to adopt in executing the Works, this will include relevant method statements;
  iii Details of the Contractor's "reasonable" estimate of the number of each class of Contractor's Personnel, Equipment, for each stage of the Works;
  iv If it is a revised programme, identification of any significant changes to the previous programme since the last submission;
  v The Contractor's proposals to overcome the effect of any delays(s) in progress of the Works.

In this respect, FIDIC 1999 required a supporting report, but only in respect of a general description of the methods which the Contractor intended to adopt, and estimates of the Contractor's expected resources for each major stage of the Works. FIDIC 2017 Contracts are therefore much more prescriptive of what is required from the Contractor.

Whilst Contractors would often provide information within their programmes additional to what the Contract required, the author welcomes the more comprehensive requirements within FIDIC 2017, to reflect the importance of the Contractor's programme as a management tool, in terms of reviewing the Contractor's intentions, his actual progress, and for assessment of claims.

Under FIDIC 1999 the Engineer (Employer under the Silver Book) did not have to carry out a review of the Contractor's programme, he only had to state if the programme was not in accordance with the Contract. If he did not respond, the Contractor could then proceed in accordance with the programme.

The Engineer (Employer under the Silver Book) under FIDIC 2017 is now required to Review (i.e. "examination and consideration by the Engineer of a Contractor's (and to what extent) it complies with the Contract and/or the Contractor's under or in connection with the Contract") the Contractor's submission, and may give a Notice to the Contractor stating the extent to which it does not comply with the Contract, ceases to reflect actual progress or is otherwise inconsistent with the Contractor's obligations.

It is unclear what a programme which "does not comply with the Contract" actually means. Obviously, if Contract dates on the programme are incorrect, then that programme is not compliant, but is a programme which does not show ALL the "ingredients" listed in Sub-Clause 8.3, a programme that "does not comply with the Contract"?

Or, is a programme where the Engineer (Employer under the Silver Book) does not agree that the Contractor has shown sufficient resources to complete the Works which form the Contract, a programme that "does not comply with the Contract"?

Essentially, the Engineer (Employer under the Silver Book) should use his own knowledge, experience and common sense to make that call.

If the Engineer (Employer under the Silver Book) does not give a Notice to the Contractor within:

- 21 days after receiving the initial programme,
- 14 days after receiving a revised programme,

the Engineer (Employer under the Silver Book) is deemed to have given a Notice of No-objection and the programme is then the Programme required by the Contract (even though it may not comply with the Contract!).

If the Engineer (Employer under the Silver Book) at any time gives a Notice to the Contractor that the programme ceases to reflect his actual progress or is otherwise inconsistent with the Contractor's obligations, the Contractor within 14 days after receiving such notice submits a revised programme.

Note that programmes are never "approved" by the Engineer (Employer under the Silver Book), the Contractor proceeds in accordance with the Programme, except to the extent that

i He has been advised of its non-compliance, or
ii It does not actually comply, with the Contract.

Note that if the Contractor submits a programme to the Engineer (Employer under the Silver Book) which shows himself completing outside the Time for Completion that is not a Notice of delay required under the contract (Clause 8.4 and 20.1) in respect of an Extension of Time, and equally, the Engineer (Employer under the Silver Book) accepting the programme is not awarding an Extension of Time.

The Engineer (Employer under the Silver Book) is merely confirming that the programme submitted by the Contractor is a fair representation of the Contractor's actual progress. FIDIC 2017 now clarifies that point.

There are specific procedures and Notices required in terms of Extensions of Time and Claims.

Advance Warning

Advance Warning is a new (and in the author's opinion welcome) provision similar to the "Early Warning" provisions in the NEC contracts, each Party being required to advise the other (and the Engineer under the Red and Yellow Books), in advance of any known or probable future events which could:

- Adversely affect the work of the Contractor's Personnel,
- Adversely affect the performance of the Works when completed,
- Increase the Contract Price, and/or
- Delay the execution of the Works or a Section.

so, it could be any combination of the above four.

Note that the above four items are all related to negative events, there is no provision for either Party to advise of positive events such as potential time, quality or cost savings or opportunities, though the Contract does include for value engineering proposals.

Note that FIDIC 1999 required the Contractor under Clause 8.3 to promptly give notice to the Engineer (Employer under the Silver Book) of any:

- Specific future events or circumstances which may adversely affect the work (note the Works is not used),
- Increase the Contract Price,
- Delay the execution of the Works.

However, it is also worth noting that the Employer or Engineer did not have a similar obligation, it was an obligation solely upon the Contractor.

The Contractor also had to submit estimates relative to these occurrences and/or a proposal under the Variation Procedure if applicable.

The Sub-Clause under the FIDIC 2017 Contracts requires each Party (i.e. the Employer and the Contractor) to advise each other and the Engineer (under the Red and Yellow Books), and the Engineer (under the Red and Yellow books) to advise the Parties. It is not for the Contractor and the Engineer to directly advise each other.

There is no prescribed format for such notifications, though Parties should adopt some form of template for such warning, which can then be sequentially numbered for the project (see Figure 7.1).

The author uses "Early Warning Notices" under the NEC Contracts as a "lessons learned" exercise on completion of the project, identifying who raised the notice, what the reason was, whether there was repetition of the same issues and how quickly they were closed out. It is hoped that FIDIC 2017 users will adopt a similar post-completion review.

The Engineer (Employer under the Silver Book) may also request the Contractor to submit a proposal to avoid or minimise the effect of such events.

Although this provision is new to the FIDIC contracts, early warnings have always been a feature of NEC contracts, and is a useful and simple risk management tool allowing risks to be notified and give the Parties and the Engineer (Employer under the Silver Book) maximum time to consider their effect quality, price and/or time, and to collaboratively work together to resolve them, without escalating immediately into the claims provisions within the Contract.

The clause does not prescribe a time scale for advising advance warnings other than "*in advance of any known or probable future events*", rather than as soon as they become

Contract: .... **Contract No:**	**ADVANCE WARNING** **AW No**......................................

Section A: Enquiry

To : Employer/Contractor

Description

This matter could:

- ☐ **Adversely affect the work of the Contractor's Personnel**
- ☐ **Adversely affect the performance of the Works when completed**
- ☐ **Increase the Contract Price**
- ☐ **Delay the execution of the Works or a Section**

Signed:................................**(Employer/Contractor)** **Date:** ....

Action by: **Date required:**

Section B: Reply

To: Employer/Contractor

Signed:................................**(Contractor/Project Manager)** **Date:** ....

Copied to:

Employer ☐ Contractor ☐ Engineer ☐ File ☐ Other .... ☐

Figure 7.1 Example of Advance Warning Notice.

aware of them, and also there does not appear to be any remedy should either Party fail to give advance warning.

The NEC contracts reflect any failure to give early warning within the Contractor's entitlement to compensation, so it is likely that there may be some amendments to this clause to specifically penalise the Contractor for not giving advance warning of something which later becomes a Claim! There may also be some amendments removing the Employer's responsibility to give advance warnings.

The author believes that a Contractor who diligently complies with what the Contract states and submits advance warnings will certainly put himself in a better position if a claim arises, and especially if it escalates into a dispute, but the advance warning provision should be seen as a claim and dispute avoidance process rarer than stocking up evidence for when the claim or dispute inevitably arises!

As with the NEC contracts, the advance warning procedure obliges people to be "proactive", notifying and dealing with risks as soon as the parties become aware of them, rather than "reactive" waiting to see what effect they have, then trying to deal with them when it is often too late. It is hoped that encouraging the early identification of problems by both parties, puts the emphasis on joint solution finding rather than blame assignment and contractual entitlement after the event.

It is hoped that FIDIC practitioners will fully embrace this new provision within its contracts.

The author has encountered situations where Employers and Engineers appear hostile to the receipt of early/advance warnings from Contractors. Sometimes they are viewed as the first stage toward a claim. That may be so, but not always, and in effect could prevent a claim occurring, or at least lessen its effect.

Statistically speaking one would expect an equal number of early warnings to be issued by the Employer and the Contractor. Typically, though, the ones issued by the Contractor frequently outnumber the ones issued by the Employer.

Employers and Engineers are often criticised for seeing advance warnings as something the Contractor has to do, and in fact most advance warnings are actually issued by the Contractor. However, the Employer and the Contractor are obliged to give advance warnings, so it is critical that they play their part in the process.

As an example, if the Employer or Engineer becomes aware that it will be late in delivering some design information to the Contractor, the Employer should issue the advance warning as soon as becomes aware that the information will not be delivered to the Contractor, not wait and subsequently blaming the Contractor for not giving an advance warning stating that it has not received the information!

The nature and format of advance warnings can have an impact of how they are received and what reaction is prompted. Contractual parties are often keen to identify each other's failings, but the obligation here potentially requires parties to identify and record their own failures. Common sense is therefore needed in how these are identified.

The "matters" which require an advance warning to be given are essentially the three criteria for success in any project i.e. price, time and quality. Therefore, the contract recognises the importance of managing "matters" that may impact on these.

A schedule of Advance Warnings should be kept (see Figure 7.2), identifying and numbering each Advance Warning, with additional information about each one.

The author has found the schedule useful for tracking advance warnings as the project progresses, but also at the end of each project to review the schedule and each

Contract: .... Contract No:				SCHEDULE OF ADVANCE WARNINGS			
Ref.	**Date notified**	**Origin? Employer or Contractor**	**Description of event**	**Relevant Clauses**	**Action with Employer or Contractor**	**Date Resolved**	**Notes**
1							
2							
3							
4							
5							
6							
7							
8							
9							
10							

Figure 7.2 Example of Schedule of Advance Warnings.

advance warning notified during the life of the project as "lessons learned" to consider in drafting Scopes for future contracts.

Extension of Time for Completion generally

Time is central to any contract, as stated above, the Contractor under the FIDIC contracts must complete the Works within the Time for Completion.

There are two types of delays which can occur on construction projects:

- Delay in activities for which there is a programme float available e.g. they can be delayed without impacting on the completion date.
- Delays in activities that will impact the completion date, these are sometimes referred to as critical delays.

Construction contracts tend to provide three different categories for delay:

Delays resulting from neutral causes

Neutral events are not the fault of any party involved in the project. The Contractor might be entitled to an Extension of Time.

Delays that are, directly or indirectly, the fault of the Employer or his agents e.g. Engineer/Employer's Representative

Where the progress of the works is materially affected by matters for which the Employer is responsible, the Contractor may be entitled to an Extension of Time, but also to claim cost incurred, and also possibly profit.

Delays that are the fault of the Contractor

Where a delay impacts on the Completion Date and is caused by the Contractor, the Contractor will be liable to pay liquidated damages/delay damages to the Employer.

The granting of an Extension of Time relieves the Contractor from liability of damages such as delay damages up to the extended Time for Completion. For the Employer, the early establishment of an extended contract completion date prevents the contract period becoming "at large" (see below).

The conditions and events that the Contractor is entitled to a claim for extension of time will be documented within the Contract, and these events will have been acknowledged by the Employer as matters affecting regular progress.

Contracts also require notification of delaying events to be given as close a time as possible to the event, a failure to provide timely notification can affect the Contractor's claim for an Extension of Time.

Note that there are several methods or "prospective" and "retrospective" delay analysis which may be used by the Contractor in formulating his claim and the Engineer/Employer's Representative in assessing entitlement, but it is outside the scope of this book to review or evaluate them.

The "prospective" approach is favoured as it considers the critical path at any particular point in time as viewed contemporaneously, whilst the "retrospective" approach considers subsequent events to determine whether the material event actually delayed completion.

Briefly some methods are:

"As-Planned v As-Built"

This takes the original "as-planned" programme and compares it with the final "as-built" programme. This method simply illustrates the impact of <u>all</u> delays rather than <u>each</u> individual delay, or delays caused by each party.

"Impacted As-Planned"

This is an "additive" method which takes the original as-planned baseline programme and impacts the effect of Employer (excusable delays) and Contractor delays to calculate project completion date/duration. It can also be used to calculate Contractor entitlement, and used widely, fairly simple to do and understand, but becomes complicated/unreliable when sequences and/or durations change.

"Windows Analysis"

This method breaks the as-planned programme down into time slices or "windows" (weeks/months/phases) and analyses the progress and the effect of any delays within those windows as the project progresses, and any delays develop. It is felt to be the most accurate method of analysing delays to a project, but it involves a lot of analysis.

"Collapsed As-Built"

This is a "subtractive" or "retrospective" method which takes the "as-built" programme and works backwards, subtracting delays caused by the Employer, to show when the project could have been completed but for those delays.

The delays caused by the Employer are the difference between the "as-built" and the "collapsed as-built" programme.

This is fairly simple to construct and analyse, but it is a retrospective method and will often make subjective assumptions that the as-built programme including sequence and durations is a <u>fair reflection</u> of how the project was built and in retrospect that they were the <u>correct</u> sequence and durations.

In theory, an extension of time should be granted as the delaying event occurs and even before i.e. prospectively, to maintain a feasible <u>contract programme</u> for co-ordination of the project. In practice, the responsible parties for assessing and granting of the Extension of Time are seldom given the authority to grant extensions at an early stage and many contracts extend well beyond their contract completion date without even the beginning of an extension of time process.

This reluctance to accept responsibility for delays and grant extensions of time is often disastrous to the project and to the Employer.

What is "Time at Large"?

Before examining the Extension of Time provisions under the FIDIC 2017 Contracts it is worth considering the principle of "Time at Large".

The requirement within the FIDIC contracts is for the Contractor to commence the execution of the Works on, or as soon as reasonably practicable after the Commencement Date, and to complete the whole of the Works within the Time for Completion (some construction contracts express the requirement to complete the Works by a specific date).

The term "Time at Large" is not a legal term but describes the situation where there is no identified Date for Completion or Time for Completion, either by absence from the contract terms or arising from events during the execution of the Works and/or the operation of law.

The term "Time at Large" is usually used in construction contracts in the situation where liquidated damages/delay damages are an issue. If time is at large then liquidated damages/delay damages cannot be applied, because there is no time or date fixed from which the liquidated damages/delay damages can be calculated. In some situations, the date for completion may be relevant to termination and the issue whether or not there has been a breach of contract by failure to complete.

Time can be made at large in any of four situations:

i No Time or Date for Completion has been fixed within the Contract by which performance must take place or be completed.

If no Time or Date has been fixed by the Contract then the objective intention of the parties must be ascertained. If the Date or Time is not fixed by a course of dealing or negotiation between the parties, a term will be implied that the Contractor's obligation is to complete within a reasonable time.

ii The Time or Date for performance has been fixed under the Contract, but has ceased to apply either by agreement or by an act of prevention (which includes instructed additional work) or by a breach of contract by the Employer with no corresponding entitlement to extension of time.

Many events may prevent completion on time including adverse weather conditions; labour or material shortages; strikes or local disturbances; lack of information or plain human error; and interference, hindrance or prevention by the Employer.

Standard forms of contract apportion liability for delay through extension of time clauses. If the Contractor is entitled to an Extension of Time for an expressly stated event, then the Employer carries the risk, otherwise it lies with the Contractor.

Which Party carries the risk of the additional cost for the delay will depend upon the other terms of the Contract.

The wording of the Extension of Time clause will determine when, and the extent to which, the risk of delay passes from the Contractor to Employer. If the Contractor is prevented from completing the Works, and there is no remedy under the contract, then time becomes at large.

iii The Employer has waived the obligation for the Contractor to complete by the Works by the specified Time or Date.

An alternative situation is that the Employer is faced with a breach of contract by the Contractor which would entitle to Employer to terminate the employment of the Contractor and/or to bring to an end the primary obligations of the Parties to perform, but instead the Employer elects to continue with the performance of the Contract.

iv The Employer has interfered in the certification process to prevent proper administration of the Contract.

Extension of Time under the FIDIC 2017 Contracts

Sub-Clause 8.5 defines the Contractor's entitlement to Extension of Time:

i *"The Contractor shall be entitled..." which* is not stated as being subject to anyone's opinion.
ii *"... subject to Sub-Clause 20.2 ..."* which may affect the entitlement to an extension of time.
iii *"... to Extension of Time... if and to the extent ... that completion ... is ... delayed..."*: so the extension should typically be calculated by reference to the delay in completion attributable to a listed cause. It may have disrupted progress but may not itself have been the cause of any delay in completion. For example, a listed cause may only delay works which are not on the critical path and thus do not delay "completion for the purposes of Sub-Clause 10.1".
iv *"... completion for the purposes of Sub-Clause 10.1 ..."*: its first sentence (sub-paragraph (i)) defines the extent of the work which is to be completed within the Time for Completion, and which is relevant when considering extensions of time. The work must include the matters described in Sub-Clause 8.2 (a) and (b) but may exclude, as permitted in Sub-Clause 10.1 (a), "any minor outstanding work and defects which will not substantially affect the use of the Works or Section of their intended purpose".

There are five grounds for Extension of Time under Clause 8.5:

a Variations

Note that there is now no requirement to comply with Clause 20.2 in terms of Claims linked to Variations. Variations on their own do not necessarily give the Contractor entitlement to an Extension of Time, the timing, size, long lead in times, etc. will determine that, rather than the fact that a Variation exists

b A cause of delay giving rise to an entitlement to an Extension of Time under the contract under a Sub-Clause of the Conditions. (These are scheduled in Chapter 10, Figure 10.1.)

See also the other chapters of this book giving details of the relevant Extension of Time Clauses within the FIDIC 2017 Contracts.

c Exceptionally adverse climactic conditions at the Site, by reference to climatic data, made available by the Employer under Sub-Clause 2.5 (note that this is excluded from the Silver Book).

Note that the reference to exceptionally adverse climactic conditions is <u>at the Site</u>, so bad weather in another location, which then delays, for example, transportation/shipping of materials would not constitute an Extension of Time.

NB: In previous editions of the FIDIC contracts the words "exceptionally adverse climactic conditions" have been subject to Parties' own interpretations and opinions, which in some cases have led to disputes.

Additional words have now been added within FIDIC 2017 allowing the Employer to make available information on climatic data within the Site Data to give an element of objectivity rather than subjectivity.

d Unforeseeable shortages in availability of personnel or goods, or Employer-supplied Materials caused by epidemic or government actions (this is new to the FIDIC 2017 Contracts, but is excluded from the Silver Book).

Employer-supplied Materials have been added to this item.

e Any delay, impediment or prevention caused by the Employer, the Employer's Personnel or the Employer's other contractors on the Site.

This could be quite subjective, but highlights that the Employer or his agents should not unnecessarily impede the Contractor's progress.

Under the Red Book, the Contractor is also entitled to an Extension of Time if the measured quantity of work is greater than the estimated quantity of the same item within the Bill of Quantities, or other Schedule by more than 10% AND the increase in quantities causes a delay to Completion.

Concurrent delay

The FIDIC 1999 Contracts (and those before them) were silent on the issue of concurrent delays, so it was left to common practice and any relevant legislation to deal with it.

The FIDIC 2017 Contracts have not prescribed a mechanism for dealing with concurrent delay, but there is now a statement within the Sub-Clause that where there is concurrent delay the Contractor's entitlement to any Extension of Time is assessed within the rules and procedures stated in the Special Provisions, where there are rules and procedures stated, due regard must be taken of all relevant circumstances.

Within the guidance on Special Provisions there is reference to the Society of Construction Law Delay and Disruption Protocol (SCL Delay and Disruption Protocol 2nd Edition: February 2017).

This is a new provision and although it does not deal with the assessment of concurrent delay it is welcomed, bearing in mind the various issues and reported outcomes with concurrent delay.

Sometimes delays exist, not in isolation but at the same time, in the case of the FIDIC contracts one an Employer Risk Event, one a Contractor Risk Event and with equal effect, each of which is an effective cause of delay to Completion (i.e. both on the critical path) so each on its own could delay completion by the same amount, these are termed "concurrent" delays.

True concurrent delay is actually rare, in that both the conditions "at the same time" and "with equal effect" in the previous paragraph have to apply.

An example would be where an Engineer was late in providing design information to the Contractor, but at the same time exceptionally bad weather prevented the Contractor from progressing with the Works, or the Contractor has insufficient resources on Site. Which delay would take precedence?

Unfortunately, there is no readily identifiable definition, along with a failure by international courts to give any clear guidance on the most suitable or appropriate method for considering an Extension of Time award when there is a current delay when one is the responsibility of the Contractor and the other his employer.

It is not particularly surprising that concurrent or simultaneous delays cause considerable confusion and even the term itself can often be understood differently in different quarters.

Notwithstanding that there is no clear definition of its meaning the problems may start much earlier in determining the inherent reasons for the cause of the concurrent delays, which often create their own factual problems such as the event in question is not critical.

There are quite a few methods for trying to determine an extension of time, each having varying degrees of success in the courts.

With regard to the effect on entitlement to recovery of costs for prolongation, where the Employer Delay and the Contractor Delay to Completion are concurrent and, as a result of that delay the Contractor incurs additional costs, then the Contractor should only be able to recover costs if it is able to separate the additional costs caused by the Employer Delay from those caused by the Contractor Delay.

If it would have incurred the additional costs in any event as a result of Contractor Delay, the Contractor will not be entitled to recover those additional costs.

Some examples of methods for trying to determine an extension of time where there is concurrent delay, each having varying degrees of success in the courts.

The more common ones used are:

a Apportionment;
b The "but for" test;
c The dominant cause approach.

Taking each in turn to consider and review their applicability:

Apportionment

Where there are two competing causes of delay of equal or relative effect a fairly natural response would be to suggest that the overrun and its consequences should be "apportioned" between the Contractor and the Employer on the basis of their relative cause and effect.

This would appear to be a reasonable approach but this method has attracted little support in jurisdictions where there is a tendency to apply principles of cause and effect in an "all or nothing" way. Consequently, the courts tend to link a single event to a cause and the claimant would either win or lose.

Some jurisdictions involve concepts of fairness and reasonableness, and arguably the apportionment approach is dependent on the court's interpretation of the facts and the application of difficult concepts of cause and effect.

The "but for" test

This method tends to attract the most support from the contracting fraternity as it tends to support the claimant (the Contractor). It is based on a simple concept that the overrun would not have occurred "but for" a default by the Employer or its representative.

It has, however, been given unsympathetic support in the courts, and this method can therefore generally be discounted.

The Dominant Cause approach

This became a popular method during the 1980s and gained popularity as the most appropriate method of calculating and dealing with concurrent delays.

Keating on Building Contracts defined it as:

> It there are two causes, one the contractual responsibility of the Defendant and the other the contractual responsibility of the Plaintiff, the Plaintiff succeeds if he establishes that the cause for which the Defendant is responsible is the effective, dominant cause. Which cause is dominant is a question of fact, which is not solved by the mere point of order in time, but is to be decided by applying common sense standards.

The dominant cause principle is based on the parties having intended that in the event of a delay one of them must be responsible.

Where you have competing delays and with the absence of apportionment one claim either for an extension of time or a cross-claim for liquidated damages must succeed.

Finally, there are other examples:

i "If there are two concurrent delays, one which is a relevant event and the other not, then the Contractor is entitled to an extension of time for the period of delay caused by the relevant delay notwithstanding the concurrent effect of the other event." – Often referred to as the "Malmaison approach" following a case involving a hotel in Manchester, UK.

ii *"In the event of a concurrent delay, a breach will always override a neutral event" (Lord Justice Devlin)*, so in this case, a delay caused by an Engineer not providing information for the Contractor to progress the Works, will override a delay caused by exceptionally adverse weather conditions

Delays caused by authorities

This Sub-Clause gives the Contractor a right to an Extension of Time for delays caused by public authorities or private utility entities, provided that the Contractor has diligently followed procedures laid down by those public authorities.

Rate of progress

If the Contractor's progress is too slow to complete the Works or a Section within the Time for Completion or progress has fallen behind the current programme, the Engineer (Employer under the Silver Book) can instruct the Contractor to submit a revised programme and proposals to rectify matters.

Unless the Engineer (Employer under the Silver Book) states otherwise, the Contractor must adopt the revised programme and proposals, including any increases in working hours and/or resources.

There is an additional paragraph within the FIDIC 2017 Contracts, allowing the Engineer (Employer under the Silver Book) to issue instructions for the Contractor to apply revised methods, including acceleration to reduce delays resulting from causes which could entitle the Contractor to an Extension of Time under Sub-Clause 8.5.

Delay damages

Delay damages are defined under Clause 1.1.28 (Clause 1.1.25 under the Silver Book) as "damages for which the Contractor shall be liable under Sub-Clause 8.8 [Delay Damages] for failure to comply with Sub-Clause 8.2 [Time for Completion]".

The FIDIC 2017 Contracts provide for the levying of delay damages by the Employer in the event that the Contractor does not complete the Works, or a Section, within the Time for Completion.

Before examining the FIDIC 2017 provisions, it is worth considering what these damages are, how are they calculated and how they are applied in the event that the Contractor is late in completing the Works.

Delay damages within the FIDIC contracts are essentially the same as "liquidated damages" under other contracts.

The general definition of liquidated damages is "fixed monetary compensation for a loss, detriment, or injury to a person or a person's rights or property, awarded by a court judgment or by a contract stipulation regarding breach of contract", in the case of the FIDIC contracts, the Contractor failing to complete the Works on time, and the damages are not awarded by a court but are a contractual obligation.

Generally, contracts that involve the promise of performance have a liquidated damages/delay damages provision. The purpose of this is to establish a predetermined sum that must be paid if a defaulting party fails to perform as promised.

Liquidated damages/delay damages within most countries' legislations should be assessed as a "genuine estimate of likely loss"…….and <u>not</u> constitute a penalty, so the Employer can only receive what he estimated he would have lost at tender stage when he set the amount. Should the Employer <u>lose</u> more or less then he is limited to the amount set in the contract.

A penalty is a sum that is disproportionate to the actual harm i.e. a fine awarded by a court.

It is basically to put the Employer, at least financially, in the position that he would have been had the delay <u>not</u> occurred and serves as a deterrent to the Contractor not to complete late (some say to motivate the Contractor to complete on time).

These damages are determined by the Employer prior to inviting tenders, and serve as protection for both parties that will enter the contract, whether they are an Employer/Contractor, Employer/Consultant, etc.

A typical example would be an Employer who requires a commercial building to be constructed by a certain date, and if it is not, then he may be liable for costs in renting existing premises.

The Employer would then set the liquidated damages/delay damages at an amount, say per day, in order to recover those damages from the Contractor.

There are several reasons for liquidated damages/delay damages provisions:

i They allow the Employer to set, and the Contractor to pay an amount for late completion through the Contract rather than an Employer having to sue the Contractor and have an amount awarded by a court with the associated time and legal fees.

ii They provide a cap on the Contractor's liability for late completion, so they know the risk at tender stage and may decide to build amounts into their Tender Price as a form of insurance should it occur. Note that it is also possible to insure against having to pay liquidated damages, essentially it is possible to insure against any risk, but the premiums will be set according to the risk.

iii They establish some predictability involving costs, so that parties can balance the cost of anticipated performance against the cost of a breach.

For example, Contractors who believe they may be late in completing the works may weigh up the risk of paying liquidated damages against paying for additional resources to complete on time.

Whilst this may seem a negative view, as the Contractor is required to complete on time, it is commercial reality.

Note that if the Employer chooses not to include liquidated damages within the contract, he may still be able to claim unliquidated damages from the Contractor.

Unliquidated damages are not fixed in advance, they remain at large and are yet to be proven, so if the Employer wishes to levy them against the Contractor he will have to prove actual loss at the time and probably take legal action against the Contractor to levy them as they cannot be deducted from monies otherwise due to the Contractor.

Note that an Employer inserting words such as "Zero" or "Nil" within the relevant part of the contract will only be able to claim for whatever the delay is multiplied by "Zero" or "Nil" as liquidated damages/delay damages, not to claim unliquidated damages as an alternative.

Sub-Clause 8.8 covers delay damages, previously called in FIDIC contracts "Liquidated Damages", which are paid by the Contractor in the event that he fails to complete the Works, or a Section within the Time for Completion.

The amount of delay damages is stated in the Contract Data and is applicable for every day between the relevant Time for Completion and the relevant Date of Completion of the Works or Section.

Note as stated above that under some jurisdictions, the amount of delay damages is required to be limited to "a genuine pre-estimate of likely loss".

Notwithstanding this, the total amount due for delay damages must not exceed the maximum amount stated in the Contract Data, unless the delay is due to fraud, gross negligence, deliberate default or reckless misconduct of the Contractor in which case the maximum amount would not apply.

So, if the Contractor is guilty of one these types of misbehaviour it will not be able to "take advantage" of the cap on delay damages.

Note the Employer should take legal advice in respect of this provision as the meaning of terms like "gross negligence" can be different under differing legislations.

The Employer may terminate the Contract under Sub-Clause 15.2.1(c) if the Contractor is sufficiently late to exceed that figure.

This provides some clarity in terms of how late the Contractor could be before the Employer can terminate, rather than the Employer having to make a subjective judgement.

The delay damages are the only damages due from the Contractor for the Contractor's failure to complete the Works or a Section within the Time for Completion.

Employer's suspension

The Engineer (Employer under the Silver Book) may under Sub-Clause 8.9 instruct the Contractor to suspend the progress of all or part of the Works stating the date and cause of the suspension, during which time the Contractor must protect, store and secure the Works or the relevant part of the Works.

Note that the Engineer (Employer under the Silver Book) may, but does not have to, notify the Contractor of the reason for the suspension.

If the Contractor suffers delay and/or cost resulting from complying with the Engineer's (Employer's under the Silver Book) instruction to suspend the carrying out of the Works, he can claim an Extension of Time and/or payment of Cost plus Profit. Formerly, FIDIC 1999 only allowed only for an Extension of Time and Cost.

The Contractor is not entitled to the Extension of Time, or to payment of the Cost plus Profit as a consequence of the Contractor's faulty or defective (design, if any) workmanship, Plant or Materials, and/or any deterioration, loss or damage caused by the Contractor's failure to protect, store or secure.

The Contractor can recover the value of any Plant and/or Materials which have not been delivered to Site if the work on the Plant and/or Materials has been suspended for more than 28 days, the Plant and/or Materials were scheduled to have been completed and ready for delivery to the Site during the suspension period, the Contractor provides the Engineer (Employer under the Silver Book) with reasonable evidence that the Plant and/or Materials comply with the Contract and the Contractor has marked the Plant and/or Materials as the Employer's property in accordance with the Engineer's (Employer under the Silver Book) instructions.

If the suspension has continued for more than 84 days, the Contractor can request the Engineer's (Employer's under the Silver Book) permission to proceed.

If the Engineer (Employer under the Silver Book) does not give permission, the Contractor and Engineer could agree to a further suspension, in which case the Parties may agree the Extension of Time and/or Cost plus Profit (if the Contractor incurs Cost), after giving a (second) Notice to the Engineer (Employer under the Silver Book), treat the suspension as an omission of the affected part of the Works.

If the suspension affects the whole of the Works, the Contractor may give a Notice of termination.

After a Notice is received from the Engineer (Employer under the Silver Book) to proceed with the Works, the Contractor and Engineer (Employer under the Silver Book) jointly inspect the Works and the Contractor makes good any deterioration or defect.

8 Suspension and Termination

Suspension and Termination are covered in the FIDIC 2017 Contracts by:

- *Conditions of Contract for Construction* (The "Red Book")
  - Clause 15 – Termination by Employer
  - Clause 16 – Suspension and Termination by Contractor
- *Conditions of Contract for Plant and Design-Build* (The "Yellow Book")
  - Clause 15 – Termination by Employer
  - Clause 16 – Suspension and Termination by Contractor
- *Conditions of Contract for EPC/Turnkey Project* (The "Silver Book")
  - Clause 15 – Termination by Employer
  - Clause 16 – Suspension and Termination by Contractor

Suspension or Termination?

"Suspension" is one Party (in the case of Clause 16, the Contractor) temporarily halting the execution of the Works due to a breach by the other Party (in this case the Employer), the Works then being restarted and resumed upon the ceasing of that breach.

"Termination" is the permanent ending of the contract by either Party. Note that certain Clauses and Sub-Clauses will survive after termination to give the Parties certain rights and obligations after termination has taken place.

FIDIC gives rights of termination for the Employer for certain breaches by the Contractor, and rights of suspension and/or termination for the Contractor for certain breaches by the Employer.

Termination by the Employer

The FIDIC contracts allow the Employer to terminate the Contract for a default by the Contractor, the Engineer (Employer under the Silver Book) giving a Notice ("Notice to Correct") to the Contractor, requiring him to make good/remedy the default within a specified time.

Sub-Clause 15.1 states that the Notice is required to describe the Contractor's failure; state the relevant Sub-Clause and/or provisions of the Contract under which the Contractor has the obligation and to which it has failed; and specify the time within which the Contractor is required to remedy the failure, and the work and/or other action required to remedy it.

After receiving the Notice, the Contractor is required to respond immediately by giving a Notice to the Engineer (Employer under the Silver Book) describing the measures he will take to remedy the failure, stating the date on which such measures will be commenced in order to comply with the time specified within the Notice.

Note that the term "Notice" has a strict definition under the Contract it must be in writing and defined as a Notice, and both Parties should ensure that they strictly comply in that respect.

So, the Engineer (Employer under the Silver Book) lays down the rules under which the Contractor is to put matters right, and how he expects him to do it, but it is then up to the Contractor to respond with what he is actually going to do and when he will do it, to suit the Engineer's (Employer's under the Silver Book) timeline.

As stated in the Preface to this book, it is not proposed to reference-specific court cases, as FIDIC operates under a multitude of different legislations worldwide, but the timeline for the Contractor to make the correction should be reasonable, dependent on what the work is and the relevant conditions at the time.

It is hoped that with all matters of termination, which is a very serious action and can be very complex, that the Employer and the Contractor take a sensible approach in deciding whether to terminate.

The Employer is not required to terminate, it is an option available to him and essentially it should be a material breach of the Contractor's obligations, where the Employer has no other options available to him.

It is often a matter of debate as to what constitutes a "material breach" under these provisions, and generally where it is stated within a Contract.

Clearly each case has to be taken on its merits, dependent on the project, the actual breach, the timing of that breach and the effect on the Parties, this matter being the subject of many discussions, debates and even disputes.

A Dispute Avoidance/Adjudication Board (DAAB) (see Chapter 10) may also take the Employer's decision to terminate into account where it was judged to be premature or wrongful.

Note that the Employer (and during the termination process it is the Employer not the Engineer which has some of the actions during the termination process) is entitled to give a Notice to the Contractor of his intention to terminate the Contract if the Contractor:

a Fails to comply with:

  i A Notice to Correct (see above);
  ii A binding agreement, or final and binding determination, under Sub-Clause 3.7 (Agreement or Determination) (this is a new reason not previously included within FIDIC 1999); or
  iii A decision of the DAAB (whether binding or final and binding) and such failure constitutes a material breach of the Contractor's obligations under the Contract (this is a new provision under FIDIC 2017);

  as stated above, the failure has to constitute a "material breach" of the Contractor's obligations.

b Abandons the Works or demonstrates an intention not to continue performance of any of its obligations under the Contract. "Demonstrating an intention" could be quite subjective, and could be real or perceived, but there may be indications that the Contractor could be about to leave Site.

c Without reasonable excuse (what is a "reasonable excuse"?) fails to proceed with the Works in accordance with Clause 8, or if there is a maximum amount of delay damages stated in the Contract Data, failure to comply, such that the Employer would be entitled to delay damages that exceed this maximum amount (the second part is a new reason not previously included within FIDIC 1999).

Note that the provision states that the Employer "would be entitled to Delay Damages", there is no requirement for the Employer to have actually deducted those delay damages.

It is not clear what will happen if the Contractor claims an Extension of Time, and it is granted by the DAAB or Arbitrator after termination so that the Delay Damages would be reduced below the maximum amount. Would the termination then be retrospectively judged as wrongful or unlawful?

d Without reasonable excuse (again, what is a "reasonable excuse"?) fails to comply with a Notice of rejection given by the Engineer (Employer under the Silver Book) under Sub-Clause 7.5 [Defects and Rejection] or an Engineer's (Employer under the Silver Book) instruction under Sub-Clause 7.6 [Remedial Work], within 28 days after receiving it. Again, common sense should prevail.

e Fails to comply with Sub-Clause 4.2 i.e. the Contractor did not obtain the required Performance Security.

f Subcontracts the whole of, or any part of, the Works in breach of Sub-Clause 5.1 (Sub-Clause 4.4 under the Yellow and Silver Books), which includes subcontracting works over the percentage values stated within the Contract Data, or any part of the Works for which subcontracting is not permitted, fails to get the Engineer's (Employer's under the Silver Book) consent in writing (for Subcontractors for which consent is required) or assigns the Contract without the required agreement (see Sub-Clause 1.7);

g Becomes bankrupt or insolvent, or if the Contractor is a Joint Venture (JV):

  i Any of these matters apply to a member of the JV, and
  ii The other member(s) do not promptly confirm to the Employer that, in accordance with Sub-Clause 1.14 (a) [Joint and Several Liability], such member's obligations under the Contract shall be fulfilled in accordance with the Contract; or

h Is found, based on reasonable evidence, to have engaged in corrupt, fraudulent, collusive or coercive practice at any time in relation to the Works or to the Contract.

Note that (h) was, in previous FIDIC contracts, only a breach by the Contractor in terms of termination, which is surprisingly common as a "one sided breach" in other standard forms of contract, but under the FIDIC 2017 Contracts it can now also be a breach by the Employer allowing the Contractor to terminate under Clause 16.

Unless the Contractor remedies the matter described in the Notice within 14 days of receiving the Notice (in which case the termination does not take place), the Employer may by giving a second Notice to the Contractor immediately terminate the Contract.

However, in the case of sub-paragraph (f), (g) or (h) above, the Employer may by giving a Notice under Sub-Clause 15.2.1 immediately terminate the Contract and the date of termination is the date the Contractor receives this Notice. These are the only three provisions where ONE notice is required, rather than TWO, the principle being that the Contractor could remedy the other matters and put them right, but not (f), (g) or (h), which cannot be remedied, at least in the short term.

The date of termination is the date the Contractor receives this second Notice. After termination of the Contract the Contractor is required to:

a Comply immediately with any reasonable instructions included in a Notice given by the Employer (note the Employer gives the instructions, not the Engineer, even under the Red and Yellow Books):

 i For the assignment of any subcontract (either to the Employer or to another Contractor),
 ii For the protection of life or property or for the safety of the Works.

 This is firmer and more prescriptive than the FIDIC 1999 Contracts which only required the Contractor to "use his best efforts to comply with any reasonable instruction included in the notice".

b Deliver to the Engineer (Employer under the Silver Book):

 i Any Goods required by the Employer,
 ii All Contractor's Documents,
 iii All other design documents compiled by or for the Contractor to the extent, if any, that the Contractor is responsible for the design of part of the Permanent Works under Sub-Clause 4.1 [Contractor's General Obligations].

c Leave the Site and, if the Contractor does not do so, the Employer has the right to expel the Contractor from the Site.

Completion of the Works

After termination, the Employer may complete the Works and/or arrange for anyone else to do so.

Obviously that is entirely up to the Employer, once the original Contractor has been terminated. The Employer may also use any Goods and Contractor's Documents made by or on behalf of the Contractor to complete the Works, obviously the matter of "title", particularly to Goods would have to be considered.

After completion of the Works (not upon termination, so it could be a considerable time later), the Employer gives another Notice to the Contractor that the Contractor's Equipment and Temporary Works can be released to the Contractor at or near the Site. The Contractor then promptly arranges their removal, at the risk and cost of the Contractor.

Note that in many cases the initial Contractor's Equipment and Temporary Works are removed at termination possibly by the hire companies who own them, and the second Contractor provides his own.

However, if the Contractor has failed to make any payment due to the Employer which he is required to make under the termination provisions, these items may be sold by the Employer in order to recover this payment. Any balance of monies is then paid to the Contractor.

Valuation after termination for Contractor's Default

After termination of the Contract, the Engineer (Employer under the Silver Book) agrees or determines the value of the Permanent Works, Goods and Contractor's Documents, and any other sums due to the Contractor for work executed in accordance with the Contract (note that the date of termination is the date of commencement of the time limit for agreement under Sub-Clause 3.7.3 (Sub-Clause 3.5.3 under the Silver Book)).

For anyone who has carried out this valuation exercise in practice, it is a considerable task to value the Works, Goods and other matters at the time of termination!

Note that the valuation does not include the value of any Contractor's Documents, Materials, Plant and Permanent Works that do not comply with the Contract.

Payment after termination for Contractor's Default

The Employer may withhold payment to the Contractor of all amounts agreed or determined until all the costs, losses and damages (if any) described in the following list have been established.

After termination of the Contract, the Employer is entitled to payment by the Contractor of:

- a Any additional costs of execution of the Works by the Employer himself or using another Contractor, and all other costs reasonably incurred by the Employer (including costs incurred in clearing, cleaning and reinstating the Site), after allowing for any sum due to the Contractor under Sub-Clause 15.3 (see above). This is often very difficult to calculate, and the subject of many disputes!
- b Any losses and damages suffered by the Employer in completing the Works. Again, this is often very difficult to calculate, and the subject of many disputes!
- c Delay damages, if the Works or a Section has not been taken over under Sub-Clause 10.1 [Taking Over the Works and Sections] at the time of termination, and if the date of termination under Sub-Clause 15.2 [Termination for Contractor's Default] occurs after the date corresponding to the Time for Completion of the Works or Section, delay damages are required to be paid for every day that has elapsed between these two dates.

Again, as stated above, this is a considerable task requiring quite specialist knowledge to calculate the cost of someone else completing the Works including clearing the Site, any losses and damages suffered by the Employer. The calculation of delay damages however is reasonably straightforward.

Termination for Employer's Convenience

This is a new provision within FIDIC 2017, allowing the Employer at any time and without giving a reason, to terminate the Contract which is essentially where no fault exists on the part of the Contractor.

Typically, an Employer may wish to terminate for his own convenience because he no longer needs the project; the market for which the building or facility was to be used has changed; or there may have been severe budget cutbacks or even a change of government, for example, on a publicly funded project,

However, it is worth mentioning that, although the Employer may terminate for convenience, which normally means that he no longer requires the Works to be carried out, he can then appoint another Contractor to carry out the Works, which does not seem equitable, even though the first Contractor is paid for loss of profit.

Depending on the legal jurisdiction applicable, this action may not be allowable under law where no breach (material or otherwise) has been committed by the Contractor, so it is strongly advised that the Employer seeks legal advice on this provision when drafting the tender documents dependent on the location of the project and the relevant legislation.

After giving a Notice to terminate for convenience under this Sub-Clause, the Employer immediately:

a Has no further right to use any of the Contractor's Documents, which must be returned to the Contractor, except any for which the Contractor has received payment from the Employer and therefore they belong to the Employer, or for which payment is already due under a Payment Certificate. This will include any design documentation.

b Has no right to allow the continued use or shared use (if any) of any Contractor's Equipment, Temporary Works, access arrangements and/or other of the Contractor's facilities or services, if Sub-Clause 4.6 [Co-operation] applies.

c Makes arrangements to return or repay the Performance Security to the Contractor.

Termination under this Sub-Clause takes effect 28 days after the later of the dates on which the Contractor receives this Notice, or the Employer returns the Performance Security.

Note that <u>until</u> the Contractor has received payment of the amount due after Termination for Employer's Convenience under this Sub-Clause, the Employer must <u>NOT</u> execute (any part of) the Works or even arrange for (any part of) the Works to be executed by any other entities.

After the termination, the Contractor proceeds in accordance with Sub-Clause 16.3 [Contractor's Obligations after Termination].

Valuation after termination for Employer's Convenience

After termination the Contractor submits detailed supporting particulars (as reasonably required by the Engineer) (Employer under the Silver Book) of:

a The value of work done, which includes:

    i The matters described in sub-paragraphs (a) to (e) of Sub-Clause 18.5 [Optional Termination] i.e. amounts payable for work executed for which a price is stated in the Contract, the Cost of Plant and Materials for which the Contractor is liable to accept delivery, and other Cost or liability reasonably incurred by the Contractor in expectation of completing the Works, the Cost of removal of Temporary Works and Contractor's Equipment, the cost of repatriation of the Contractor's staff and labour <u>wholly in connection with the Works</u> at the time of termination.

    ii Any additions and/or deductions, and the balance due (if any), by reference to the matters described in sub-paragraphs (a) and (b) of Sub-Clause 14.13 [Issue of FPC] i.e. any amounts due to the Contractor under Sub-Clause 3.5 [Agreement or Determination], or under the Contract.

b The amount of any loss of profit or other losses and damages suffered by the Contractor as a result of this termination. Note that the entitlement is not just to loss of profit, but any other losses or damages resulting from the termination (which can be a considerable amount and a considerable task to calculate). Note that the entitlement and calculation of losses and damages, and any limitations, may also be different under differing legislation.

The Engineer (Employer under the Silver Book) then agrees or determines the matters described above, and the date the Engineer (Employer under the Silver Book) receives the Contractor's particulars is the date of commencement of the time limit for agreement under Sub-Clause 3.7.3.

The Engineer (Employer under the Silver Book) must issue a Payment Certificate for the amount agreed or determined, without the need for the Contractor to submit a Statement.

Payment after termination for Employer's Convenience

The Employer is required to pay the Contractor the amount certified in the Payment Certificate within 112 days after the Engineer (Employer under the Silver Book) receives the Contractor's submission.

Suspension by Contractor

If:

a The Engineer fails to certify a payment as required (in terms of timing and amount),
   (Note that (a) is omitted from the Silver Book)
b The Employer fails to provide "reasonable evidence" in accordance with Sub-Clause 2.4 [Employer's Financial Arrangements], this can be quite an onerous requirement for Employers, and is often omitted from the Contract anyway,
c The Employer fails to comply with Sub-Clause 14.7 [Payment] (again in terms of timing and amount),
d The Employer fails to comply with:
   i A binding agreement, or final and binding determination under Sub-Clause 3.7 [Agreement or Determination], or
   ii A decision of the DAAB under Sub-Clause 21.4 [Obtaining DAAB's Decision] (whether binding or final and binding),

and the failure constitutes a "material breach" (again, what is a "material breach"?) of the Employer's obligations under the Contract, the Contractor may, not less than 21 days after giving a Notice to the Employer (not then Engineer under the Red and Yellow Books), wholly or partially suspend work (or reduce the rate of work) until the Employer has remedied the default. Note that if the Employer remedies the matter, suspension does not take place.

The Contractor is also still entitled to financing charges under Sub-Clause 14.8 [Delayed Payment] and to termination under Sub-Clause 16.2 [Termination by Contractor] because the payment is late.

In some countries, failure of payment to be certified on time and/or the Employer to pay on time will allow the Contractor to suspend the Works under a statutory entitlement by giving a notice, which may be shorter than the 21 days (in many cases 7 days) in the Contract. It is important to note that rights of statute will normally prevail over rights within the Contract.

If the Employer subsequently remedies the default as described in the above Notice before the Contractor gives a Notice of termination under Sub-Clause 16.2, the Contractor is required to resume normal working as soon as is reasonably practicable.

If the Contractor suffers delay and/or incurs Cost as a result of suspending work (or reducing the rate of work) in accordance with this Sub-Clause, the Contractor shall be entitled to an Extension of Time and/or Cost plus Profit.

Termination by Contractor

Notice

The FIDIC 2017 Contracts now include several new grounds under which the Contractor may terminate the Contract (see below).

The Contractor is entitled to give a Notice to the Employer of the Contractor's intention to terminate the Contract or, in the case of sub-paragraph (g) (ii), (h), (i) or (j) below a Notice of termination, if:

a The Contractor does not receive the "reasonable evidence" within 42 days after giving a Notice under Sub-Clause 16.1 [Suspension by Contractor] in respect of a failure to comply with Sub-Clause 2.4 [Employer's Financial Arrangements] (again as stated above, this can be quite an onerous requirement for Employers, and is often omitted from the Contract anyway).

b The Engineer fails, within 56 days after receiving a Statement and supporting documents, to issue the relevant Payment Certificate.

Note (b) is omitted from the Silver Book.

c The Contractor does not receive the amount due (in full) under any Payment Certificate within 42 days after the expiry of the time stated in Sub-Clause 14.7 [Payment] (this previously only applied to the non-payment of Interim Payment Certificates, but now relates to a failure by an Employer to pay the amount due under any Payment Certificate i.e. Advance, Interim or Final Payment).

d The Employer fails to comply with:

  i A binding agreement, or final and binding determination under Sub-Clause 3.7 [Agreement or Determination] (this is a new provision under the FIDIC 2017 Contracts); or

  ii A decision of the DAAB and the failure constitutes a "material breach" of the Employer's obligations under the Contract (this is a new provision under FIDIC 2017, note the reference to a "material breach" of the Employer's obligations under the Contract). What constitutes a "material breach" is likely to be the subject of many differences of opinion and even disputes.

e The Employer substantially fails to perform, and such failure constitutes a "material breach" of the Employer's obligations under the Contract. Note that again the reference to a "material breach" which is a serious and/or major breach, in the

FIDIC 1999 Contracts, the relevant clause merely stated, "*the Employer substantially fails to perform his obligations under the Contract*".

f The Contractor does not receive a Notice of the Commencement Date under Sub-Clause 8.1 [Commencement of Works] within 84 days after receiving the Letter of Acceptance (this is a new provision under the FIDIC 2017 Contracts and is similar to the provision within the "FIDIC Pink Book").

This protects the Contractor from fluctuations in the rates and prices where there is an extended delay to the commencement of the Works, although the Contractor could be entitled to damages for breach of contract anyway. Note that the Contractor could claim for loss of profit on the entire project!

g The Employer:

 i Fails to comply with Sub-Clause 1.6 [Contract Agreement] in issuing or signing the Contract Agreement, or
 ii Assigns the Contract without the required agreement under Sub-Clause 1.7 [Assignment].

h A prolonged suspension (more than 84 days and the work is not resumed) affects the whole of the Works as described in sub-paragraph (b) of Sub-Clause 8.12.

i The Employer becomes bankrupt or insolvent; goes into liquidation, administration, reorganisation, winding-up or dissolution; becomes subject to the appointment of a liquidator, receiver, administrator, manager or trustee; enters into a composition or arrangement with the Employer's creditors; or any act is done or any event occurs which is analogous to or has a similar effect to any of these acts or events under applicable Laws (the definitions have been widened from the FIDIC 1999 Contracts).

j The Employer is found, based on "reasonable evidence", to have engaged in corrupt, fraudulent, collusive or coercive practice at any time in relation to the Works or to the Contract.

 Note, under FIDIC 1999, only the Employer could terminate for corrupt practices engaged in by the Contractor, but now under FIDIC 2017, the equivalent entitlement exists for the Contractor where the Employer has engaged in such practices.

Note that in some countries, extended failure of payment to be certified on time and/or the Employer to pay on time will allow the Contractor to terminate the Contract under a statutory entitlement by giving a notice, which may be shorter than the period in the Contract. It is important to note that rights of statute will normally prevail over rights of Contract.

Termination

Unless the Employer remedies the matter described in a Notice within 14 days of receiving the Notice (in which case termination would not take place), the Contractor may by giving a <u>second</u> Notice to the Employer immediately terminate the Contract.

The date of termination is then the date the Employer receives this second Notice.

However, in the case of sub-paragraph (g) (ii), (h), (i) or (j) of Sub-Clause 16.2.1 [Notice], by giving a Notice under Sub-Clause 16.2.1 the Contractor may terminate the Contract immediately, without giving a second Notice, and the date of termination is the date the Employer receives this Notice.

These are the only three provisions where ONE notice is required, rather than TWO, the principle being that the Contractor could remedy the other matters and put them right, but not (g), (h), (i) or (j) which cannot be remedied, at least in the short term.

If the Contractor suffers delay and/or incurs Cost during the above period of 14 days, the Contractor is entitled subject to an Extension of Time and/or payment of Cost plus Profit.

Contractor's obligations after termination

After termination of the Contract under Sub-Clause 15.5 [Termination for Employer's Convenience], Sub-Clause 16.2 [Termination by Contractor] or Sub-Clause 18.5 [Optional Termination], the Contractor is required to promptly:

a Cease all further work, except for any work which may have been instructed by the Engineer for the protection of life or property or for the safety of the Works;

   Note that whilst it is the subject of an instruction, the Contractor should not abandon a dangerous structure because he never received an instruction!

   If the Contractor incurs Cost as a result of carrying out such instructed work the Contractor shall be entitled subject to be paid such Cost plus Profit.

b Deliver to the Engineer all Contractor's Documents, Plant, Materials and other work for which the Contractor has received payment;

c Remove all other Goods from the Site, except as necessary for safety, and leave the Site.

Payment after termination by Contractor

After termination by the Contractor, the Employer is required to promptly:

a Pay the Contractor in accordance with Sub-Clause 18.5 [Optional Termination];

b Pay the Contractor any loss of profit or other losses and damages suffered by the Contractor, as a result of the termination.

Contract termination

Under Sub-Clause 1.16, subject to any mandatory requirements under the governing law of the Contract, termination of the Contract under any Sub-Clause of these Conditions requires no action of any kind by either Party, other than as stated in the Sub-Clause.

Note that under many legislations, the Contract can only be terminated by approval or by an order of a Court. Parties should seek legal advice in that respect.

9 Care of the Works and Indemnities, Exceptional Events and Insurance

1 Care of the Works and Indemnities is covered in the FIDIC 2017 Contracts by:
- *Conditions of Contract for Construction* (The "Red Book")
  - Clause 17 – Care of the Works and Indemnities
- *Conditions of Contract for Plant and Design-Build* (The "Yellow Book")
  - Clause 17 – Care of the Works and Indemnities
- *Conditions of Contract for EPC/Turnkey Project* (The "Silver Book")
  - Clause 17 – Care of the Works and Indemnities

2 Exceptional Events is covered in the FIDIC 2017 Contracts by:
- *Conditions of Contract for Construction* (The "Red Book")
  - Clause 18 – Exceptional Events
- *Conditions of Contract for Plant and Design-Build* (The "Yellow Book")
  - Clause 18 – Exceptional Events
- *Conditions of Contract for EPC/Turnkey Project* (The "Silver Book")
  - Clause 18 – Exceptional Events

3 Insurance is covered in the FIDIC 2017 Contracts by:
- *Conditions of Contract for Construction* (The "Red Book")
  - Clause 19 – Insurance
- *Conditions of Contract for Plant and Design-Build* (The "Yellow Book")
  - Clause 19 – Insurance
- *Conditions of Contract for EPC/Turnkey Project* (The "Silver Book")
  - Clause 19 – Insurance

In the FIDIC 1999 Contracts, Clause 17 was titled "Risk and Responsibility", but it is now entitled "Care of the Works and Indemnities", and although the Sub-Clauses within the FIDIC 2017 Contracts have been re-arranged, and clarification has been added, the Clause itself is broadly similar to the Clause in the previous FIDIC contracts.

Note that there is no definition in the Contract of the term "Care of the Works", but it can be readily seen from the relevant Sub-Clauses as the documents which form the Contract, as to what it covers.

Responsibility for care of the Works

The Contractor is fully responsible for the care of the Works, Goods and Contractor's Documents, including rectifying any loss or damage during the period from the Commencement Date until the issue of the Taking-Over Certificate for the Works (as stated within the Taking-Over Certificate or when it is deemed to be issued), at which time that responsibility passes to the Employer. The exception is if the Contract is terminated when the responsibility runs from the Commencement Date until the date of termination.

Note that the published Errata (see Appendix to this Book) has changed "Date of Completion of the Works" with "issue of the Taking-Over Certificate for the Works".

The Contract has separate Sub-Clauses for "Responsibility for Care of the Works" and "Liability for Care of the Works", "responsibility" and "liability" often being seen as having the same meaning as each other.

As stated above, the Contractor is fully responsible for the care of the Works from the Commencement Date until the issue of the Taking-Over Certificate for the Works. During that period, he has possession of the Site and therefore is in control of what happens on the Site.

However, liability covers the period after the Taking-Over Certificate has been issued, when the Contractor no longer has possession and control of the Site, but he still may be liable for matters that arise after that time, or that arose before that time but possibly materialise after that time.

Note that when the Contractor is "responsible" he has to rectify the loss or damage at his own cost, whereas the Contract is silent on what the Contractor has to do when he is "liable", other than taking liability. Possibly the Employer will claim for any losses for which the Contractor is liable.

Note also that if something occurs after the issue of the Taking-Over Certificate (or earlier termination), for which the Contractor was responsible prior to the Date of Completion, and obviously dependent on what the issue is, the Contractor would likely still be liable, so the Contractor does not absolve himself of all liability once the Date of Completion arises.

The Contractor would obviously still be liable for correction of Defects occurring after the Date of Completion.

Responsibility for a Section or Part passes to the Employer if a Taking-Over Certificate is issued for that Section or Part.

If the Contract is terminated in accordance with these Conditions or otherwise, the Contractor shall cease to be responsible for the care of the Works from the date of termination.

Liability for care of the Works

There are exceptions to the Contractor's liability, in that it holds no liability for loss or damage to the Works, Goods or Contractor's Documents caused by any of the following events, essentially because he did not cause them or have any control of them:

a Interference with any right of way, light, air, water or other easement which is the unavoidable result of the execution of the Works. This is an important provision, as there may be an inevitable consequence, whether temporary or permanent of the Contractor having to execute the Works.

b Use or occupation by the Employer of any part of the Permanent Works, except as specified in the Contract. The Employer becomes liable for care of the Works if he occupies it, as the provisions state, unless states otherwise.

c A fault, error, defect or omission in any element of the design of the Works by the Employer, which an "experienced Contractor" exercising due care would not have discovered when examining the Site and the Specification and Drawings before submitting the Tender, other than design carried out by the Contractor (under the Red Book, the Contractor may have partial responsibility for design of the Works, but under the Yellow and Silver books, the Contractor has full responsibility for the design of the Works). Note the reference to "an experienced Contractor".

d Any operation of the "forces of nature" (other than those allocated to the Contractor in the Contract Data) which is Unforeseeable, or against which an experienced Contractor could not "reasonably" have been expected to have taken adequate preventative precautions. By using the word "reasonably" and "forces of nature", without being specific, there is an element of subjectivity.

e Any of the events or circumstances listed as Exceptional Events (See Clause 18) (below).

f Any act or default of the Employer's Personnel or the Employer's other Contractors.

If any of the six events above results in damage to the Works, Goods or Contractor's Documents the Contractor must immediately give a Notice to the Engineer (the Employer under the Silver Book), and the Contractor rectifies any loss and/or damage as instructed by the Engineer (the Employer under the Silver Book).

The instruction is then deemed to be a Variation; therefore, the Contractor is entitled to be paid for rectifying any loss.

If the loss or damage to the Works or Goods or Contractor's Documents results from a combination of any of the six events above, and is a cause for which the Contractor is liable or partially liable, and the Contractor suffers a delay and/or incurs Cost from rectifying the loss and/or damage, the Contractor is only entitled to a proportion of the Extension of Time and/or payment of Cost plus Profit to the extent that for any of the above events he may have contributed to such delays and/or Cost, so there may be a partial recovery for the Contractor.

Intellectual and Industrial Property Rights

Whenever a Party receives a claim, but fails to give Notice to the other Party of the claim within 28 days of receiving the Notice, the first Party is deemed to have waived any right to indemnity under the Sub-Clause.

There is then an indemnity on both sides.

The Employer is required to indemnify the Contractor against any claim alleging an infringement which is:

a An "unavoidable" result of the Contractor's compliance with the Specification and Drawings and/or any Variation;

b A result of any Works being used by the Employer for a purpose other than that indicated by, or reasonably to be inferred from (which is very subjective), the Contract, or in conjunction with anything not supplied by the Contractor, unless such use was disclosed to the Contractor before the Base Date or is stated in the Contract (quite a complex provision!);

The Contractor is required to indemnify the Employer against any other claim (including legal fees and expenses), alleging an infringement which arises out of or in relation to the Contractor's execution of the Works, or the use of Contractor's Equipment.

If a Party is entitled to be indemnified, the indemnifying Party (the other Party) may assume overall responsibility for negotiating the settlement of the claim, and/or any litigation or arbitration which may arise from it.

The other Party, at the request and cost of the indemnifying Party, assists in contesting the claim.

The other Party must not make any admission which might be prejudicial to the indemnifying Party, unless the indemnifying Party failed to promptly assume overall responsibility for the conduct of any negotiations, litigation or arbitration after being requested to do so by the other Party.

Indemnities by Contractor

The Contractor is required to indemnify the Employer and his agents against all third-party claims, damages, losses and expenses (including legal fees and expenses) in respect of:

a Bodily injury, sickness, disease or death of any person arising in the course of the Contractor's execution of the Works;
b Damage to or loss of any property, real or personal (other than the Works), to the extent that the damage or loss arises in the course of the Contractor's execution of the Works, and is attributable to any negligence, wilful act or breach of the Contract by the Contractor.

Note that within this Sub-Clause, the Contractor's indemnities strictly relate to injury or death to persons and damage to property <u>other than the Works</u>.

This is similar to FIDIC 1999, but the Contractor's and the Employer's indemnities are now limited to "third-party" claims, thereby preventing the Parties from claiming against each other, so the reference to third parties is critical.

Clearly, the reference to third-party claims is intended to exclude any claims for damage to the Employer's property, but one could envisage a claim coming from a third party in respect of something of which the Employer has ownership but the third party has liability.

Note the new provision under the FIDIC 2017 Contracts that if the Contractor is responsible for the design of part of the Permanent Works (under the Red Book he <u>may</u> be, under the Yellow and Silver Books he <u>will</u> be), he indemnifies the Employer against all acts, errors or omissions by the Contractor in carrying out the Contractor's design obligations that result in the Works (or Section or Part or major item of Plant, if any), when completed, not being fit for the purpose(s) for which they are intended. That is a particularly onerous risk for Contractors (see also Chapter 5).

In considering indemnities provided by both Parties one should also consider the critically important Sub-Clause 1.15 [Limitation of Liability], which has been moved from the old FIDIC 1999 Contracts Clause 17.

Note that neither Party is liable to the other Party for loss of use of any Works, loss of profit and loss of any contract, or for any indirect or consequential loss or damage which may be suffered by the other Party in connection with the Contract, other than under:

a Sub-Clause 8.8 [Delay Damages]
b Sub-paragraph (c) of Sub-Clause 13.3.1 [Variation by Instruction]
c Sub-Clause 15.7 [Payment after Termination for Employer's Convenience]

d Sub-Clause 16.4 [Payment after Termination by Contractor]
e Sub-Clause 17.3 [Intellectual and Industrial Property Rights]
f The first paragraph of Sub-Clause 17.4 [Indemnities by Contractor]
g Sub-Clause 17.5 [Indemnities by Employer]

The total liability of the Contractor to the Employer under or in connection with the Contract, other than:

i Under Sub-Clause 2.6 [Employer-Supplied Materials and Employer's Equipment]
ii Under Sub-Clause 4.19 [Temporary Utilities]
iii Under Sub-Clause 17.3 [Intellectual and Industrial Property Rights]
iv Under the first paragraph of Sub-Clause 17.4 [Indemnities by Contractor]

must not exceed the sum stated in the Contract Data or (if a sum is not so stated) the Accepted Contract Amount.

Note that the Sub-Clause does not limit liability in any case of fraud, gross negligence, deliberate default or reckless misconduct by the defaulting Party, which clearly is a sensible exclusion (note that "gross negligence" has been added to the FIDIC 2017 Contracts).

It is advised that Parties take legal advice as to what constitutes, for example, "gross negligence" as this can be interpreted differently under differing legislations.

Within the Yellow and Silver Books the Contractor has full design responsibility anyway, and Sub-Clause 17.4 is worded accordingly.

Indemnities by Employer

The Employer's indemnities have been expanded within the FIDIC 2017 Contracts, and now cover the Contractor against all third-party claims, damages, losses and expenses in respect of:

a Bodily injury, sickness, disease or death, or loss of or damage to any property other than the Works, which is attributable to any negligence, wilful act or breach of the Contract by the Employer, the Employer's Personnel or any of their respective agents (this indemnity has been expanded from FIDIC 1999); and
b Damage to or loss of any property, real or personal (other than the Works), to the extent that such damage or loss arises out of any event described under sub-paragraphs (a) to (f) of Sub-Clause 17.2 [Liability for Care of the Works] (again this indemnity has been expanded from FIDIC 1999).

Shared indemnities

The Contractor's liability to indemnify the Employer is reduced proportionately to the extent that any event described under sub-paragraphs (a) to (f) of Sub-Clause 17.2 [Liability for Care of the Works] the Employer may have contributed to the damage, loss or injury.

The Employer's liability to indemnify the Contractor is reduced proportionately to the extent that any event for which the Contractor is responsible under Sub-Clause 17.1 [Responsibility for Care of the Works] and/or under Sub-Clause 17.3 [Intellectual and Industrial Property Rights] may have contributed to the said damage, loss or injury.

This is a new provision under the FIDIC 2017 Contracts, so the contributory effect by both Parties is taken into account in assessing liability and any ensuing loss. This would probably be taken into account anyway under previous FIDIC contracts, but it is important to spell it out. Notice that liability is "reduced proportionally" which is the correct thing to do, but the "amount of proportionality" could be the subject of debate.

Note again that under Clause 1.15 of FIDIC 2017 (see above), neither Party is liable to the other for loss of use of any of the Works, loss of profit, loss of contract or any indirect or consequential loss or damage other than for the items listed under (a) to (g), and liability is capped for items (i) to (iv).

Exceptional Events

In FIDIC 1999, Clause 19 was titled "Force Majeure", but its equivalent within FIDIC 2017 Clause 18 is now entitled "Exceptional Events" though the Clause itself is broadly similar to the FIDIC 1999 Clause.

Force Majeure is a well-known internationally known legal/contractual term, so it is not clear why FIDIC has changed the name of the Clause, particularly as the definition of an "Exceptional Event" is virtually the same as "Force Majeure".

Note that the Conditions of Contract for Design, Build and Operate Projects 2008 ("Gold Book") adopted the term "Exceptional Events" in lieu of "Force Majeure" so there is precedence within the FIDIC contracts.

The author believes that the use of this new term does not provide any clarity as to the provision, and in effect could cause confusion to FIDIC 2017 users.

The author has discussed this with various practitioners, and it has been stated that the term "Force Majeure" is subject to various long held and in some cases incorrect perceptions as to what it actually means, so practitioners will now have to <u>read the Contract</u> to interpret what an "Exceptional Event" is, which is always advisable anyway, but also the concept of Force Majeure is not recognised within some jurisdictions.

Note that the NEC contracts (of which the author has a great deal of experience) do not use the term "Force Majeure" either, but Sub-Clause 60.1(19), within the compensation event provisions, refers to an event which

- Stops the Contractor completing the whole of the Works, or
- Stops the Contractor completing the whole of the Works by the date for planned Completion shown on the Accepted Programme,

and which

- Neither Party could prevent,
- An experienced contractor would have judged at the Contract Date to have such a small chance of occurring that it would have been unreasonable to have allowed for it, and
- Is not one of the other compensation events stated in this contract.

So, the change within the FIDIC 2017 Contracts is not without precedent in other published contracts.

"Exceptional Event" is defined within the FIDIC 2017 Contracts as an event which:

i Is beyond a Party's (the Employer's or Contractor's not the Engineer or Employer's Representative) control;
ii The Party could not reasonably have provided against it before entering into the Contract. In the Contractor's case when he could have priced and/or programmed for it;
iii Having arisen, the Party could not reasonably have avoided or overcome it;
iv Is not substantially attributable to the other Party.

It may comprise (note this is not an exhaustive list) any of the following events which are listed in the Contract:

a War, hostilities, invasion, act of foreign enemies;
b Rebellion, terrorism, revolution, insurrection, military or usurped power, or civil war;
c Riot, commotion or disorder by persons other than the Contractor's Personnel and other employees of the Contractor and Subcontractors;
d Strike or lockout not solely involving the Contractor's Personnel and other employees of the Contractor and Subcontractors;
   It is curious that the FIDIC 1999 Contracts previously included "strikes and lockouts" with riots, commotion and disorder," but it is now a separate entry.
   One could say that "riots, commotion and disorder" are considered violent acts, whilst a strike or lockout may not be, but note that "riots, commotion and disorder" are related to "persons other than the Contractor's Personnel and other employees of the Contractor and Subcontractors", whilst "strikes or lockouts" are not solely involving the Contractor's Personnel and other employees of the Contractor and Subcontractors, so there is a subtle difference.
e Encountering munitions of war, explosive materials, ionising radiation or contamination by radio-activity, except as may be attributable to the Contractor's use of such munitions, explosives, radiation or radio-activity; or
f Natural catastrophes such as earthquake, tsunami, volcanic activity, hurricane or typhoon.

NB: Should a pandemic, for example COVID-19, be expressly considered as an "Exceptional Event"?

As this book is being written in the midst of a COVID-19 lockdown, it is worth reviewing the provisions within the Contract.

The Contract states, against the list (a) to (f) above, that the list is not exhaustive, provided that the conditions within the definition are met i.e. beyond a Party's control, the Party could not reasonably have provided against it before entering into the Contract, the Party could not reasonably have avoided or overcome it, and it is not substantially attributable to the other Party. In that sense COVID-19 would fall under the definition.

As with most issues within the Contract, one must also consider the law in respect of the pandemic, for example, is there a government lockdown, curfew, etc., which prevents the Contractor from working, or does it just make it restrictive and more difficult for the Contractor to work?

If the Contractor feels that the pandemic is to be classified as an "Exceptional Event" then he must clearly follow the requirements of the Contract.

The Parties should first discuss the pandemic, and its effect on the progress of the Works. All too often Parties quote the Contract and rights and obligations, before they have even spoken to each other about the issues and how to achieve a mutually agreeable outcome.

- i The Contractor should give a Notice to the Employer within the required time scale.
- ii Use all reasonable endeavours to minimise any delay.
- iii If the Exceptional Event is continuing for a prolonged period, the affected Party gives further Notices every 28 days after giving the first Notice.
- iv The Employer should then respond as soon as possible with any Extension of Time and/or payment of Cost.
- v There is the option of termination if progress with the Works is prevented for a continuous period of 84 days, but that would need to be discussed between the Parties.

Note also the provision for Extension of Time under Clause 8.5 item (d), "Unforeseeable shortages in the availability of personnel or Goods (or Employer-Supplied Materials, if any) caused by epidemic or governmental actions".

Note also "Adjustments for Changes in Laws" Sub-Clause 13.6, which could give rise to an Extension of Time and/or payment of Cost.

Again, the Contractor must clearly follow the requirements of the Contract in terms of Notices and the necessary procedures.

As stated below, (a) to (e) will give rise to an entitlement to an Extension of Time, (b) to (e) of that Sub-Clause, if it occurs in the Country, will also give rise to payment of Cost.

Notice of an Exceptional Event

If a Party (the "affected Party") is prevented from performing any (any, not necessarily all) of his obligations under the Contract due to an Exceptional Event, then that Party gives a Notice to the other Party specifying those obligations.

This Notice is given <u>within</u> 14 days after the Party <u>became aware</u>, or <u>should have become aware</u> ("should have become aware" is always a very subjective decision and it is not unusual for one Party to assert when the other Party <u>should</u> have become aware!) of not being able to perform those obligations, the Party then being excused performance of the prevented obligations from the date the performance is prevented by the Exceptional Event.

If the Notice is received by the other Party <u>after</u> the period of 14 days, the Party is only excused performance <u>from the date on which the Notice is received by the other Party</u>.

Note that the obligation of either Party to make periodic payments to the other Party is not excused by an Exceptional Event.

Duty to minimise delay

Each Party must at all times use all reasonable endeavours to minimise any delay in the performance of the Contract as a result of an Exceptional Event. The duty to mitigate is a common requirement and does not require the Party to put matters right, but to contain the matter as far as they are able, so it does not get worse. It amplifies the fact that there is no entitlement for the affected Party to "relax" once he has issued the Notice!

If the Exceptional Event has a continuing effect (i.e. more than 28 days) as it is quite likely to, the affected Party gives further Notices describing the effect every 28 days after giving the first Notice.

The Party immediately gives a Notice to the other Party when the affected Party ceases to be affected by the Exceptional Event. It is critical that both Parties are aware of when it no longer becomes a problem.

If the affected Party fails to do so, the other Party may give a Notice to the affected Party, stating that the other Party considers that the affected Party's performance is no longer prevented by the Exceptional Event, with reasons, so that will close the matter out.

Consequences of an Exceptional Event

If the Contractor is the affected Party and suffers delay and/or incurs Cost by reason of the Exceptional Event, the Contractor is entitled to an Extension of Time, and/or if the Exceptional Event is of the kind described in sub-paragraphs (a) to (e) of Sub-Clause 18.1 [Exceptional Events] i.e. excepting natural catastrophes, and, in the case of sub-paragraphs (b) to (e) of that Sub-Clause i.e. war, hostilities, and natural catastrophes, and occurs in the Country, payment of Cost.

Optional termination

If the execution of "substantially all the Works in progress" is prevented for a <u>continuous</u> period of 84 days by reason of an Exceptional Event of which Notice has been given, or for <u>multiple</u> periods which total more than 140 days due to the <u>same</u> Exceptional Event, then either Party may give to the other Party a Notice of termination of the Contract.

The word "substantially" obviously needs to be considered on a case-by-case basis. Is it major items, high value items and how substantial is "substantial"!

The date of termination is then seven days after the Notice is received by the other Party.

(See also Chapter 8 on Termination).

After the date of termination, the Contractor submits detailed supporting particulars (as reasonably required by the Engineer (Employer's Representative under the Silver Book)) of the value of the work done, which includes:

- a The amounts payable for any work carried out for which a price is stated in the Contract;
- b The Cost of Plant and Materials ordered for the Works <u>which have been delivered to the Contractor</u>, or of which the Contractor <u>is liable to accept delivery</u>. This Plant and Materials become the property of the Employer only when paid for by the Employer;
- c Any other Cost or liability which, under the circumstances, was reasonably incurred by the Contractor in the expectation of completing the Works (this could be quite subjective and difficult to calculate);

d The Cost of removal of Temporary Works and Contractor's Equipment from the Site and the return of these items to the Contractor's place of business in the Contractor's country (or to any other destination(s) at no greater cost);
e The Cost of repatriation of the Contractor's staff and labour employed wholly in connection with the Works at the date of termination (note the term "wholly in connection" with the Works).

The Engineer (Employer's Representative under the Silver Book) then agrees or determines the value of work done (and for the purpose of Sub-Clause 3.7.3 [Time limits] (Sub-Clause 3.5.3 under the Silver Book)), and the date the Engineer (Employer's Representative) receives the Contractor's particulars under this Sub-Clause is the date of commencement of the time limit for agreement under Sub-Clause 3.7.3. (Sub-Clause 3.5.3 under the Silver Book).

The Engineer (Employer under the Silver Book) then issues a Payment Certificate, for the amount agreed or determined, without the need for the Contractor to submit a Statement as he would have done for normal payments.

Release from performance under the law

If any event arises outside the control of the Parties (including, but not limited to, an Exceptional Event) which:

a Makes it impossible or unlawful for either Party or both Parties to fulfil their contractual obligations, or
b Entitles, under the law of the Contract, the Parties to be released from further performance of the Contract, and if the Parties are unable to agree on an amendment to the Contract that would permit the continued performance of the Contract (this needs legal advice), then after either Party gives a Notice to the other Party of such event:

i The Parties will then be discharged from further performance.
ii The amount payable by the Employer to the Contractor is the same as would have been payable under Sub-Clause 18.5 [Optional Termination], and such amount is certified by the Engineer, as if the Contract had been terminated under that Sub-Clause.

Obviously, as stated above, legal advice needs to be taken in terms of "the law governing the Contract" and the provisions within it.

Insurance

Insurance is now covered within the FIDIC contracts by Clause 19.

As with much of the FIDIC 2017 Contracts, the requirements within Clause 19 and the Contract Data are now considerably wordier and more prescriptive, which is good in the sense of clarity, but would again lead to the perception that the Contract appears administratively burdensome.

Under the FIDIC 1999 Contracts and previously in Clause 18 it referred to the "Insuring Party", allowing for flexibility in the sense that provisions referred to whichever Party was required to insure.

Under the FIDIC 2017 Contracts under Clause 19, the FIDIC drafters have discarded the term "Insuring Party", and almost all the insurance obligations are on the Contractor, so if there are exceptions to that principle, then the Contract will have to be amended.

The principle under FIDIC 2017 is that the Employer is required to add "annexed memoranda" to the Letter of Acceptance at the time of appointment, as stated on the Letter of Acceptance template.

There is clearly a concern that even where the Parties essentially remain within the outline of the Clause 19 terms, many typical insurance policies may well not match the now much more specific requirements within Clause 19.

Items that would need to be considered within any amendments include:

- Joint names insurance cover extending to all parties for their Site interests, particularly Subcontractors of any tier and other contractors of the Employer, as may be applicable;
- The Works forming a part of a larger project, all at or near the Site;
- The presence of significant existing property of the Employer at or close to the Site.

In summary the more extensive and prescriptive nature of Clause 19 and the associated Contract Data is a positive development if the outcome is that Parties will consider the requirements carefully and take the necessary professional insurance advice...but often they do not!

Insurance generally

Insurance is the principle of many paying in for the few to be compensated should an insured event or series of events occur, which cause a loss.

It is basically a contract between the "insured" who offers to pay an agreed sum, a premium, to the "insurer" who warrants to pay out an agreed sum should a particular and specified event occur e.g. damage to property, death or injury, etc. In effect, it is possible to insure against anything, which holds an element of risk.

The required insurances under any construction contract can normally be grouped under the following headings.......

Loss or damage to Works, Plant or Materials

Insurance of the Works is often covered by a Contractor's All Risks (CAR) policy.

CAR insurance normally covers damage to property, such as damage to buildings, infrastructure or other structures being worked on. It also covers liability for third-party claims for injury and death or damage to third-party property.

This insurance may be taken out in the joint names of the Contractor and the Employer. Other interested parties, such as funders, often ask to be added as a joint name. The theory is that if damage occurs to the insured property then, regardless of fault, insurance funds will be available to allow for reinstatement.

The effect of Joint Names Insurance is that each party has its own rights under the policy and can therefore claim against the insurer. Each insured should comply with the duties of disclosure and notification.

The minimum amount of cover for all insurances is stated in the Contract Data; however, the Contractor is liable for whatever the amount of any claim therefore it must consider the minimum value in the Contract Data purely as a guide.

This insurance covers loss or damage to the Works and Plant or Materials intended for the Works, including any unfixed Plant and Materials on Site.

Loss or damage to Equipment

Again, the CAR policy should cover this. Any reference to "replacement cost" means the cost of replacement with Equipment of similar age and condition rather than "new for old".

Public liability

Loss or damage to property (except the Works, Plant or Materials) or bodily injury or death not to an employee of the Contractor arising in connection with the Contractor providing the Works.

This requires the Contractor to indemnify the Employer against any loss, expense, claim, etc., in respect of any personal injury or death caused by carrying out of the work, other than their own employees. This includes the liability toward members of the public who may be affected by the construction work, although they have no part in it. In the case where a party makes a claim directly against the Employer due to a death or injury the Contractor should either take on that claim, or alternatively the Employer can sue the Contractor to recover any monies.

Employer's liability

Death of, or bodily injury to, employees of the Contractor.

In many countries, employers of staff in any company have a statutory obligation to provide cover for their employees in the event of death, injury or damage caused during the course of their employment. "Worker's Compensation" can also form part of this insurance.

Professional Indemnity Insurance

Particularly where the Contractor is designing parts of the Works, it would be advisable for the Employer to include a requirement for the Contractor to provided Professional Indemnity (PI) Insurance. Particularly if, in providing the design, the Contractor makes a mistake, is found in some way to be negligent or gives inaccurate advice, then the Contractor will be liable to the Employer in event that the Employer incurs a loss as a result.

Insurance under the FIDIC 2017 Contracts

The Contractor is required to effect and maintain all insurances for which it is responsible with insurers and in terms, both the insurer and the terms of which are subject to consent by the Employer.

The terms of the insurances are required to be consistent with terms (if any) agreed by both Parties before the date of the Letter of Acceptance.

Note that the insurances required to be provided under this Clause are the minimum required by the Employer, and the Contractor may, at the Contractor's own cost, add such other insurances that he may deem prudent.

Whenever required by the Employer, the Contractor is required to produce the insurance policies which it is required to effect and maintain under the Contract. As each premium is paid, the Contractor then submits either a copy of each receipt or confirmation from the insurers that the premium has been fully paid.

If the Contractor fails to effect and keep in force any of the insurances required under the Contract then the Employer may effect and keep in force such insurances and pay any premium as may be necessary and recover the same from the Contractor. This is a standard provision in most construction contracts.

If either the Contractor or the Employer fails to comply with any condition of the insurances effected under the Contract, the Party failing to comply indemnifies the other against all direct losses and claims arising from such failure.

The Contractor is also responsible for:

a Notifying their insurers of any changes in the nature, extent or programme for the execution of the Works;
b The adequacy and validity of the insurances in accordance with the Contract at all times during the performance of the Contract.

The permitted deductible (excess) limits allowed in any policy shall not exceed the amounts stated in the Contract Data (if not stated, the amounts agreed with the Employer).

Where there is a shared liability the loss is borne by each Party in proportion to each Party's liability, provided the non-recovery from insurers has not been caused by a breach of this Clause by the Contractor or the Employer.

In the event that non-recovery from insurers has been caused by such a breach, the defaulting Party is required to bear the loss suffered.

Insurance to be provided by the Contractor

The Contractor must provide the following insurances:

The Works

The Contractor is required to insure and to maintain the insurance policies in the joint names of the Contractor and the Employer from the Commencement Date until the date of the issue of the Taking-Over Certificate for the Works:

a The Works and the Contractor's Documents, together with Materials and Plant which are intended for incorporation into the Works. The insurance cover is also required to extend to include loss and damage of any part of the Works as a consequence of failure of elements defectively designed or constructed with defective material or workmanship.
b An additional amount of 15% of the replacement value (or any other amount specified in the Contract Data) to cover any additional costs related to the rectification of loss or damage, including professional fees, and the cost of demolition and removal of debris.

The insurance cover is required to cover the Employer and the Contractor against all loss or damage from whatever cause arising until the issue of the Taking-Over Certificate for the Works.

From that point, the insurance is required to continue until the date of the issue of the Performance Certificate in respect of any incomplete work for loss or damage arising from any cause occurring before the date of the issue of the Taking-Over Certificate for the Works, and for any loss or damage occasioned by the Contractor in the course of any operation carried out by the Contractor for the purpose of complying with the Contractor's obligations under Clause 11 [Defects after Taking Over] and under the Yellow and Silver Books, Clause 12 [Tests after Completion].

Note the published Contract included "Clause 12" within the Red Book but this has been changed by the Errata (see Appendix).

However, the insurance cover provided by the Contractor for the Works may exclude any of the following:

- i The cost of making good any part of the Works which is defective, or otherwise does not comply with the Contract, provided that it does not exclude the cost of making good any loss or damage to any other part of the Works attributable to such defect or non-compliance;
- ii Indirect or consequential loss or damage including any reductions in the Contract Price for delay;
- iii Wear and tear, shortages and theft;
- iv Unless otherwise stated in the Contract Data, the risks arising from Exceptional Events (See Clause 18).

Goods

The Contractor is required to insure, in the joint names of the Contractor and the Employer, the Goods and other things brought to Site by the Contractor to the extent specified and/or amount stated in the Contract Data (if not specified or stated, for their full replacement value including delivery to Site).

The Contractor is required to maintain this insurance from the time the Goods are delivered to the Site until they are no longer required for the Works.

Liability for breach of professional duty

Where the Contractor is designing parts of the Works, it would be advisable for the Employer to include a requirement for the Contractor to provide PI Insurance.

Particularly if, in providing the design, the Contractor makes a mistake, is found in some way to be negligent or gives inaccurate advice, then the Contractor will be liable to the Employer in event that the Employer incurs a loss as a result.

This loss can be very significant where the design has to be corrected, parts of the structure have to be taken down and reinstated, a facility has to be closed down whilst the remedial measures take place and there is also likely to be significant also legal costs.

PI claims can arise where there is negligence, misrepresentation or inaccurate advice, which does not give rise to bodily injury, property damage or personal injury, but it does give rise to some financial loss.

Additional coverage for breach of warranty, intellectual property, personal injury, security and cost of contract can be added.

Note that PI Insurance policies are effected on a "claims made" basis, meaning that the policy only covers claims made during the policy period when the policy is "live" (and the premiums are being paid), so claims which may relate to events occurring before the coverage was active may not be covered.

However, these policies may have a retroactive or retrospective date which can operate to provide cover for claims made during the policy period, but which relate to an incident which occurs after the retroactive date.

The Engineer under the Red and Yellow Books, on behalf of the Employer, should ensure that the Contractor has taken out and maintained the required insurances for the full period of its liability and has certificates and other policy documents to prove it.

Note that the Contractor's liability is likely to extend many years beyond the completion of the Works, dependent on the relevant law this typically could be 6, 10, 12 years or more.

Under the FIDIC 2017 Contracts, if the Contractor is responsible for the design of any part of the Permanent Works under Red Book Sub-Clause 4.1 [Contractor's General Obligations], and/or any other design under the Contract, and of course he is fully responsible for design under the Yellow and Silver Books, also under the Red, Yellow and Silver Books, he has to provide certain indemnities under Clause 17 [Care of the Works and Indemnities]:

a The Contractor must take out and maintain PI Insurance against his liability arising out of any act, error or omission in carrying out his design obligations.

  The insurance must be maintained in an amount not less than that stated in the Contract Data. Note that if no amount is stated, the insurance must be to an amount agreed by the Employer.

  Also,

b If it is stated in the Contract Data, PI Insurance to indemnify the Contractor against liability arising out of any act, error or omission by the Contractor in carrying out those design obligations results in the Works (or a Section/Part or major item of Plant), when completed, not being fit for the purpose(s) for which they are intended under Sub-Clause 4.1 [Contractor's General Obligations] (see Chapter 5).

The Contractor is required to maintain PI Insurance for the period specified in the Contract Data, as stated above, this period could be many years after the Contractor completed the works.

Injury to persons and damage to property

This is Public liability Insurance.

The Contractor must insure, in the joint names of the Contractor and the Employer, against liabilities for death or injury to any person, or loss of or damage to any property other than the Works arising out of the performance of the Contract and occurring before the issue of the Performance Certificate, other than loss or damage caused by an Exceptional Event.

The insurance policy must include a "cross liability clause" such that the insurance shall apply to the Contractor and the Employer as separate insured parties.

The insurance is required to be effected before the Contractor begins any work on the Site and must remain in force until the issue of the Performance Certificate and must be for not less than the amount stated in the Contract Data (if not stated, the amount as agreed with the Employer).

Note that some construction contracts have specific and separate provision for insuring neighbouring property which may suffer from collapse, subsidence, vibration, etc., due to damage by, for example, piling, deep excavations or demolition works during the execution of the Works, and where not caused by the negligence of the Contractor.

The FIDIC contracts do not have this as a specific requirement, though it could be covered here.

The Employer should consider its liabilities in respect of neighbouring owners who may be affected by the Works and if necessary either instruct the Contractor to provide such insurance or take it out itself.

Such a policy would normally exclude:

i Damage caused by the negligence of the Contractor or its Subcontractors;
ii Damage which can be reasonably foreseen e.g. cracking may have already occurred to a neighbouring building prior to the works commencing. In this respect, it is important that a condition survey with photographic evidence be carried out before construction commences.
iii Where the damage is caused by any Excepted Risks identified in the contract. These risks will normally be held by the Employer.

Injury to employees

The Contractor must effect and maintain insurance against liability for claims, damages, losses and expenses (including legal fees and expenses) arising out of the execution of the Works in respect of injury, sickness, disease or death of any person employed by the Contractor, or any of the Contractor's other personnel.

The Employer and the Engineer must also be indemnified under the policy of insurance, except that this insurance may exclude losses and claims to the extent that they arise from any act or neglect of the Employer or of the Employer's Personnel.

The insurance must be maintained in full force and effect during the whole time that the Contractor's Personnel are assisting in the execution of the Works.

For any person employed by a Subcontractor, the insurance may be effected by the Subcontractor, but the Contractor is responsible for the Subcontractor's compliance with this Sub-Clause.

Other insurances required by Laws and by local practice

The Contractor is required to provide all other insurances required by the Laws of the countries where (any part of) the Works are being carried out at the Contractor's own cost.

Other insurances required by local practice (if any) are detailed in the Contract Data, and the Contractor provides the insurances in compliance with the details given at the Contractor's own cost.

10 Employer's and Contractor's Claims

Employer's and Contractor's Claims are covered in the FIDIC 2017 Contracts by:

- *Conditions of Contract for Construction* (The "Red Book")
  - Clause 20 – Employer's and Contractor's Claims
- *Conditions of Contract for Plant and Design-Build* (The "Yellow Book")
  - Clause 20 – Employer's and Contractor's Claims
- *Conditions of Contract for EPC/Turnkey Project* (The "Silver Book")
  - Clause 20 – Employer's and Contractor's Claims

Before examining the details of the FIDIC 2017 Contracts in terms of Employer's and Contractor's Claims, let us first consider the matter of Claims generally.

What is a Claim?

If an Employer, an agent of the Employer (e.g. Engineer/Employer's Representative), delays and/or disrupts a Contractor, or causes him to incur additional expense over and above what he would have naturally incurred in carrying out the work, or in some cases, an event happens that is not directly caused by the Employer or his agent, but is outside the control of the Contractor, then normally a Claim arises, for which contracts provide a direct remedy in the form of Extensions of Time and/or payment of Cost and in some cases Cost plus Profit.

It is important, particularly in dealing with claims, to appreciate the distinction between a "*claim under the contract*" and a "*claim for breach of contract*".

A claim under the contract arises when some event occurs (which may or may not be a breach of contract) for which the contract provides a specific remedy. The remedy is usually the award of an Extension of Time (EOT) to the Contractor and/or payment of a sum of money, based on Cost, and sometimes profit in addition.

Often the same event will give rise to claims both under and in breach of contract.

But the consequences of the two are different.

For example, consequential damages may be recovered for a breach, but under the contract only such remedies as are provided within the contract can be recovered. A claim under a contract is a way of enforcing its provisions.

A breach of contract may have two principal consequences. First, every breach entitles the innocent party to sue for damages. Second, if the breach is sufficiently serious it gives the innocent party an option to treat the party in breach as having repudiated the whole contract, and termination could arise.

In such a case the innocent party may bring the contract to an end by accepting the repudiation, or he may at his option treat the contract as subsisting, when it will continue to bind both parties.

The subject of claims is often an emotive subject, tending to polarise the two sides and making them think badly of each other.

There are those who are prepared to exploit a situation and take advantage of any loophole, and there are also those who prepare contracts in haste and later attempt to defend those documents.

There may also be concern as to the impartiality of, for example, the Contract Administrator, whether they be an Engineer or an Employer's Representative, not only with his "dual" role in representing the Employer and administering the Contract, but particularly where the Engineer/Employer's Representative is also the Employer or a salaried employee of the Employer.

By legal definition a contract is an agreement freely entered into by the two parties.

To tackle the problems of claims and to remove the apparent stigma attached to them it is necessary for both sides to have a better understanding of the basis and principles of contracts and, therefore, of the rights and obligations of the contracting parties.

It is necessary for professional representatives to have an understanding of the business side of industry in order to appreciate the circumstances in which the Contractor tenders, the risks which he can properly and contractually be expected to undertake and the financial consequences if things go wrong. A contract states the obligations to be undertaken, and no-one is entitled to expect more than compliance with those obligations.

Types of Claim

A claim is a process where the Contractor considers that certain matters are changed or fall outside the original basis of contract. It is a request for payment by the Contractor in respect of some occurrence which he considers was not envisaged in the contract and for which agreement has not been reached.

The Engineer or Employer's Representative acts independent of both Employer and Contractor and is empowered by the terms of the contract to settle claims and certain disputes between them. He alone must discharge those responsibilities in an independent and impartial way.

Claims made by the Contractor fall into <u>three</u> general types:

i Contractual claims – within the contract and can be dealt with by the Engineer/ Employer's Representative under provisions within the contract.

  An example would be a delay caused by a Variation, for which the Contractor would notify a Claim.

ii Extra-contractual claims – outside the contract and can only be negotiated by Employer and Contractor i.e. there is no express provision within the contract, and they are therefore founded on a breach of contract and settlement is in terms of damages.

  An example would be negligent design by the Contractor, for which the Employer would submit a claim.

iii Ex-gratia claims – no basis in contract or common law and is usually an occurrence which was unforeseen by either party and was significant. The payments are totally discretional and are the province of the Employer.

An example would be where an exceptional event occurs such as a pandemic or severe inflation, or where the Contractor encounters severe hardship, and the Employer decides to pay the Contractor an amount, for which he would not be entitled to be paid under the Contract.

Contractual claims

A contractual claim must be made under a specific clause of the contract in order to establish that it is in accordance with and under the provisions of some condition or requirement of the Contract.

The basis of all claims is, almost without exception, change e.g. new instructions, changed circumstances, new requirements, changes in work content, etc. It must be said that disputes could, in fact, be avoided given sufficient information and with good communication between the parties involved.

There are basically two important aspects of claims:

i Contractual principal and validity;
ii Quantification of time and/or payment.

and generally contractual disputes fall into three categories which relate to:

- What is to be done;
- The time in which it is, or has had, to be done;
- What is to be paid.

A general review of claims management

Well-conceived, well-presented realistic submissions, based on a clear understanding of the contract, and forwarded as soon as possible after the circumstances giving rise to the claim, will speed the resolution including any consideration of interim payments. The converse will, of course, apply.

The Contractor should be prepared to provide all written evidence and records that the Contract Administrator may reasonably require to establish the validity of the claim. If the claim has ultimately to be referred to arbitration the respondent will have the right to demand such evidence. This applies in particular to the Contractor's pricing notes, which are often considered by the Contractor to be confidential, but which would nevertheless have to be produced at the discovery stage of an arbitration or legal action. The Contractor may, not unreasonably, require that any confidential information submitted in support of a claim is used only for that purpose and is not disclosed to third parties.

Whilst claims are never welcomed by the recipient, a well-structured claim will be better received than a sketchy estimate of alleged monies lost.

There is essentially no right or wrong way to prepare and present a claim for loss and expense; however, there are certain guidelines, which enable a claim to be presented in the best possible light.

First, it must be factual, there is nothing worse than a Contract Administrator analysing a claim and finding that even a minority of elements are untrue, as it will prejudice his views against the majority of items that are factual.

Claims generate a mass of correspondence, time sheets, invoices and other paperwork, all of which must be included in order that the claim can "stand alone" as a complete document, rather than having to refer to other information stored elsewhere. Therefore, in order for the recipient to quickly analyse the basis of a claim, it is advisable to compile a concise and brief Claim Document, with all the necessary backup information included within separate appendices.

The Claim Document should be set out as follows:

i Project Details

It is worthwhile to begin the claim by listing the Project Details i.e. Parties to the Contract, Name and Type of Project, Value, Original Commencement and Completion Dates. This sets the scene for the claim to follow.

ii Problems

List the matters that brought about the claim e.g. Delays in receipt of Design Information, delays caused by others, etc., including the efforts which you as the Contractor made to prevent these delays. Contrast what was included in the Tender Documents and therefore priced for, against what actually happened, and the consequences.

It is good policy to include within this Section a letter list in date order showing the letters sent to or from the Employer, Contract Administrator or other party and the relevant responses. A brief summary of the contents can be included within this list, and with the actual letters included in a separate volume.

Claim calculations

The "Heads of Claim" can then be addressed, and all calculations included.

Typical items for which the Contractor may seek reimbursement are.....

Prolongation of Preliminaries Costs

This may include Site Management, Site Accommodation, Plant & Scaffolding, which may be chargeable on a Monthly or Weekly basis.

Disruption

This may include additional labour costs to carry out the work in a different sequence or under different conditions to those tendered for. A delay may also extend the contract into a winter period with its associated disruption, for example groundwork externally, without the delay, may have been completed before the winter. In order to prove these costs, the Contractor will need to show how his original tender was calculated, and relate this to his actual work sequence and associated costs. It should be noted that many Contractors might seek to recover additional costs due to their own failings via this section which they are clearly not entitled to do!

Fluctuations in cost of Labour, Materials or Plant

The Contractor will have made assessments of likely increases over the Contract Period, but if this period is extended he is entitled to recover any additional costs he has incurred.

N.B: particularly during a recession there may also have been decreases!

Increased Insurances, Bond, etc.

If the Contract Period is extended, the Insurance risk and cover period will also be extended. Bonds normally apply for a block period such as a year or six months, so care must be taken in calculating and checking this element.

Prolonged Head Office Costs

In addition to the Direct Site Costs incurred in Item (i) there may also be Head Office Costs such as Project Managers, Quantity Surveyors and Secretarial Staff including a proportion of Rent, Heating and Lighting Costs. This is a very difficult cost to assess, particularly when the Head Office supports a number of Projects, so if prolonged head office costs have been incurred, and there is NO OTHER WAY to calculate them, a theoretical calculation such as Hudson's Formula may be used.......

$$\frac{\text{Percentage Head Office Overheads}}{100} \times \frac{\text{Contract Sum}}{\text{Contract Period}} \times \text{Delay in Weeks}$$

or more rarely the Emden Formula......

$$\frac{\text{Total Overheads}}{\text{Total Turnover}} \times \frac{\text{Contract Sum}}{\text{Contract Period}} \times \text{Delay in Weeks}$$

or even more rarely the Eichleay Formula......

$$\frac{\text{Contract Invoices}}{\text{Total Invoices}} \times \frac{\text{Total Overheads} \times \text{Contract Sum}}{\text{Contract Period}} \times \text{Delay in Weeks}$$

Finance Charges

The Contractor may also be involved with financing elements of the project, such as later release of retention due to delays.

Cost of claim preparation

The cost of preparing a claim is not a legitimate head of claim and must be covered by the Contractor's Overheads. The reason is that under many forms of contract the Contractor is not required to make a detailed loss and expense claim, merely to furnish such information as may be required for the Contract Administrator to assess loss and expense. In practice, however, a well-prepared and well-documented claim will repay the Contractor in terms of recovered amount and will also speed up the settlement of claims.

Record keeping

It is vital that comprehensive records are kept so that the Contractor can substantiate claims and also that the Contract Administrator can defend them in the event that they arise.

These can include (as stated above):

i Correspondence (letters, emails, etc.)
ii Instructions

iii Progress Reports
iv Programmes (Planned versus Actual)
v Requests for Information
vi Technical Queries
vii Site Diaries and Daily Reports
viii Site Meeting Minutes
ix Photographic Evidence
x Meteorological Records

In the event that the claim proceeds to adjudication, arbitration and/or litigation, then proper records are absolutely vital to prove a case.

No records usually means no success!

Consideration of claim by the Engineer/Employer's Representative

Notwithstanding what the contract states:

- If the claim is acceptable in principle the Contractor should be notified accordingly and as soon as possible.
- Similarly, if the claim is unacceptable the Contractor should be notified as soon as possible.
- If further information and/or elucidation required the Contractor should be told promptly.
- There should be an arrangement for the keeping of contemporaneous records. These are not necessarily admissible but an agreed statement of fact taken at the time.
- If the claim is accepted, payment should be certified promptly of such monies as can be properly substantiated.
- If the full amount is not certified in full, state this with reasons and add what further information would justify reconsideration if this is appropriate.
- Ensure that all information required for verification is obtained from the Contractor and dealt with promptly to ensure that the certification can be issued.

Claims under the FIDIC 2017 Contracts

The FIDIC 2017 Contracts provide for various types of claims:

Contractor's Claims against the Employer

- Claims which allow the Contractor an EOT;
- Claims which allow the Contractor an EOT plus Cost;
- Claims which allow the Contractor an EOT plus Cost plus Profit.

Employer's Claims against the Contractor

- Claims which allow the Employer to claim costs from the Contractor and/or an extension to the Defects Notification Period.

See Figure 10.1 for relevant Sub-Clauses entitling the Contractor to claim an EOT only, or an EOT plus "Cost" or an EOT plus "Cost Plus Profit".

See Figure 10.2 for relevant Sub-Clauses entitling the Employer to claim costs from the Contractor and/or an extension to the Defects Notification Period.

FIDIC Red Book 2017	FIDIC Yellow Book 2017	FIDIC Silver Book 2017
Clauses giving the Contractor right to claim Extension of Time Only		
Clause 8.5 – Extensions of time	Clause 8.5 – Extensions of time	Clause 8.5 – Extensions of time
Variations	Variations	Variations
Cause of delay covered by subaclauses	Cause of delay covered by sub-clauses	Cause of delay covered by sub-clauses
Exceptionally adverse climactic conditions	Exceptionally adverse climactic conditions	
Unforeseeable shortages	Unforeseeable shortages	
Any delay, impediment etc caused by the Employer or its agents.	Any delay, impediment etc caused by the Employer or its agents.	Any delay, impediment etc caused by the Employer or its agents.
Measurement		
Clauses giving the Contractor right to claim Extension of Time + Cost		
Clause 4.12 – Unforeseeable physical conditions	Clause 4.12 – Unforeseeable physical conditions	
Clause 4.15 – Access Route	Clause 4.15 – Access Route	Clause 4.15 – Access Route
Clause 4.23 – Archaeological and Geological Findings	Clause 4.23 – Archaeological and Geological Findings	Clause 4.23 – Archaeological and Geological Findings
Clause 13.6 – Adjustments for Changes in Laws	Clause 13.6 – Adjustments for Changes in Laws	Clause 13.6 – Adjustments for Changes in Laws
Clause 18.4 – Consequences of an Exceptional Event	Clause 18.4 – Consequences of an Exceptional Event	Clause 18.4 – Consequences of an Exceptional Event
Clauses giving the Contractor right to claim Extension of Time + Cost + Profit		
Clause 1.9 – Delayed drawings or instructions	Clause 1.9 – Errors in the Employer's Requirements	
Clause 1.13 – Compliance with Laws	Clause 1.13 – Compliance with Laws	Clause 1.12 – Compliance with Laws
Clause 2.1 – Right of Access to the Site	Clause 2.1 – Right of Access to the Site	Clause 2.1 – Right of Access to the Site
Clause 4.6 – Co-operation	Clause 4.6 – Co-operation	Clause 4.6 – Co-operation
Clause 4.7 – Setting out	Clause 4.7 – Setting out	

Figure 10.1 (Continued).

Clause 7.4 – Testing by the Contractor	Clause 7.4 – Testing by the Contractor	Clause 7.4 – Testing by the Contractor
Clause 7.6 – Remedial Work	Clause 7.6 – Remedial Work	Clause 7.6 – Remedial Work
Clause 8.10 – Consequences of Employer's Suspension	Clause 8.10 – Consequences of Employer's Suspension	Clause 8.10 – Consequences of Employer's Suspension
Clause 8.12 – Prolonged Suspension	Clause 8.12 – Prolonged Suspension	Clause 8.12 – Prolonged Suspension
Clause 10.2 – Taking Over Parts	Clause 10.2 – Taking Over Parts	
Clause 10.3 – Interference with Tests on Completion	Clause 10.3 – Interference with Tests on Completion	Clause 10.3 – Interference with Tests on Completion
Clause 11.7 – Right of Access after Taking Over	Clause 11.7 – Right of Access after Taking Over	Clause 11.7 – Right of Access after Taking Over
Clause 11.8 – Contractor to Search	Clause 11.8 – Contractor to Search	Clause 11.8 – Contractor to Search
	Clause 12.2 – Delayed tests	Clause 12.2 – Delayed tests
	Clause 12.4 – Failure to pass tests	Clause 12.4 – Failure to pass tests
Clause 13.3 – Variation Procedure	Clause 13.3 – Variation Procedure	Clause 13.3 – Variation Procedure
Clause 16.1 – Suspension by Contractor	Clause 16.1 – Suspension by Contractor	Clause 16.1 – Suspension by Contractor
Clause 16.2 – Termination by Contractor	Clause 16.2 – Termination by Contractor	Clause 16.2 – Termination by Contractor
Clause 16.3 – Contractor's Obligations after Termination	Clause 16.3 – Contractor's Obligations after Termination	Clause 16.3 – Contractor's Obligations after Termination
Clause 16.4 – Contractor's Obligations after Termination	Clause 16.4 – Contractor's Obligations after Termination	Clause 16.4 – Contractor's Obligations after Termination
Clause 17.2 – Liability for Care of the Works	Clause 17.2 – Liability for Care of the Works	Clause 17.2 – Liability for Care of the Works

Figure 10.1 Sub-Clauses entitling the Contractor to make a claim

FIDIC Red Book 2017	FIDIC Yellow Book 2017	FIDIC Silver Book 2017
Clauses giving the Contractor right to claim Extension of Time Only		
Clause 8.5 – Extensions of time	Clause 8.5 – Extensions of time	Clause 8.5 – Extensions of time
Variations	Variations	Variations
Cause of delay covered by subaclauses	Cause of delay covered by sub-clauses	Cause of delay covered by sub-clauses
Exceptionally adverse climactic conditions	Exceptionally adverse climactic conditions	
Unforeseeable shortages	Unforeseeable shortages	
Any delay, impediment etc caused by the Employer or its agents.	Any delay, impediment etc caused by the Employer or its agents.	Any delay, impediment etc caused by the Employer or its agents.
Measurement		
Clauses giving the Contractor right to claim Extension of Time + Cost		
Clause 4.12 – Unforeseeable physical conditions	Clause 4.12 – Unforeseeable physical conditions	
Clause 4.15 – Access Route	Clause 4.15 – Access Route	Clause 4.15 – Access Route
Clause 4.23 – Archaeological and Geological Findings	Clause 4.23 – Archaeological and Geological Findings	Clause 4.23 – Archaeological and Geological Findings
Clause 13.6 – Adjustments for Changes in Laws	Clause 13.6 – Adjustments for Changes in Laws	Clause 13.6 – Adjustments for Changes in Laws
Clause 18.4 – Consequences of an Exceptional Event	Clause 18.4 – Consequences of an Exceptional Event	Clause 18.4 – Consequences of an Exceptional Event
Clause 1.9 – Delayed drawings or instructions	Clause 1.9 – Errors in the Employer's Requirements	
Clause 1.13 – Compliance with Laws	Clause 1.13 – Compliance with Laws	Clause 1.12 – Compliance with Laws
Clause 2.1 – Right of Access to the Site	Clause 2.1 – Right of Access to the Site	Clause 2.1 – Right of Access to the Site
Clause 4.6 – Co–operation	Clause 4.6 – Co–operation	Clause 4.6 – Co-operation
Clause 4.7 – Setting out	Clause 4.7 – Setting out	

Figure 10.2 (Continued).

Clause 7.4 – Testing by the Contractor	Clause 7.4 – Testing by the Contractor	Clause 7.4 – Testing by the Contractor
Clause 7.6 – Remedial Work	Clause 7.6 – Remedial Work	Clause 7.6 – Remedial Work
Clause 8.10 – Consequences of Employer's Suspension	Clause 8.10 – Consequences of Employer's Suspension	Clause 8.10 – Consequences of Employer's Suspension
Clause 8.12 – Prolonged Suspension	Clause 8.12 – Prolonged Suspension	Clause 8.12 – Prolonged Suspension
Clause 10.2 – Taking Over Parts	Clause 10.2 – Taking Over Parts	
Clause 10.3 – Interference with Tests on Completion	Clause 10.3 – Interference with Tests on Completion	Clause 10.3 – Interference with Tests on Completion
Clause 11.7 – Right of Access after Taking Over	Clause 11.7 – Right of Access after Taking Over	Clause 11.7 – Right of Access after Taking Over
Clause 11.8 – Contractor to Search	Clause 11.8 – Contractor to Search	Clause 11.8 – Contractor to Search
	Clause 12.2 – Delayed tests	Clause 12.2 – Delayed tests
	Clause 12.4 – Failure to pass tests	Clause 12.4 – Failure to pass tests
Clause 13.3 – Variation Procedure	Clause 13.3 – Variation Procedure	Clause 13.3 – Variation Procedure
Clause 16.1 – Suspension by Contractor	Clause 16.1 – Suspension by Contractor	Clause 16.1 – Suspension by Contractor
Clause 16.2 – Termination by Contractor	Clause 16.2 – Termination by Contractor	Clause 16.2 – Termination by Contractor
Clause 16.3 – Contractor's Obligations after Termination	Clause 16.3 – Contractor's Obligations after Termination	Clause 16.3 – Contractor's Obligations after Termination
Clause 16.4 – Contractor's Obligations after Termination	Clause 16.4 – Contractor's Obligations after Termination	Clause 16.4 – Contractor's Obligations after Termination
Clause 17.2 – Liability for Care of the Works	Clause 17.2 – Liability for Care of the Works	Clause 17.2 – Liability for Care of the Works

Figure 10.2 Sub-Clauses entitling the Employer to make a claim

Clause 20 of FIDIC 1999 was entitled "Claims, Disputes and Arbitration", but has now been split into two Clauses:

- Clause 20 – Employer's and Contractor's Claims;
- Clause 21 – Disputes and Arbitration (Clause will be dealt with within Chapter 11).

This is a sensible change as the previous Clause 20 in the FIDIC 1999 Contracts almost implied that if the Contractor submitted a claim, it was the natural step to escalate it into a dispute within the same Clause!

Employer's Claims were also previously covered under Clause 2.5 of the FIDIC 1999 Contracts, now Contractor's Claims <u>and</u> Employer's Claims are dealt with under the same provisions.

Hence, Clause 20 covering the submission of claims by the Employer and by the Contractor is now significantly longer than in the FIDIC 1999 Contracts.

Employer's and Contractor's Claims

Three definitions should be particularly borne in mind when dealing Claims under the FIDIC 2017 Contracts:

i Under Sub-Clause 1.1.6 (Sub-Clause 1.1.5 Yellow Book/Sub-Clause 1.1.3 Silver Book) "Claim" means "a request or assertion by one Party to the other Party for an entitlement or relief under any Clause of these Conditions or otherwise in connection with, or arising out of, the Contract or the execution of the Works".

All claims, whether they be the Contractor claiming from the Employer, or the Employer claiming from the Contractor, are now dealt with in exactly the same way and under the same Clauses.

It is critical that in assessing claims there is a full understanding of what constitutes "Cost" as many claims refer to payment of "Cost".

ii Under Sub-Clause 1.1.19 (Sub-Clause 1.1.16 Silver Book) "Cost" means "*all expenditure reasonably incurred (or to be incurred) by the Contractor in performing the Contract, whether on or off the Site, including taxes, overheads and similar charges, but does not include profit. Where the Contractor is entitled under a Sub-Clause of these Conditions to payment of Cost, it shall be added to the Contract Price*".

It is also critical that in assessing claims there is a full understanding of what constitutes "Profit" as many claims refer to payment of "Cost Plus Profit".

iii Under Sub-Clause 1.1.20 (Sub-Clause 1.1.17 Silver Book) "Cost Plus Profit" means "*Cost plus the applicable percentage for profit stated in the Contract Data (if not stated, five percent (5%)). Such percentage shall only be added to Cost, and Cost Plus Profit shall only be added to the Contract Price, where the Contractor is entitled under a Sub-Clause of these Conditions to payment of Cost Plus Profit*".

This percentage may be taken from the Contractor's tender or may be set by the Employer.

Previously in the FIDIC 1999 Contracts, certain clauses referred to "Cost Plus Profit", but never stated the amount or the percentage that could be claimed as profit, leaving it to the Contractor and the Engineer or Employer (or the Employer's Representative) to agree an amount or percentage either on a claim-by-claim basis, or more often, a pre-agreed percentage to be applicable to all Claims.

The Contractor is also now expressly entitled to recover lost profit where Works are omitted, or where the Contract is terminated for convenience (although the margin of profit is not stipulated).

Previously, under the FIDIC 1999 Contracts, claims by the Employer against the Contractor were dealt with under Clause 2.5, where the Employer had to notify its claim "*as soon as practicable after the Employer became aware of the event or circumstances giving rise to the claim*", whilst claims by the Contractor against the Employer were dealt with under Clause 20.1, but now all claims by either Party are dealt with under Clauses 20.1 and 20.2.

So, the Employer also now has to give notice of a claim within 28 days after it became aware, or should have become aware, of the event or circumstance resulting in a claim for payment (or reduction of the contract price) or an extension of the Defects Notification Period.

In addition, both the Employer and the Contractor must submit a fully detailed claim within 84 days. In the FIDIC 1999 Contracts, the time frame for the Contractor to submit a fully detailed claim was 42 days.

Under Clause 20.1, a Claim is stated as being:

a The Employer considering that it is entitled to any additional payment from the Contractor or reduction in the Contract Price and/or to an extension of the Defect Notification Period.

  As stated previously, within the FIDIC 1999 Contracts, Employer's Claims were dealt with separately under Sub-Clause 2.5.

b The Contractor considers that it is entitled to any additional payment from the Employer and/or EOT.

c Either Party considering that it is entitled to any other relief against the other Party. As this would exclude any additional payment and/or EOT, it could include any form of entitlement or relaxation under any part of the Contract.

Sub-Clause 20.2 deals with (a) or (b), in the case of (c), where either Party has disagreed with the other, the matter should be referred to the Engineer for Agreement or Determination.

If a Claim arises under (c), where the other Party or the Engineer has disagreed with the Claim, then a Notice should be issued by the Claiming Party, a Dispute shall not be deemed to have arisen but the claiming Party may, by giving a Notice, refer the Claim to the Engineer and the Engineer must make a determination under Sub-Clause 3.7.

Notice of a Claim

If either Party (Employer or Contractor) considers himself entitled to a claim as detailed above, the Party gives a Notice of Claim to the Engineer (the other Party under the Silver Book) not later than 28 days after the claiming Party became aware, or should have become aware ("should have become aware" is very subjective but stresses that the claiming Party should notify as soon as possible!), of the matter which gave rise to the cost, loss, delay or extension of the Defects Notification Period which are the subject of the Claim. The words "should have become aware" are obviously subject to opinion and possible debate!

If the claiming Party does not give notice within 28 days, it loses its right to a Claim and the Claim will be time barred (see "Initial Response" below).

It is likely that there may be some "creative drafting" in the 28-day time bar applying to the Employer as well as to the Contractor so that the claiming Party can comply with the time limit!

Note that FIDIC 1999 also required the Contractor to give notice of a Claim, whereas FIDIC 2017 requires the claiming Party to give a Notice of Claim to the Engineer, which is more prescriptive.

Notice is now a defined term Sub-Clause 1.1.56 (Sub-Clause 1.1.48 Silver Book) defining "Notice" as "a written communication identified as a Notice and issued in accordance with Sub-Clause 1.3" (Notices and Other Communications).

Sub-Clause 1.3 requires that a Notice must be:

i A paper-original signed by the Contractor's Representative, the Engineer or the authorised representative of the Employer; or
ii An electronic original generated from any of the systems of electronic transmission stated in the Contract Data or acceptable to the Engineer (Employer under the Silver Book).

and if it is a Notice, it must be identified as a Notice, <u>not just an email or other communication purporting to be a Notice</u>.

So, the Notice of Claim is now more formal than previously under FIDIC 1999, where previously Contractors would notify informally through emails, wording within letters or Meeting Minutes.

The new deadline of 28 days for the Employer to issue a Notice of Claim, now the same time scale as the Contractor always had, is calculated from the point in time when the Employer should have become aware and imposes strict obligations on the Employer which did not exist before, but is likely to be welcomed by Contractors!

Some Employers have amended this time scale within the Contract, on the basis that they believe that the Employer's Claims tend to be more complex than the Contractor's Claims, but then maybe it is just Employers not being used to preparing and notifying claims to deadlines!

Initial Response

If the Engineer (the other Party under the Silver Book) considers the Notice of Claim to be outside the 28-day period and therefore out of time, he must notify the claiming Party within 14 days of receiving the Notice, with reasons, or the Notice of Claim will be deemed valid, so for the 28-day time bar for a Notice of Claim time bar to be effective, the Engineer (the other Party under the Silver Book) must give that Notice to the claiming Party within 14 days of receiving the Party's Notice of Claim or the lapse of the 84 days for the fully detailed Claim (see below).

Notwithstanding this deemed acceptance, the other (non-claiming) Party may, in turn, give a subsequent Notice disagreeing with the deemed validity, in which case, the

Engineer (the other Party under the Silver Book) is required to review the issue within his Determination.

Also, if the Engineer (the other Party under the Silver Book) issues its Notice deeming the Notice of Claim invalid i.e. out of time, the claiming Party may include in its fully detailed Claim of its disagreement or justification of the late submission.

Even if a 14-day Notice has been issued, the Engineer (the other Party under the Silver Book) is required to agree or determine the substance of the Claim pursuant to Sub-Clause 3.7 and include a Determination on the validity of the Notice.

The FIDIC 1999 Contracts had a simple time bar, in that a claim notified outside the 28-day period was time barred i.e. *"if the Contractor fails to give notice of a claim within such period of 28 days, the Time for Completion shall not be extended, the Contractor shall not be entitled to additional payment, and the Employer shall be discharged form all liability in connection with the claim"*, so the Employer absolves himself of all liability simply because the notice of a claim was outside the 28-day period.

The FIDIC 2017 Contracts, however, although having the same time bar, have "softened" the approach, allowing a form of appeal against the time bar. The claiming Party can either argue that the Notice of Claim or the fully detailed Claim was served within their time limits or submit a justification for why it was late. The Engineer may consider prejudice to the other Party and prior knowledge by the other Party.

If the Engineer (the other Party under the Silver Book) does not confirm that, it is deemed to be a valid submission, however the other Party may disagree with that, and in that case issues a Notice to the Engineer stating that fact.

Various opinions have been published about the effectiveness and enforceability of time bars in contracts such as FIDIC, commentators particularly debating whether the clause is a condition precedent to the Party being able to recover time and money, and whether a Party can benefit from its own breach of contract to the detriment of the injured Party.

For example, if the Employer does not provide something which it is to provide under the Contract, can the Employer prevent the Contractor from receiving any remedy because the Contractor failed to notify the Engineer (the other Party under the Silver Book) within the eight-week limit, and possibly, if Completion is delayed the Contractor has to pay delay damages?

There are those within the industry, notably within the contracting fraternity, that consider this time barring clause to be totally unfair and unduly harsh and an obstruction to good relationships between the contracting parties since such a clause promotes a Notice being submitted at the merest hint of a potential claim.

It is also stated that in some jurisdictions such a fatal Notice clause may be contrary to the intent of the relevant Law. If that is the case, then that would have to be considered and legal advice sought.

The author, whilst not being a lawyer, is of the opinion that apart from the legal issue in the previous paragraph, the time bar is correct and enforceable as the parties are clear as to what the terms of their agreement are at the time the Contract is formed, it is also clear what happens if the Contractor does not notify a claim in the correct format within the time stated.

It is advisable that Claimants properly respect the time limits within the Contract and comply with them, rather than to question whether they are ultimately enforceable, or whether Claimants can appeal against them.

Considering that the content requirements for the Notice of Claim is quite brief and straightforward, it would hard to justify why anyone would issue a Notice of Claim late, possibly apart from the usual difficulty of identifying the start of the period the delay/cost occurred. Having said that it is likely that Contractors and Employers are likely to provide very scant detail at the Notice of Claim stage, just to "get their foot in the door" in time.

The fully detailed Claim is a different matter as it probably requires a significant amount of evidence, including a statement of the contractual/legal basis of the Claim.

No specific deadline is set out in Clause 20 in respect of other claims than mentioned above but if there is a disagreement in respect of such other Claim, the claiming party may refer the claim to the Engineer (the other Party under the Silver Book), which shall be done "*as soon as practicable after the claiming party becomes aware of the disagreement*".

These new provisions make it very important to distinguish between Claims for additional payment, reduction in Contract Price and Extension of the Defects Notification Period, and other claims.

Under FIDIC 2017, the provision applies to the claiming Party so it could be the Contractor or the Employer, which is unusual in that the rules of notifying claims are the same for the Employer claiming from the Contractor as the Contractor submitting a contractual claim to the Employer, but also as stated the time bar has been "softened" in that if the claiming Party disagrees or he/she believes there are circumstances which may justify the late notice they detail these circumstances within their fully detailed claim, and the Engineer is required to proceed with determination of the Claim, but taking into account the points raised by the claiming Party.

Note that under FIDIC 1999 Clause 2.5 the Employer only had to raise its claim "*as soon as practicable after the Employer became aware of the event or circumstances giving rise to the claim*". There was no right of the Contractor to be discharged of liability of the Employer failed to raise its claim as soon as practicable, so effectively the claims provisions of FIDIC 1999 were biased in favour of the Employer. FIDIC 2017 now provides equality, in that the 20-day rule applies to either claiming Party.

Contemporaneous records

The claiming Party as with previous FIDIC contracts is required to keep contemporaneous records as necessary to substantiate its claim, the term "contemporaneous" means records that are prepared or generated at the same time, or immediately after the event or circumstance giving rise to the Claim.

These can include, where appropriate:

i Correspondence (letters, emails, etc.)
ii Instructions
iii Progress Reports
iv Programmes (Planned versus Actual)
v Requests for Information
vi Technical Queries
vii Site Diaries and Daily Reports
viii Site Meeting Minutes

ix Photographic Evidence
x Meteorological Records

The Engineer has the right to monitor and inspect these records. If additional records are required to be kept a Notice may be issued for them.

It is likely that a professional Contractor will have the majority of such records in place in any event as a matter of course, and may only need to focus those records upon the actual claim submitted.

The Engineer may also instruct the Contractor to keep specific records and monitor those being kept by the Contractor, although these specific requirements should not be considered as an admission of liability by the Engineer or the Employer.

It is vital that the Contractor and the Engineer have some documents as to what is required and both parties are proactive towards concluding the claim as quickly as possible and for the benefit of all the parties.

Note that if the Engineer chooses to monitor, inspect or instruct, this shall not imply acceptance of the accuracy or completeness of the Contractor's contemporary records.

Fully Detailed Claim

The claiming Party is required to submit a fully detailed Claim to the Engineer (the Employer's Representative under the Silver Book) within 84 days of when it became aware or should have become aware of the event giving rise to the Claim.

The detailed Claim includes:

a A detailed description of the event or circumstance giving rise to the Claim;
b A statement of the contractual and/or legal basis of the Claim;
c Contemporary records on which the Claim relies;
d Detailed supporting particulars of the amount of additional payment (or reduction in the contract price if the Employer is the claiming party), EOT or extension of the Defects Notification Period claimed.

If fully detailed particulars are not provided within the 84-day period, the Notice of Claim will lapse and will no longer be valid.

When the Engineer (the Employer's Representative under the Silver Book) receives the detailed Claim, he is required to give a Determination under Clause 3.7.

Within either:

i 84 days (previously under FIDIC 1999 it was 42 days) after the claiming Party became aware, or should have become aware, of the event or circumstance giving rise to the Claim,

or

ii Such other period as may be proposed by the claiming Party and agreed by the Engineer (the Employer's Representative under the Silver Book), the claiming Party shall submit to the Engineer a fully detailed Claim.

Again, as stated above, the words "should have become aware" may give rise to possible debate!

If within this time limit the claiming Party fails to submit the statement under sub-paragraph (b) above, the Notice of Claim shall be deemed to have lapsed, it shall no longer be considered as a valid Notice and the Engineer (the Employer's Representative under the Silver Book) shall, within 14 days after this time limit has expired, give a Notice to the claiming Party accordingly.

Under the FIDIC 1999 Contracts the period was 42 days, but there was no time bar, so there are serious consequences for not complying with the time limit for submitting a fully detailed Claim in the 2017 Red Book!

However, as with the initial 28-day time limit discussed above, the FIDIC 2017 Contracts provide further provisions to give the provision some balance.

After receiving a fully detailed Claim, or an interim or final fully detailed Claim, the Engineer (the Employer's Representative under the Silver Book) is required to agree or determine:

a Any additional payment to which the claiming Party is entitled or the reduction of the Contract Price (in the case of the Employer as the claiming Party); and/or
b Any Extension of the Time for Completion, or the extension of the Defects Notification Period, to which the claiming Party is entitled under the Contract.

If the Engineer (the Employer's Representative under the Silver Book) has given a Notice under Sub-Clause 20.2.2 [Engineer's Initial Response] and/or under Sub-Clause 20.2.4 [Fully detailed Claim], the Claim shall nevertheless be agreed or determined in accordance with this Sub-Clause 20.2.

The agreement or determination of the Claim includes whether or not the Notice of Claim is to be treated as a valid Notice taking account of the details (if any) included in the fully detailed claim of the claiming Party's disagreement with such Notice(s) or why late submission is justified (as the case may be). The circumstances which may be taken into account (but shall not be binding) may include:

- Whether or to what extent the other Party would be prejudiced by acceptance of the late submission;
- In the case of the time limit under Sub-Clause 20.2.1 [Notice of Claim], any evidence of the other Party's prior knowledge of the event or circumstance giving rise to the Claim, which the claiming Party may include in its supporting particulars; and/or
- In the case of the time limit under Sub-Clause 20.2.4 [Fully detailed Claim], any evidence of the other Party's prior knowledge of the contractual and/or other legal basis of the Claim, which the claiming Party may include in its supporting particulars.

If, having received the fully detailed Claim under Sub-Clause 20.2.4 [Fully detailed Claim], or in the case of a Claim under Sub-Clause 20.2.6 [Claims of continuing effect] an interim or final fully detailed Claim (as the case may be), the Engineer (the Employer's Representative under the Silver Book) requires necessary additional particulars:

i He promptly gives a Notice to the claiming Party, describing the additional particulars, and the reasons for requiring them.

ii He gives his response on the contractual or other legal basis of the Claim, by giving a Notice to the claiming Party, within the time limit for agreement under Sub-Clause 3.7.3 [Time limits].
iii As soon as practicable after receiving the Notice under sub-paragraph (i) above, the claiming Party submits the additional particulars, and the Engineer (the Employer's Representative under the Silver Book) proceeds to agree or determine the matters.

If the matters giving rise to a Claim have a continuing effect, which is quite common, for example, with long-term issues with availability of Goods caused by epidemic or governmental actions:

a The fully detailed Claim is considered as interim, but note that the Claim must still be dealt with in the normal way, even though it may be interim.
b The Engineer (the Employer's Representative under the Silver Book) gives his response on the contractual or other legal basis of the Claim, by giving a Notice to the claiming Party, within the time limit under the Contract.
c After submitting the first interim fully detailed Claim the claiming Party is required to submit further interim fully detailed Claims at <u>monthly</u> intervals, giving the accumulated amount of additional payment claimed, and/or EOT claimed or extension of the Defects Notification Period (in the case of the Employer as the claiming Party).
d The claiming Party is required to submit a final fully detailed Claim within 28 days after the <u>end</u> of the effects resulting from the event or circumstance, or within such other period as may be proposed by the claiming Party and agreed by the Engineer.
   This final fully detailed Claim is required to give the total amount of additional payment claimed (or the reduction of the Contract Price, in the case of the Employer as the claiming Party), and/or EOT claimed (in the case of the Contractor as the claiming Party) or extension of the Defects Notification Period (in the case of the Employer as the claiming Party).

After receiving the Notice of Claim, and until the Claim is agreed or determined, in each Payment Certificate the Engineer (the Employer's Representative under the Silver Book) must include such amounts as due to the claiming Party under the relevant provision of the Contract.

The Employer is only entitled to claim any payment from the Contractor and/or to extend the Defects Notification Period or set off against or make any deduction from any amount due to the Contractor, by complying with this Sub-Clause 20.2.

The requirements of Sub-Clause 20.2 are in addition to those of any other Sub-Clause which may apply to the Claim.

If the claiming Party fails to comply with this or any other Sub-Clause in relation to the Claim, any additional payment and/or any EOT (in the case of the Contractor as the claiming Party) or extension of the Defects Notification Period (in the case of the Employer as the claiming Party) shall take account of the extent (if any) to which the failure has prevented or prejudiced proper investigation of the Claim by the Engineer (the Employer's Representative under the Silver Book).

Agreement or Determination

Previously under FIDIC 1999, Clause 3.5 (Determinations) was very brief, requiring the Engineer simply to proceed to agree or determine any matter, including consulting with each Party, but if agreement could not be reached he was required to make a fair determination and whether there was agreement or the Engineer had to determine he issued a Notice to both Parties once it had been completed.

FIDIC 2017 Sub-Clause 3.7 (Agreement or Determination) (Sub-Clause 3.5 Silver Book) is much more comprehensive and prescriptive with a number of new provisions, and provides for various steps (with time limits) to be taken in agreeing or determining various matters, with the outcomes of either:

- Agreement,
- Determination in the absence of agreement,
- Effect of the agreement or determination, or
- Dissatisfaction with the determination.

These "matters" are not just Claims, they are <u>any</u> matters for which the Engineer (the Employer's Representative under the Silver Book) must try to reach an agreement or to make a fair determination.

It includes Employer's Claims, Contractor's Claims, measurement of the Works where applicable, the evaluation of Variations including dayworks, where actual progress differs from planned progress affecting payment in accordance with a Schedule of Payments, the amount that should be paid to the Contractor in respect of Plant and/or Materials when shipped or delivered to Site, under Sub-Clause 14.5 [Plant and Materials intended for the Works], payment issues including where the Engineer (the Employer's Representative under the Silver Book) states that the Engineer (the Employer's Representative under the Silver Book) has undervalued the Works, and also whether work is defective.

An important point stated within this Sub-Clause is that the Engineer (the Employer's Representative under the Silver Book) is required to act neutrally between the Parties and not be biased towards, or on behalf of the Employer.

The Engineer (the Employer's Representative under the Silver Book) is also required to consult with each Party in an attempt to reach a fair agreement. The Engineer (the Employer's Representative under the Silver Book) is also required to provide a record of that consultation.

The Engineer (the Employer's Representative under the Silver Book) also gives the Parties a "Notice of the Parties' Agreement".

In the absence of an agreement within the time scales of the Contract, the Engineer (the Employer's Representative under the Silver Book) gives a notice to that effect, and makes a fair determination, referred to as an Engineer's (the Employer's Representative under the Silver Book) Determination.

When agreeing or determining any matter or Claim, the Engineer (the Employer's Representative under the Silver Book) must follow the following steps under the Contract.

Consultation to reach agreement

This is a new provision under the FIDIC 2017 Contracts in that the Engineer (the Employer's Representative under the Silver Book) is first required to consult with BOTH

Parties to try to create discussion between them in an effort to reach agreement within the time limits laid down by the Contract.

The discussion could be with each Party separately or both Parties at the same time. The Engineer (the Employer's Representative under the Silver Book) provides both Parties with a record of the consultation he has had with them.

This is an important point as the Engineer (the Employer's Representative under the Silver Book) must seek agreement between the Parties, and must not act unilaterally.

If agreement is achieved within the time limit laid down by the Contract, the Engineer (the Employer's Representative under the Silver Book) confirms that agreement by giving a Notice to both Parties describing and quantifying the agreement, both Parties then sign the agreement.

If:

a No agreement is achieved within the time limit for agreement under the Contract; or
b Both Parties advise the Engineer (the Employer's Representative under the Silver Book) that no agreement can be achieved within this time limit whichever is the earlier, the Engineer (the Employer's Representative under the Silver Book) gives a Notice to the Parties and immediately proceeds to carry out an Engineer's (the Employer's Representative under the Silver Book) Determination.

If agreement cannot be reached between the Parties, the Engineer (the Employer's Representative under the Silver Book) makes a determination, and in doing so, take due regard of all relevant circumstances.

Within the time limit for determination, the Engineer (the Employer's Representative under the Silver Book) gives a Notice to both Parties of his determination. The Notice must state that it is a "Notice of the Engineer's (the Employer's Representative under the Silver Book) Determination" and describe the determination in detail with reasons and detailed supporting particulars.

Time limits

If agreement is achieved between the Parties, the Engineer (the Employer's Representative under the Silver Book) gives a Notice of agreement within 42 days or other time limit proposed by the Engineer (the Employer's Representative under the Silver Book) and agreed by both Parties, after:

a If a matter is not a Claim, i.e. the Employer considering that it is entitled to any additional payment from the Contractor or reduction in the Contract Price and/or to an extension of the Defect Notification Period, the date of commencement of the time limit for agreement as stated in the applicable Sub-Clause of these Conditions.
b If it is a Claim where the Contractor is considering that it is entitled to any additional payment from the Employer and/or an EOT, the date the Engineer receives a Notice under Sub-Clause 20.1 from the claiming Party.
c In the case of a Claim where either Party considers that it is entitled to any other relief against the other Party e.g. any form of entitlement or relaxation under any

part of the Contract, the date the Engineer (the Employer's Representative under the Silver Book) receives a fully detailed Claim under Sub-Clause 20.2.4 [Fully detailed Claim], or in the case of a Claim under Sub-Clause 20.2.6 [Claims of continuing effect], an interim or final fully detailed Claim.

The Engineer (the Employer's Representative under the Silver Book) is required to give his Notice of determination within 42 days or within such other time limit as may be proposed by the Engineer (the Employer's Representative under the Silver Book) and agreed by both Parties (after the date corresponding to his obligation to proceed).

If the Engineer (the Employer's Representative under the Silver Book) does not give the Notice of agreement or determination within the relevant time limit:

i In the case of a Claim, the Engineer (the Employer's Representative under the Silver Book) is deemed to have given a determination rejecting the Claim; or
ii In the case of a matter to be agreed or determined, the matter is deemed to be a Dispute which may be referred by either Party to the Dispute Avoidance/Adjudication Board (DAAB) for its decision under Sub-Clause 21.4, without the need for a Notice of Dissatisfaction.

The above provision obviously applies to Claims raised by the Employer as well as by the Contractor.

The time in which the Engineer (the Employer's Representative under the Silver Book) must make a determination is now 84 days.

i 42 days for the Engineer to give a "Notice of the Parties Agreement" if agreement is achieved, and if no agreement is reached;
ii A further 42 days for the Engineer to give Notice of the Engineer's Determination.

Previously under the FIDIC 1999 Contracts there was an initial 42-day period for the Engineer (the Employer's Representative under the Silver Book) to respond to a claim or any further particulars supporting a previous claim with approval, or with disapproval and detailed comments. But then to make a fair determination under Sub-Clause 3.5 for which there was no express time limit.

Note that time limits can be amended/extended if proposed by the Engineer (the Employer's Representative under the Silver Book) and accepted by both Parties.

Effect of the agreement or determination

Each agreement or determination is binding on both Parties (and the Engineer (the Employer's Representative under the Silver Book)) unless and until it is corrected (see below), or in the case of a determination, it is revised under Clause 21 (Disputes and Arbitration).

If an agreement or determination concerns the payment of an amount from one Party to the other Party, the Contractor includes the amount in the next Statement and the Engineer (the Employer's Representative under the Silver Book) includes it in the Payment Certificate that follows that Statement.

If, within 14 days after giving or receiving the Engineer's (the Employer's Representative under the Silver Book) Notice of agreement or determination, any error of a typographical or clerical or arithmetical nature is found:

a By the Engineer (the Employer's Representative under the Silver Book), then he immediately advises the Parties; or
b By a Party, then that Party gives a Notice to the Engineer (the Employer's Representative under the Silver Book), clearly identifying the error. If the Engineer does not agree he immediately advises the Parties.

The Engineer (the Employer's Representative under the Silver Book) within seven days of finding the error, or receiving a Notice, gives a Notice to both Parties of the corrected agreement or determination. The corrected agreement or determination is then treated as the agreement or determination.

Dissatisfaction with the Engineer's (the Employer's Representative under the Silver Book) determination

If either Party is dissatisfied with a determination of the Engineer (the Employer's Representative under the Silver Book):

a The dissatisfied Party gives a Notice of Dissatisfaction to the other Party, with a copy to the Engineer (the Employer's Representative under the Silver Book).
b The Notice of Dissatisfaction must clearly state that it is a "Notice of Dissatisfaction with the Engineer's (the Employer's Representative under the Silver Book) Determination" and set out the reason(s) for dissatisfaction.
c The Notice of Dissatisfaction is given within 28 days after receiving the Engineer's (the Employer's Representative under the Silver Book) Notice of the determination, or his Notice of the corrected determination or, in the case of a deemed determination rejecting the Claim, within 28 days after the time limit for determination.
d Either Party may proceed under Sub-Clause 21.4 (Obtaining DAAB's Decision).

If no Notice of Dissatisfaction is given by either Party within the period of 28 days, the determination of the Engineer (the Employer's Representative under the Silver Book) is deemed to stand and to have been accepted by both Parties and is final and binding on them.

If the dissatisfied Party is dissatisfied with only part(s) of the Engineer's (the Employer's Representative under the Silver Book) determination:

i The specific part must be clearly identified in the Notice of Dissatisfaction;
ii The specific part(s), and any other parts of the determination that are affected by such part(s) or rely on such part(s) for completeness, is deemed to be severable (able to be separated) from the remainder of the determination; and
iii The remainder of the determination becomes final and binding on both Parties as if the Notice of Dissatisfaction had not been given.

In the event that a Party fails to comply with an agreement of the Parties, or a final and binding determination of the Engineer (the Employer's Representative under the Silver Book), the other Party may refer the failure directly to arbitration in which case the first and the third paragraphs of Sub-Clause 21.7 [Failure to Comply with DAAB's Decision] apply in the same manner as these paragraphs apply to a final and binding decision of the DAAB.

How can claims and their escalation into disputes be avoided?

Claims are unfortunately considered as a "necessary evil" when drafting and managing construction contracts, but there are some basic things that can be done to prevent claims and/or their escalation into disputes:

Tender Documents

It is generally found that an appropriate procurement strategy, together with a competent, robust and unambiguous tender document, is the key to avoiding or minimising contractual problems at a later date.

The tender Documents should clearly set out the respective responsibilities of the Parties to the contract in clear and unambiguous terms with a clear definition of risks and which party is carrying which risk and how.

Any physical, financial or time constraints should also be set down with any required parameters.

Changes during construction

The incidence of claims generally lies in the effects of change. This may be due to unforeseen conditions or unexpected events or imposed change such as change in the Works to be carried out i.e. Variations.

Conditions of Contract recognise that unforeseen conditions can arise and have a contractual mechanism to deal with the effects of that situation. Imposed changes are generated by the issue of Variations, but too often only the direct cost issues are considered, and little attention is given to the delay and disruption effects of the changed work. This can often be mitigated by discussion with the Contractor prior to the issue of the Variation.

It is important that when a Contractor notifies a claim that he is required to state the Clause/Sub-Clause in the Contract under which he considers he has an entitlement.

If the Contract Administrator (Engineer or Employer's Representative) agrees then this establishes the procedures for the evaluation of that claim. If this cannot be established then it is possibly a claim that is founded in a breach of contract, is outside the contract and therefore cannot be administered by the Contract Administrator.

Timely submission and assessment of claims

During the construction of the Works the Contractor may register the possibility of a claim (if he is not to prejudice his rights or prejudice its consideration) though he may subsequently not pursue it.

Claims should not be put aside to be dealt with later, in the hope that they may go away, the facts tend to be become obscured (particularly with the dispersal of staff) and attitudes tend to harden. In short claims do not improve with storage.

Communication

Communication and poor claims administration are still seen as significant problems in most disputes. It is important that the procedures and spirit required by the contract are followed and that the parties demonstrate that they are actively seeking resolution of the matters in dispute.

The keeping of contemporaneous records

A pre-requisite to problem avoidance is for the parties to understand the contract and be clear as to their respective obligations and responsibilities. It has also been said that every dispute should be considered at the time as though it were going to arbitration or litigation.

Probably the most usual reason why many potentially valid claims are abandoned is the claimant's failure to maintain adequate records of events or costs. Without proper relevant and contemporaneous records, the evidence needed to establish the claim is not available, the claim is doomed and the proceedings become fractious.

It is vital that comprehensive records are kept so that the Contractor can substantiate claims and also that the Contract Administrator/Employer can defend them in the event that they arise.

These can include:

1 Correspondence
2 Instructions
3 Programmes (Planned versus Actual)
4 Site Diaries
5 Site Meeting Minutes
6 Photographic Evidence
7 Meteorological Records

Good records are fundamental to the resolution of claims. As soon as the Contractor becomes aware that he has, or may have, a claim under the contract, he must ensure that full and detailed records relating to the history of the claim are prepared and maintained throughout the period during which conditions giving rise to the claims remain in existence.

These records should be submitted to the Contract Administrator for his agreement. Where the Contract Administrator refuses or fails to check the Contractor's records that fact should be recorded by the Contractor.

A Contract Administrator's refusal to agree or accept records, usually where he does not admit the existence of the claim does not help the Employer if there is a valid claim.

If the claim is ultimately referred to Arbitration, the Arbitrator can only consider the evidence before him. Where a Contractor has maintained accurate and contemporaneous records, which the Contract Administrator has refused to check or agree, it may be difficult for the Contract Administrator to produce convincing evidence to refute the claim. Where, as is often the case, the Contract Administrator maintains a separate set of records a problem will arise, if and to the extent, that they are different to the Contractor's.

In such a situation, the Arbitrator can only weigh the evidence before him and decide which is the more convincing. The fact that the Contract Administrator refused to consider the Contractor's records would, in all probability, weigh in the Contractor's favour.

There is no logical reason not to agree a set of contemporaneous records of events or any contentious issue. These records can be qualified if the claim at that point in time is not admitted or for any relevant reason.

In the event that the claim proceeds to a formal dispute resolution process, be it DAAB, arbitration and/or legal proceedings, then proper records are **absolutely vital** to prove a case.

The principle is Records, Records, Records! And that is why the author makes no apology in mentioning the keeping of contemporaneous records three times within this chapter!

Agreeing cost and price

The issue of agreeing cost is quite clear, in that the Contractor must provide whatever information and records are necessary to reach that agreement. What is more problematical is that Contractors are often reluctant to provide proof of overhead and profit levels included in the contract. There is no reason for the pricing notes not to be made available and their confidentiality recognised. If the dispute goes to arbitration the pricing notes would be subject to discovery anyway.

In establishing rates for changes or additional works the contract requires contract rates, or rates based on contract rates, should be used wherever possible. It is insufficient, however, to use rates where the item description is the same without taking into account the circumstances under which the changed work was carried out.

All too often the disruptive effects of changed or altered works are not recognised. It is appropriate to analyse the constituent elements of the relevant rates and adjust those elements to reflect the new requirements (including differences in location, timing, sequence, etc., and both quantity-related and time-related costs).

Cost information is a matter of record in most cases, particularly in relation to labour and materials. Plant costs can cause difficulty particularly with contractor-owned plant or plant on internal company hire. The Contractor may seek to charge daywork rates or internal hire rates, both of which may be inappropriate.

Summary

Claims are often seen as an inevitable consequence of working in the construction industry.

They are **not** inevitable and can be avoided by some simple rules:

i Clearly state in all tender and contract documentation what is required of all parties.

Claims often arise because the parties are not aware of their responsibilities.

ii Communicate effectively with all parties.

Claims usually escalate when the parties do not speak to each other!

iii Put everything in writing, in the event that a claim arises you have documentary evidence to support or dismiss it. Every claim submitted deserves some time spent in analysing and constructively responding to it. Rarely is a claim totally fictitious or unwarranted, but the onus is upon the claimant to prove a loss, not for the recipient to disprove it.

Claims usually fail because the claimant does not have factual evidence of what he asserts within his claims or he did not give proper written notification at the time of the problem.

11 The Dispute Avoidance/Adjudication Board and the Resolution of Disputes

The Dispute Avoidance/Adjudication Board and the Resolution of Disputes are covered in the FIDIC 2017 Contracts by:

- *Conditions of Contract for Construction* (The "Red Book")
  - Clause 21 – Disputes and Arbitration
- *Conditions of Contract for Plant and Design-Build* (The "Yellow Book")
  - Clause 21 – Disputes and Arbitration
- *Conditions of Contract for EPC/Turnkey Project* (The "Silver Book")
  - Clause 21 – Disputes and Arbitration

What is a Dispute?

"Dispute" is a defined term under Sub-Clause 1.1.29 (Sub-Clause 1.1.26 in the Silver Book, of the FIDIC 2017 Contracts), and is *any situation where*:

- One Party makes a claim against the other Party (which may be a Claim, as defined in these Conditions, or a matter to be determined by the Engineer (Employer's Representative under the Silver Book) under these Conditions, or otherwise);
- The other Party (or the Engineer under Sub-Clause 3.7.2 [Engineer's Determination] (Employer's Representative's Determination under the Silver Book) rejects the claim in whole or in part; and
- The first Party does not acquiesce (by giving a NOD (Notice of Dissatisfaction) under Sub-Clause 3.7.5 [Dissatisfaction with Engineer's determination] (Employer's Representative's Determination under the Silver Book) or otherwise;

  provided however that a failure by the other Party (or the Engineer/Employer's Representative under the Silver Book) to oppose or respond to the claim, in whole or in part, may constitute a rejection if, in the circumstances, the DAAB or the arbitrator(s), as the case may be, deem it reasonable for it to do so.

While an attempt has clearly been made to narrow the scope of the term "Dispute", this definition is not particularly straightforward and therefore may be a source of confusion and possible debate in the future.

Under FIDIC 1999, the "Dispute Adjudication Board" (DAB) was only put in place if/when a dispute arose between the Parties.

Under FIDIC 2017 the Parties are jointly required to appoint a Dispute Avoidance/Adjudication Board (DAAB) at the start of the Contract, which must visit the Site on a regular basis, and will remain in place for the duration of the Contract.

Sub-Clause 1.1.22 (Sub-Clause 1.1.20 under the Silver Book) defines "DAAB" or "Dispute Avoidance/Adjudication Board" as "*the sole member or three members (as the case may be) so named in the Contract, or appointed under Sub-Clause 21.1 [Constitution of the DAAB] or Sub-Clause 21.2 [Failure to Appoint DAAB Member(s)]*".

Sub-Clause 1.1.23 (Sub-Clause 1.1.20 under the Silver Book) defines "DAAB Agreement" as

> the agreement signed or deemed to have been signed by both Parties and the sole member or each of the three members (as the case may be) of the DAAB in accordance with Sub-Clause 21.1 [Constitution of the DAAB] or Sub-Clause 21.2 [Failure to Appoint DAAB Member(s)], incorporating by reference the General Conditions of Dispute Avoidance/Adjudication Agreement contained in the Appendix to these General Conditions with such amendments as are agreed.

There is a new provision under Clause 21.4.1, in that the reference of a Dispute to the DAAB shall, unless prohibited by law, be deemed to interrupt the running of any applicable statute of limitation or prescription period.

"Notice of Dissatisfaction" or "NOD" is defined as the Notice one Party may give to the other Party if it is dissatisfied, either with an Engineer's determination (Employer's Representative's determination under the Silver Book) or with a DAAB's decision.

Note that a Party may draft a NOD in respect of parts of the DAAB's decision. The part or parts that the Party is dissatisfied with should be clearly identified in the NOD.

As stated in Chapter 8, Clause 20 of FIDIC 1999 was entitled "Claims, Disputes and Arbitration", but has now been split into two Clauses:

- Clause 20 – Employer's and Contractor's Claims
- Clause 21 – Disputes and Arbitration

Previously, the FIDIC 1999 Contracts only had 20 clauses, Clause 20 being entitled "Claims, Disputes and Arbitration". FIDIC 2017 now has 21 clauses, the new Clause 21 being entitled "Disputes and Arbitration", Claims being included within the new contracts under Clause 20 entitled "Employer's and Contractor's Claims".

The FIDIC 1999 Contracts provided for a DAB ("Dispute Adjudication Board"), FIDIC 2017 providing for a DAAB ("Dispute Avoidance/Adjudication Board") which can be comprised of a sole member or three members named in the Contract or appointed under Clause 21.1.

Disputes and Arbitration under the FIDIC 2017 Contracts

FIDIC 1999 provided for a DAB to be appointed by the date stated in the Appendix to Tender, but would usually be appointed on an "ad hoc" only when a dispute arose under the Contract and only dealing with the dispute.

It is generally felt within the industry, and by FIDIC, that disputes can be dealt with better, and with greater consistency if the Board is a "standing board" which is in place throughout the Contract and essentially acts as part of the "project team", and this is how the FIDIC 2017 Contracts have been drafted in terms of the DAAB, unless the Parties agree otherwise.

There is a debate as to whether the provision of a "standing DAAB" actually does aid dispute avoidance by engaging with the parties and carrying out regular visits, but it is felt that it certainly assists with dispute avoidance.

Many Contracts are now including provisions for avoidance as well as resolution of disputes, and to this end a new provision has been added within the FIDIC 2017 Contracts, to allow the Parties, should they agree, to request the DAAB to provide assistance in resolving any matter that has not formally escalated into a Dispute. If the Parties enable these provisions, they are not bound to act on any advice given by the DAAB. This provision previously existed within the Conditions for Design, Build and Operate Projects (Gold Book) 2008.

The General Conditions of the Dispute Avoidance/Adjudication Agreement and the DAAB Procedural Rules have also been amended by including express provisions in respect of the DAAB's role in assisting the Parties to avoid Disputes.

Note that it is beyond the scope of those books to examine or review the General Conditions of the Dispute Avoidance/Adjudication Agreement.

This is why the DAAB has been renamed from its former DAB to stress that its role is not just to resolve the dispute when it "inevitably" happens.

Disputes under FIDIC 2017 are now decided by the DAAB, the Parties jointly appointing the members of the DAAB within the time stated in the Contract Data, the default period being 28 days after the date the Contractor receives the Letter of Acceptance.

The DAAB consists of either one or three members, the default being three members, with members' names being selected from a list within the Contract Data.

The Parties are required to sign a DAAB Agreement, by which time the composition of the DAAB must have been agreed together with the remuneration terms.

If a Party fails to agree the appointment of DAAB members or failing to agree the remuneration or to sign the DAAB agreement, then an entity stated in the Contract Data (if nothing is stated, the President of FIDIC) to appoint the DAAB member(s) and set the fees, and a DAAB agreement will be deemed to have been signed by the Parties.

Each Party is jointly and severally liable for paying half the DAAB fees, so if the Contractor pays the fees in full, he includes it in a Statement and recovers the difference from the Employer, if the Employer pays it in full, a deduction for the difference is made in the Statement from amounts otherwise due to the Contractor.

Obtaining the DAAB's decision

Either Party (Employer or Contractor) may refer a Dispute to the DAAB for its decision.

The reference of a Dispute to the DAAB is required:

a If Sub-Clause 3.7 [Agreement or Determination] (Sub-Clause 3.5 under the Silver Book) applies to the subject in Dispute, be made within 42 days of giving or receiving a Notice of Dissatisfaction.

   If the Dispute is not referred to the DAAB within this period of 42 days, the Notice of Dissatisfaction is deemed to have lapsed and is no longer valid.

b State that the reference is given under Sub-Clause 21.4.1.

c Set out the referring Party's case relating to the Dispute.
d Be in writing, with copies to the other Party and the Engineer (note there in no Engineer under the Silver Book so the copies just go to the other Party).
e For a DAAB of three persons, be deemed to have been received by the DAAB on the date it is received by the designated chairperson of the DAAB.

Both Parties are required to make available to the DAAB all information, access to the Site and appropriate facilities, as the DAAB may require for the purposes of deciding on the Dispute.

Whilst the DAAB deliberates on the matter, the Parties are required to continue to perform their obligations under the Contract (this is a normal requirement).

The DAAB gives its decision on the outcome of the dispute within:

a 84 days after receiving the reference (this is the same as FIDIC 1999); or
b Such period as may be proposed by the DAAB and agreed by both Parties.

However, if at the end of this period, the due date(s) for payment of any DAAB member's invoice(s) has passed but the invoice(s) remains unpaid, the DAAB is not obliged to give its decision until the outstanding invoice(s) have been paid in full, the DAAB then gives its decision as soon as practicable after payment has been received.

The decision is given in writing to both Parties with a copy to the Engineer (again there is no Engineer under the Silver Book so the decision just goes to the Parties), and must include reasons for the decision.

The decision is binding on both Parties, who are required to comply with it, whether or not a Party gives a Notice of Dissatisfaction. The Employer is responsible for ensuring that the Engineer (under the Red and Yellow books) complies with the DAAB decision.

Note that if the decision of the DAAB requires a payment of an amount by one Party (Employer/Contractor) to the other Party (Contractor/Employer), the amount is immediately due and payable without the requirement for any certification or Notice.

The DAAB proceeding is not deemed to be an arbitration, and the DAAB is not acting as arbitrator(s).

If either Party is dissatisfied with the DAAB's decision:

a The dissatisfied Party may (if they wish to take the matter further) give a Notice of Dissatisfaction to the other Party, with a copy to the DAAB.
b The Notice of Dissatisfaction must clearly state that it is a "Notice of Dissatisfaction with the DAAB's decision" and set out the reason(s) for dissatisfaction.
c This Notice of Dissatisfaction must be given within 28 days after receiving the DAAB's decision.

If the DAAB fails to give its decision within 84 days, then either Party may, within 28 days after the period has expired, submit a Notice of Dissatisfaction to the other Party.

Neither Party is entitled to commence arbitration on a Dispute unless and until a Notice of Dissatisfaction in respect of that Dispute has been given.

If the DAAB has given its decision, and no Notice of Dissatisfaction with the DAAB's decision has been given by either Party within 28 days after receiving the DAAB's decision, then the decision becomes final and binding on both Parties.

Amicable Settlement

If a Notice of Dissatisfaction with the DAAB's decision has been given, both Parties must attempt to settle the Dispute amicably before the commencement of arbitration.

Amicable Settlement could include a meeting between the Parties, or possibly mediation/conciliation. The key is to exhaust all efforts at resolving the Dispute before it goes to arbitration if arbitration cannot be avoided altogether.

However, unless both Parties agree otherwise, arbitration may be commenced on or after the 28th day (FIDIC 1999 had 56 days) after the day on which the Notice of Dissatisfaction was given, even if no attempt at Amicable Settlement has been made.

Arbitration generally

Before considering the provisions within the FIDIC 2017 Contracts for arbitration it is worth reviewing the process of arbitration in general terms, for those not familiar with it.

Note that this is a generic summary of the arbitration process, not specifically applicable to the FIDIC contracts, the exact details of an arbitration being dependent on where the arbitration is held, and the specific process involved.

Any dispute between two or more parties can be resolved through the Courts, a process known as "litigation" or "legal proceedings".

However, litigation has many disadvantages, not least the cost and process of a full court hearing and in many cases sometimes a number of years waiting time before the matter actually reaches the Courts, so arbitration is often used as a simpler, more convenient (and usually cheaper) method of dispute resolution. But one should not enter into an arbitration process lightly!

Arbitration is a procedure for the settlement of disputes, under which the parties agree to be bound by the decision of an Arbitrator or Arbitrator(s) whose decision is, in general, final and legally binding on both parties.

The Contract should include provisions for the resolution by international arbitration of any Dispute which is not settled amicably by the Parties.

In international contracts which are covered by the FIDIC contracts, for example, international commercial arbitration has numerous advantages over litigation in national courts and is likely to be more acceptable to the Parties.

Careful consideration should be given to ensuring that the international arbitration rules that are chosen are compatible with the provisions of the Contract.

It is important that the Parties agree on the number of Arbitrators and the language of arbitration. In the absence of specific requirements as to the number of Arbitrators and the place of arbitration in the Contract, an International Court of Arbitration will make the decision.

Sometimes a professional institution recognised by the Parties and the relevant country will be used to appoint the Arbitrator(s). Such institutions are well equipped to appoint Arbitrators and have panels of Arbitrators for that purpose, but it may be prudent to check, before designating an institution in the Contract, that the institution named is prepared to appoint an Arbitrator, if required.

For major projects tendered internationally, it is desirable that the place of arbitration be situated in a country other than that of the Employer or Contractor.

This country should have ratified a bilateral or multilateral convention (such as the 1958 New York Convention on the Recognition and Enforcement of Foreign Arbitral

Awards), or both, that would facilitate the enforcement of an arbitral award in the states of the Parties.

Note, however, that in some legislations an award by an Arbitrator may still have to be ratified by a Court.

It may be considered desirable in some cases for other Parties to be joined into any arbitration between the Parties or for two or more pending arbitrations to be consolidated, thereby creating a multi-party arbitration.

Some Employers have included multi-party arbitration clauses in their contracts; however, such clauses require skilful drafting, and usually need to be prepared on a case-by-case basis.

It is not unusual that the arbitration of a complex dispute is carried out and concluded long after a project has been completed. The author has been involved with some arbitrations of that nature which extend several years after project completion.

The Arbitrator may be a lawyer, but arbitration is not the premise of lawyers; in fact, legal qualifications are not necessary if the person has technical qualifications, is an expert in the field of the dispute and has qualified as an Arbitrator.

He will make a decision according to the rules of arbitration and the applicable laws, the decision, known as an "award", is legally binding and can be enforced through the courts if required.

Arbitration is not a new concept; in fact, it has been in existence for almost as long as the law itself, the first official recognition being the Arbitration Act 1697, which largely governed disputes about the sale of livestock.

One does not issue writs in arbitration as one would with litigation, BOTH parties merely agree to enter into arbitration because it is a provision within the contract between them, and thus be bound by the decision of the Arbitrator.

Arbitrators are normally appointed by one of THREE methods:

i As a result of a pre-agreed decision between the parties or through a contract to refer a dispute to arbitration in the event of it arising;
  In this case the Arbitrator himself may be named or subject to nomination by a professional body, say The Chartered Institute of Arbitrators.
ii By Statute under certain legislation disputes are required to be settled by arbitration;
iii By Order of a Court.

The Arbitration Acts (dependent on the country where the arbitration is to be held) set out procedures in the absence of any agreement to the contrary between the parties. As arbitration is a more flexible arrangement than litigation, whatever procedure both parties agree on, then that is often sufficient in an arbitration case.

It may be surprising to hear that arbitration is common as a settlement procedure for disputes under trade agreements, maritime and insurance, consumer matters such as package holidays and property valuations, as well as construction industry disputes.

The contracting party that initiated that the dispute be referred to arbitration is usually referred to as the "Claimant", whilst the other party is referred to as the "Respondent". Should there be a Joint Agreement to go to arbitration, the arbitrator will decide who is the Claimant, and who is the Respondent. It is important that these titles are clarified as the arbitration rules repeatedly refer to them.

There are several advantages in using arbitration rather than litigation:

i Privacy
In litigation, dependent on the specific legislation, private and personal matters are debated in full view of the public in open court, and often published in the press and technical magazines, whereas arbitration is carried out in private sessions, usually at the arbitrator's office or a neutral venue.

ii Convenience
Arbitration is carried out at a time and place to suit the parties, for example evenings and weekends, not when the court is available for session.

iii Speed
The dispute is settled efficiently and usually without the normal delays and procedural matters involved in litigation, not least the waiting time before actually getting into court.

iv Simplicity
There are fewer technical and formal procedures than in the courts, the parties set their own procedures within a set of ground rules.

v Expertness
The arbitrator will be an expert on the matter in dispute, for example, an Architect where the dispute is about building design, a Quantity Surveyor or quantum expert where it is related to a claim, whereas judges and other legal professionals are experts primarily on principles of law and expert witnesses may have to be called to a court. The dispute will often not hinge on a legal problem, for example, in matters of design, who is better than a qualified experienced designer to resolve it?

vi Thoroughness
Because the Arbitrator is a skilled technical person, the dispute can often be dealt with more thoroughly and matters more rigorously aired than in a confined and formal courtroom atmosphere.

vii Expense
It can readily be seen that arbitration is generally a less expensive option than litigation, though it must not be seen as a cheap way to resolve a dispute. In some complex cases where experienced arbitrators demand significant fees, arbitration can be even more expensive than litigation.

The main disadvantages of using arbitration rather than litigation are:

i Lack of Legal Knowledge
A dispute may hinge on a difficult point of law, which the Arbitrator has insufficient legal knowledge to resolve, as he is an expert in the field relating to the dispute, but not necessarily in the law itself. He is, however, entitled to seek legal advice on a point of law, but not on the dispute itself, and also the disputing parties can refer a question of law to a court.

NB: The arbitration award itself cannot be referred to the Courts unless BOTH parties consent, or the Courts are referred to on a point of law, the resolution of which could substantially overturn an arbitrator's decision.

Further, arbitrators cannot be sued for negligence as they do not hold themselves up as "experts" in the sense of being advisory bodies and therefore liable for giving wrongful advice.

ii Precedents

Arbitration is not subject to rulings in previous cases as with the courts, each case being judged on its own merits on the day, and also to some extent subject to the knowledge and experience of the Arbitrator, therefore you have no guide as to how successful you may be with your action.

It is prudent to always seek professional advice before resorting to arbitration as whilst it is simpler than Court action, the Arbitrator's decision is final and binding therefore you must be confident that you have a strong case.

iii Decision Making

There are some perceptions that Arbitrators, being technical people who may be or have been previously Contractors or Consultants themselves, tend to be more "empathetic" than lawyers, and may wish to "split the difference" between the parties when making their awards as they have possibly been in a similar position themselves, but judges are inclined to take colder decisions and award wholly in favour of one of the parties.

The author believes that all professionals have a duty to uphold the responsibilities they are vested with, so it is not a view that he shares.

Notice of reference to Arbitration

Proceedings may begin by either party sending a written notice to the other stating that they wish to refer the dispute to arbitration, and if they cannot jointly agree on a person to act as arbitrator, the Appointer named in the contract, usually a professional body will appoint the Arbitrator. Any correspondence flowing between the two parties as from the date of appointment must then also be copied to the Arbitrator.

Within a defined period of the notification date i.e. the date the Arbitrator writes to the parties accepting the appointment, a preliminary meeting may be held between the parties at a time and venue to be directed by the Arbitrator to decide the form of the hearing which may be on a documents only basis or an oral hearing.

Documents only tends to be cleaner and quicker as there is less opportunity for the parties to debate the issue.

The role of the Arbitrator

The role of the Arbitrator is only to examine the evidence before him enabling him to resolve the dispute, and nothing more, he cannot look into other matters not in dispute.

He can normally conduct the arbitration as he wishes (within any specified procedures), though he must set the procedures in a way that is not detrimental to either party.

He must also have a regard for the rules of evidence....

For example, ...

i Hearsay evidence is not permissible.

ii The Arbitrator is not permitted to use evidence from someone not qualified to give such evidence.

iii The Arbitrator is not allowed to take into account any evidence he has discovered himself, which was not provided by either party.

iv The Arbitrator is not allowed to use any privileged evidence, for example, that which is marked "Without Prejudice".

v The rules of discovery apply where either party is entitled to have access to the file or documents of the other, provided that it relates to matters at issue.

The Hearing

At the Hearing, the parties are entitled to legal representation should they wish, though they should be made aware that the Arbitrator has the power equivalent to a judge in litigation.

The Arbitrator normally commences by introducing the parties and summarising the dispute itself.

Then the Claimant or his legal representative is invited to outline his version of the dispute and his party's involvement, emphasising any matters which he feels may be important. The Arbitrator may also invite the Claimant to call witnesses, though they must first be sworn in under oath, and both the Claimant and the Respondent must have the opportunity to cross-examine them, in addition to the Arbitrator should he so wish.

The Respondent or his legal representative is then invited to outline his version of the dispute and his party's involvement, emphasising any matters which he feels may be important. Again, witnesses may appear, with the same provisos as before.

Finally, the Respondent makes his closing speech, followed by the Claimant, following which the Arbitrator closes proceedings and advises the parties as to when he will make his award.

Inspection by the Arbitrator

The Arbitrator is entitled to inspect any property, work or material at any premises, and may invite the claimant and/or the respondent to accompany him purely for the purpose of identifying the work or materials. No-one else is allowed to accompany the Arbitrator unless he specifically invites him.

The Award

The Award is binding upon both parties and must be served in the appropriate manner with full headings, summary of issues in dispute, outline of events leading to the dispute, his decision based on the hearing and finally the monetary award.

Arbitrator's fees and expenses

In all of the above cases both parties will normally incur costs, though it is the Arbitrator's decision as to how the costs are apportioned.

Both parties are initially liable to the Arbitrator for payment of his fees and expenses, but the Arbitrator will direct the parties as to the proportion they are liable for once the decision is reached. Should the parties agree on a settlement of the dispute at any time during the Arbitration, they will jointly be liable for the Arbitrator's Costs.

Arbitration under the FIDIC 2017 Contracts

There are four routes to arbitration:

i The issue of a Notice of Dissatisfaction with the DAAB's decision, followed by an attempt at Amicable Settlement;
ii Any failure to comply with a DAAB decision;
iii Any failure to comply with an agreement or a final and binding determination of the Engineer (Employer's Representative under the Silver Book);
iv No DAAB has been put in place.

Note that the Arbitrator may take into account the extent to which a Party has failed to co-operate in constituting and appointing a DAAB, in any award dealing with costs of the arbitration.

Also, where a Party has failed to comply with a DAAB's decision, whether binding or final and binding, the arbitral tribunal will have the power to order, by an interim or provisional measure or an award, the enforcement of that decision.

The provisional measure or an award is subject to the express reservation that the rights of the parties are reserved until the merits of the dispute are resolved by an award.

As a further incentive for the Parties to cooperate with each other in the appointment of the DAAB, under Sub-Clause 21.6 [*Arbitration*] a new provision has been added in respect of any award costs, allowing the arbitral tribunal to take account of a Party's failure to co-operate in the appointment of the DAAB.

If the Dispute is not settled by the DAAB provisions, or by Amicable Settlement it is finally settled by international arbitration.

Unless the Parties agree otherwise:

a The Dispute is finally settled under the Rules of Arbitration of the International Chamber of Commerce.
b The Dispute is settled by one or three arbitrators appointed in accordance with these Rules.
c The arbitration is conducted in the ruling language defined in the Contract.

The arbitrator(s) have full power to open up, review and revise any certificate, determination (other than a final and binding determination), instruction, opinion or valuation of the Engineer (Employer/Employer's Representative under the Silver Book), and any decision of the DAAB (other than a final and binding decision) relevant to the Dispute.

In any award dealing with costs of the arbitration, the arbitrator(s) may take account of the extent (if any) to which a Party failed to cooperate with the other Party in constituting a DAAB.

Arbitration may be commenced before or after completion of the Works. The obligations of the Parties, the Engineer (under the Red and Yellow Books) and the DAAB are not altered by reason of any arbitration being conducted during the progress of the Works.

If an award requires a payment of an amount by one Party to the other Party, this amount is immediately due and payable without any further certification or Notice.

Failure to comply with the DAAB's decision

If either Party fails to comply with the DAAB's decision, then the other Party may refer the failure directly to arbitration (see above).

The arbitral tribunal may decide on the enforcement of that decision.

No DAAB in place

If there is no DAAB in place and a Dispute arises, the DAAB and Amicable Settlement provisions do not apply, and the Dispute may be referred by either Party directly to arbitration.

12 Tendering using the FIDIC 2017 Contracts

Deciding the procurement strategy

Prior to drafting the tender documents for a FIDIC 2017 Contract, the Employer must first decide on his procurement strategy.

This is the process of considering the most optimum way of achieving the Employer's objectives, taking into account the project itself, its current and future needs and also the inherent risks in taking a particular route towards achieving those objectives.

The procurement strategy should consider value for money in a holistic way, considering life cycle costing as well as the capital cost of building the project.

It is not intended to consider procurement strategy and life cycle costs in any detail within this book.

The initial briefing is the fundamental first step in establishing what is required, when it is required, what the Employer is proposing and what he can afford to pay.

Through a process of interviewing key members of the Employer's team and reviewing its business plans, specific objectives, options and constraints, a procurement strategy is developed which forms the basis for future assessments and recommendations.

Whilst it may sound obvious, it is vital that the Employer understands what he wants and can convey it clearly. He may not know what he wants, but also he can often be "persuaded" by Consultants or other third parties into accepting something that he does not want.

Deciding the procurement strategy is the foundation stone to the successful outcome to any contract, but this process is very often undervalued, or even overlooked as Employers (and their advisers), in their urgency to get projects out to tender and commenced on site, tend to favour the procurement strategy most familiar to them and the one that is easiest to comply with, rather than to spend time properly considering each project on its merits in terms of the Employer's resources and his objectives in terms of Time, Price (or Cost) and Quality, with the additional factor of Risk having to also be considered and evaluated (see Figure 12.1).

Time

Most Employers want their projects to be completed as quickly as possible so they can start using the building or facility. Some projects also have an absolute deadline e.g. international sporting events.

If the Employer sees time as being the most important factor, he may have to pay a higher cost, and quality may lower in order to achieve his time objective.

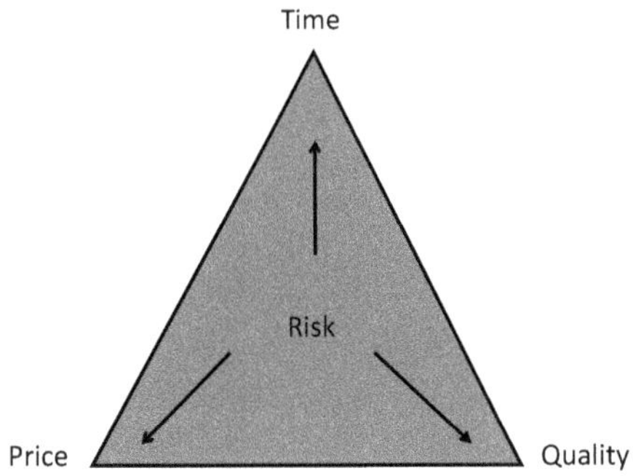

Figure 12.1 "The Time/Price/Quality Triangle".

The procurement method will have to consider timing through programme requirements, but also through provisions such as delay damages. As stated in Chapter 1, Design and Build could present substantial time savings from project inception to completion.

Price

All Employers will have some form of budget, some budgets are tighter than others, but the price they can, or want to pay, will always be a major consideration, particularly if the Employer requires a return on his investment.

The budget may be fixed, representing all that the Employer has available through its own funds or borrowings in order to fund the project, or it may have some degree of flexibility. The procurement strategy will need to address the budget and any inherent flexibility.

Quality

Finally, quality will always be a major factor. Poor quality using cheap materials or bad design could lead to significant problems and additional costs in the future as well as a building or facility which is not aesthetically pleasing, which could be particularly important where customers have a perception of a company by the appearance of its buildings and other facilities.

If time and price are more important, some Employers will lower the expected standards. It is important that the design lowers the standards, not that a "cheap" Contractor is employed who will unilaterally lower the standards himself!

It is rarely achievable for all three to be fully satisfied it is usually a balance.

Open questions that could be used in establishing the correct procurement strategy include:

- How important to the Employer is certainty of price, particularly if changes are expected? Lump sum options will give greater certainty of price.
- What resources and expertise does the Employer have?
- Which party is to be responsible for design and/or which party has the necessary design expertise? Using the FIDIC contracts, design can be assigned to the

Contractor under the Red Book (partial Contractor design), Yellow Book (full Contractor design) or Silver Book (full Contractor design).
- How important to the Employer is an early start and/or early completion?
- How clearly defined is the scope of the works, and what form does it take, for example, is it in the form of detailed drawings and specification, or performance-based criteria? Again, the FIDIC contracts can be used with detailed drawings or performance-based criteria, where the Contractor carries out the design.
- What is the likelihood of change to those defined requirements? Many Employers prefer a lump sum option for price certainly, some prefer a remeasurement option where there is likelihood of many changes.
- What is the Employer's attitude towards risk? Who is best placed to manage the inherent risks within the project?

  Lump sum contracts assign the greatest risk to the Contractor, and other types of contract assign greater risk to the Employer.

  It is a poorly acknowledged fact that the Employer always pays for risk, regardless of whether it is carried within the contract by the Employer or the Contractor. If it is the Contractor's risk he is deemed to have priced and programmed for it within his tender, if it is an Employer's risk, the Employer will pay for it if it occurs.

Risk

Whilst "Risk" is a major subject, it is worth briefly reviewing it in the light of construction projects and deciding on a procurement strategy.

The various risks in a construction project can result in:

- Failure to keep within budget,
- Failure to keep within the required or contracted time scales,
- Failure to meet the required standards for quality or fitness for purpose.

The ultimate consequences are essentially financial, and the need for risk identification and control are therefore self-evident.

The important aspect to recognise is that the risks are always there.

The issue is that a risk must be identified, and arrangements made to deal with it, and that the necessary finance has been provided either by the Contractor within the contract or by the Employer within his budget.

In very simple terms, the Employer always pays for risk.

As for other areas of commercial activity, risks or their financial consequences can be:

- "eliminated, reduced or controlled" by altering design requirements, construction methods, materials, increasing investigation, etc.
- "transferred contractually to others", for example, risks can be transferred
  - From Employer to Contractor or insurer,
  - From Contractor to Sub-Contractor or insurer,
  - From Designer to insurer.
- "retained" by the Employer and financial provision made.

It is considered good commercial practice that the consequences of risk occurring should be carried by the party who is in the best position to take that risk and minimise its effects.

For instance, if the Contractor is required to include in his pricing for all risks, some will occur and some will not; The Employer will, however, pay for all risks including those which have and will not occur.

In most forms of contract, the areas of risk to be borne by the Employer are normally stated and all other risk is borne by the Contractor.

With the exception of cost inflation or escalation, almost all other risks and uncertainties can be assessed as a change in the demand for resources. Hence, the particular importance of time-related costs and the need to define all costs correctly.

If a delay or disruption occurs, then the cost of maintaining the relevant resources for a longer time will increase and the resulting cost can be determined on the time basis.

The allowance to be made by a tendering contractor for the likely risks which are his responsibility requires judgement based on experience, but it is important that all risks and uncertainties are identified and taken into account. Each individual risk should be considered carefully, and a reasonable probability and impact identified.

It is also important to consider which elements of work account for the major elements of the total cost together with the appropriate uncertainties and assess their possible influence on the proposed tender sum. This should also be considered in relation to a programme in order to assess the delay and disruption effects on time and cost.

Contract Choice

Once these questions have been asked and answered and analysed, if using FIDIC 2017 Contracts, the correct Contract can be chosen whether it be the Red, Yellow or Silver Book.

As stated in previous chapters:

- The Red Book, or to give it its full title *Conditions of Contract for Construction*, is primarily intended for building or engineering works where the Employer bears the design responsibility; however, the works may include some Contractor-designed civil, mechanical, electrical and/or construction works.

  The basis of payment to the Contractor is through rates and prices in the Bills of Quantities, though it can be amended to provide a lump sum price or even as a cost reimbursable contract.
- The Yellow Book, full title *Conditions of Contract for Plant and Design-Build*, is intended for contracts on electrical/mechanical installations, and for the design and execution of building or engineering works where the Contractor bears the design responsibility. The Contractor designs and provides the Works in accordance with the Employer's Requirements.
- The Silver Book, full title *Conditions of Contract for EPC/Turnkey Projects*, is suitable for the provision on a turnkey basis of a process or power plant, of a factory or similar facility or of an infrastructure project or other type of development, where the Contractor takes total responsibility for the design and execution of the project, and a higher degree of certainty of final price and time is required.

Whilst many traditional contracts are based on Bills of Quantities with the Employer designing the project, and detailed bills of quantities being sent to each tendering contractor for pricing, in the past 30 years there has been a movement away from the use of traditional Bills of Quantities and towards Contractor-designed projects, with payment mechanisms such as milestone payments and activity schedules, with payment based on progress achieved, rather than purely on the quantity of work done.

A well-considered and appropriate procurement strategy, together with a clear, robust and unambiguous tender document, is the key to avoiding or at least minimising contractual problems at a later date.

Regarding tender documents, the author uses a "3 C's" approach, in that tender documents should always be:

- Clear – Unambiguous,
- Concise – Not over wordy,
- Complete – Have nothing missing.

Application of Laws

As FIDIC is an international contract, it is vital that drafters of tender documents correctly take into account the prevailing Law in the country where the contract is to be based. The FIDIC contracts provide for the Law of the Contract to be inserted into the Contract Data.

When using a FIDIC contract, the drafter, and in turn the contracting parties, must consider how the chosen law of contract and any specific local laws will affect the interpretation of the contract conditions. Will there be a conflict between the contract terms and the applicable law?

Where necessary, amendments should be made to the contract, although the law will ultimately prevail over the contract conditions.

The author has had personal experiences of the law when working in the Middle East, and also in the Caribbean, India, Sri Lanka and Malaysia.

Governing law

In some countries the local courts will uphold a governing law of the Contract, which is not the local law, as long as it does not directly conflict with the local law.

Whilst the language of the contract may be English, local courts may conduct their proceedings in a local language, so the documents referred to in the dispute may have to be translated.

In the UK, where FIDIC is not used to any great extent, the FIDIC dispute resolution procedures do not comply with the Housing Grants Construction and Regeneration Act 1996 amended by the Local Democracy, Economic Development and Construction Contract 2009 ("The Construction Act") which allows any party at any time to refer a dispute under a contract to adjudication. If the contract has not been amended to provide for adjudication, the parties have a statutory right of referral.

"Pay when paid" is also prohibited by the UK Construction Act.

"Fitness for purpose" is also something to consider within the applicable law, a fitness for purpose obligation may be implied unless the contract contains an express alternative standard. In the UK, for example, it is common practice for construction contracts, including design and build, to include a lower standard of "reasonable skill and care".

Parties using FIDIC should look carefully at the limitation on liability clauses and consider whether they wish to exclude recovery of both direct and indirect losses of profit claims and whether they are prepared to cap liability in the ways proposed.

Additionally, the parties should consider the FIDIC clause relating to adjustments for changes in legislation.

Drafters also need to consider long-term liabilities for design and/or construction defects, which could be, for example, 6, 10, 12 or 15 years, with the associated insurance premiums for Professional Indemnity (PI) insurance.

Good faith

The concept of "good faith" is often applicable to jurisdictions, including requirements and restrictions in respect of equity and unjust enrichment.

The requirement to act in good faith is not merely a requirement not to act in bad faith and not to deceive one another. Each party is instead under a legal obligation to exercise good faith in the performance of all of its contractual obligations and its dealings with the other party.

Exports/Imports

One potential situation would be where substantial parts of the Plant/Equipment/Materials are manufactured in another country and then transported to the Site.

The mechanism in the contract for dealing with any impact on the works (delays or additional costs) resulting from changes in the law applies only in relation to the "Laws" of the country where the project is located.

Changes in the law of another country which affect the Plant/Equipment/Materials being manufactured elsewhere will not be included unless bespoke drafting is included in the contract.

Delay damages/liquidated damages

The maximum level of delay damages is often capped at 10% of the total Contract Price, though the local law may not allow capped liabilities, or they may have to be capped at the total Contract Price.

The amount of delay damages stated within the contracts may be changed by the local courts, and also the agreed amount may be waived in their entirety if:

- The defaulting party can show that no loss has been suffered by a non-defaulting party;
- The level of the pre-agreed damages was excessive; or
- The obligation has been partially performed.

Contract Variations

Under some jurisdictions there may be a cap on the value of Variations in relation to the Contract Price, beyond which a contract may become subject to termination or be unenforceable.

Force Majeure

Some jurisdictions have differing definitions of Force Majeure (now termed under FIDIC 2017 as "Exceptional Events") and their effect on the execution of the Works.

In the event of Force Majeure, the affected party is relieved from performing only that part of its obligation which is impossible due to reasons beyond its control. Any other contractual obligation not affected must be performed. As soon as the cause of Force Majeure ceases, the corresponding obligation must be performed.

Termination

Under some jurisdictions, a construction contract can be terminated if its performance becomes impossible due to an external reason beyond the control of either party.

A contract may also be terminated on the death of a Contractor, provided that its heirs are unable to take over the work due to the contractor's specialist knowledge.

Tendering using FIDIC 2017 Contracts

The FIDIC contracts are based on the assumption that the project (the Works) will be competitively tendered, and that the Employer will send a Letter of Acceptance in relation to the accepted tender. Note that the Silver Book does not have a Letter of Acceptance.

The contract contains the following pro formas:

- Letter of Tender
- Letter of Acceptance.
- Contract Agreement
- Dispute Adjudication Avoidance Agreement

Annexes

The FIDIC 2017 Contracts contain the following Annexes within the contracts:

- Annex A – Example Form of Parent Company Guarantee
- Annex B – Example Form of Tender Security
- Annex C – Example Form of Performance Security – Demand Guarantee
- Annex D – Example Form of Performance Security – Surety Bond
- Annex E – Example Form of Advance Payment Guarantee
- Annex F – Example Form of Retention Money Guarantee
- Annex G – Example Form of Payment Guarantee by Employer

If we could briefly review each in turn:

Annex A – Example Form of Parent Company Guarantee

This form of guarantee is given by the Contractor's parent company, to guarantee the proper performance of a contract by one of its subsidiaries (the Contractor), who, whilst having limited financial resources itself, may be owned by a larger, more financially sound parent company.

Such a guarantee is cheaper than a performance bond, as the Contractor will normally just charge an administration fee rather than the case of performance bond, where the Contractor is actually paying a premium for an insurance policy, but it may give less certainty of redress because it is not supplied by an independent third party so it is dependent on the survival, and the ability to pay, of the parent company.

However, whilst accepting less independence, parent company guarantees for the proper performance of the contract can be more advantageous than bonds. Rather than receiving a fixed amount in compensation, the parent company is normally obliged to either complete the works in accordance with the Contractor's original obligations on behalf of its subsidiary company, or fund the completion of the contract by others, so effectively completion is guaranteed.

In addition, further recompense can be sought for time delays in completion through the normal clauses incorporated in the contract. In the case of performance bonds, the "guarantee" is that a sum of money is available to at least partly compensate the Client in the event of the Contractor's default.

Because the financial strength of the parent company may be linked to that of the Contractor, a parent company guarantee will be acceptable only if the parent company (or holding company) is financially strong and its financial resources are largely independent of those of the Contractor.

Obviously, the status of the parent company guarantee and the parent company itself must be consistent with the Law of the Contract and relevant jurisdiction.

Annex B – Example Form of Tender Security

The Tender Security is required to protect the Employer (the Beneficiary) in the event that a tendering Contractor (the Applicant) withdraws his offer without the agreement of the Employer, and has refused to accept correction of errors in his Tender, the offer has been accepted but the Contractor has failed in respect of Sub-Clause 1.6 to sign a Contract Agreement, or the offer has been accepted but the Contractor has failed in respect of Sub-Clause 4.2.1 to deliver the Performance Security.

A specified amount is then guaranteed to be paid by the Guarantor to the Employer.

Annex C – Example Form of Performance Security – Demand Guarantee

This form of guarantee is given by a Guarantor on behalf of the Contractor where the Guarantor will pay a fixed sum of money to the Employer in the event the Contractor is in breach of its obligations.

The amount of the Performance Security is stated in the Contract Data.

The sum is adjusted in the event that the Contract Price increases or decreases by more than 20%.

Annex D – Example Form of Performance Security – Surety Bond

This Surety Bond is a form of guarantee given by a Guarantor on behalf of Contractor (the Principal) where, if the Contractor defaults in any of its obligations, the Guarantor will discharge (pay for) any damages incurred by the Employer (the Beneficiary) up to the value of the Bond Amount.

The amount of the Performance Security is stated in the Contract Data.

Annex E – Example Form of Advance Payment Guarantee

This is a guarantee where the Contractor receives an Advance Payment, but fails to make the repayments as required, the Guarantor makes the payment on his behalf.

Annex F – Example Form of Retention Money Guarantee

This is a guarantee where the Contractor receives an early repayment of retention money but fails to carry out his obligations in respect of correcting defects. The Guarantor becomes liable for any costs.

Annex G – Example Form of Payment Guarantee by Employer

This is a guarantee where the Employer fails to make a payment as required under the contract. The Contractor issues a demand a written statement and receives the money from the Guarantor.

Letters of intent

It is common with a project requiring an early start on site for a Contractor to receive a letter of intent from the Employer stating that he accepts his tender and that the Employer is in the process of preparing the contract documents.

The Employer may also include in his letter a requirement to carry out some initial design work, ordering of materials, or even start, maybe some enabling work on site prior to a formal contract being drawn up. Caution must be taken when drafting the letter of intent, and also by the Contractor when he receives one.

It is common for this letter to mark the beginning of on-site works, though negotiations of minor issues have yet to be finalised, it is therefore critical this letter is not be interpreted as the "Letter of Acceptance", so drafting must be meticulous, as they often do not properly constitute an acceptance of tender and therefore do not create any contractual liability, rarely giving any express undertaking by the Employer to pay for any work, which could put the Contractor at risk if, for example, the project were to be abandoned.

The golden rule is to execute contracts as early as possible so all parties know the ENTIRE terms and conditions, and not rely on letters of intent to provide false security. If letters of intent are used they should CLEARLY state the intentions of the parties, particularly who will pay what, and how much if matters do not proceed.

The structure of the FIDIC 2017 Contracts

Within the *Conditions of Contract for Construction* (The "Red Book") the "Contract" is defined under Sub-Clause 1.1.10 as meaning:

i The Contract Agreement
ii The Letter of Acceptance
iii The Letter of Tender
iv Any Addenda referred to in the Contract Agreement
v The Conditions (the General Conditions as amended by the Particular Conditions)
vi The Specification

vii The Drawings
viii The Schedules
ix The JV (Joint Venture) Undertaking (if applicable) and,
x The further documents (if any) which are listed in the Contract Agreement or in the Letter of Acceptance.

Note that the published Contract refers to "the Contractor's Proposal", which has been deleted by the published Errata.

The documents forming the FIDIC contract are to be taken as mutually explanatory of one another, which means that they should all be read together and each document has equal weight.

However, if there is any conflict, ambiguity or discrepancy between the documents, the priority of the documents is prescribed in each of the Contracts.

Red Book (in descending order)

i The Contract Agreement
ii The Letter of Acceptance
iii The Letter of Tender
iv The Particular Conditions Part A – Contract Data
v The Particular Conditions Part B – Special Provisions
vi The General Conditions
vii The Specification
viii The Drawings
ix The Schedules
x The JV Undertaking (if the Contractor is a JV)
xi Any other documents forming part of the Contract

Note "Contract Data" (d) was formerly referred to in FIDIC 1999 Contracts as "Appendix to Tender".

The Red Book has its own set of tender documents:

- Letter of Tender
- Conditions of Contract
- Specification
- Drawings
- Schedules

The Red Book also includes a Letter of Acceptance and a Contract Agreement, either document may bring the Contract into force.

Within the *Conditions of Contract for Plant and Design-Build* (The "Yellow Book") the "Contract" is defined under Sub-Clause 1.1.9 as meaning:

i The Contract Agreement
ii The Letter of Acceptance
iii The Letter of Tender
iv Any addenda referred to in the Contract Agreement
v The Conditions (the General Conditions as amended by the Particular Conditions)
vi The Employer's Requirements
vii The Schedules

viii The Contractor's Proposal
ix The JV Undertaking (if the Contractor is a JV)
x The further documents (if any) which are listed in the Contract Agreement or in the Letter of Acceptance

Note the priority of documents within the Yellow Book (in descending order)

i The Contract Agreement
ii The Letter of Acceptance
iii The Letter of Tender
iv The Particular Conditions Part A – Contract Data
v The Particular Conditions Part B – Special Provisions
vi The General Conditions
vii The Employer's Requirements
viii The Schedules
ix The Contractor's Proposal
x The JV Undertaking (if the Contractor is a JV)
xi Any other documents forming part of the Contract

The Yellow Book has its own set of tender documents:

- Letter of Tender
- Conditions of Contract
- Employer's Requirements
- Schedules
- Contractor's Proposal

The Yellow Book includes a Letter of Acceptance and a Contract Agreement, either document may bring the Contract into force.

Within the *Conditions of Contract for EPC/Turnkey Projects* (The "Silver Book") the "Contract" is defined under Sub-Clause 1.1.7 as meaning:

i The Contract Agreement.
ii Any addenda referred to in the Contract Agreement
iii The Conditions (the General Conditions as amended by the Particular Conditions)
iv The Employer's Requirements
v The Schedules
vi The Tender
vii The JV Undertaking (if the Contractor is a JV)
viii The further documents (if any) which are listed in the Contract Agreement or in the Letter of Acceptance

Note the priority of documents within the Silver Book (in descending order)

i The Contract Agreement
ii The Particular Conditions Part A – Contract Data (formerly referred to in FIDIC 1999 Contracts as "Appendix to Tender")
iii The Particular Conditions Part B – Special Provisions
iv The General Conditions

v The Employer's Requirements
vi The Schedules
vii The Tender
viii The JV Undertaking (if the Contractor is a JV)
ix Any other documents forming part of the Contract

If either Party finds an ambiguity or discrepancy in the documents (ambiguity is something unclear, discrepancy is something missing), that Party must promptly give a Notice to the Engineer (Employer under the Silver Book), describing the ambiguity or discrepancy.

After giving or receiving a Notice, the Engineer/Employer must issue the necessary clarification or instruction to correct matters.

Note that any Special Provisions (Particular Conditions – Part B) will always overrule and supersede the equivalent provisions in the General Conditions.

In writing the Particular Conditions it is therefore essential that the changes are easily identifiable by using the same clause numbers and titles as in the General Conditions. Furthermore, it is necessary to add a statement in the tender documents for a contract that:

> Special Provisions (Particular Conditions – Part B) take precedence over the provisions found within the General Conditions, and also that the provisions of the Contract Data (Particular Conditions – Part A) take precedence over the Special Provisions (Particular Conditions – Part B).

If we consider each of the contents of the FIDIC 2017 Contracts in turn:

The Silver has its own set of tender documents:

- Conditions of Contract
- Employer's Requirements
- Tender

Whilst the Red Book and the Yellow Book include a Letter of Acceptance and a Contract Agreement, the Contract Agreement will bring an EPC Contract such as the Silver Book into force.

The Contract Agreement

The "Contract Agreement" is defined as meaning "*the agreement entered into by both Parties in accordance with Sub-Clause 1.6 [Contract Agreement]*".

The Silver Book also includes the words "*including any annexed memoranda*".

A template is included at the back of the Contracts.

This is the Agreement signed by both Parties, and refers to the documents that form the Contract, the covenant by the Contractor to execute and complete the Works and remedy any defects, in accordance with the Contract, and the covenant by the Employer to pay the Contract Price, in accordance with the Contract.

The Letter of Acceptance

The "Letter of Acceptance" is defined in the Red and Yellow Books (but not the Silver Book) as meaning the letter of formal acceptance, signed by the Employer, of the Letter of Tender, including any annexed memoranda comprising agreements between and signed by both Parties. If there is no such letter of acceptance, the expression "Letter of Acceptance" means the Contract Agreement and the date of issuing or receiving the Letter of Acceptance means the date of signing the Contract Agreement.

A template is included at the back of the Contracts.

This is the letter completed and signed by the Employer accepting the Contractor's Tender.

Note that the Letter of Tender and the Letter of Acceptance together constitute a binding contract until the Contract Agreement is executed.

The Letter of Tender

The "Letter of Tender" is defined in the Red and Yellow Books (but not the Silver Book) as meaning "*the letter of tender, signed by the Contractor, stating the Contractor's offer to the Employer for the execution of the Works*".

A template is included at the back of the Contracts.

This is the letter completed and signed by the Contractor forming the Contractor's offer.

Note that the Letter of Tender and the Letter of Acceptance together constitute a binding contract until the Contract Agreement is executed.

The Addenda referred to in the Contract Agreement

This is only relevant if there are Addenda referred, and attached, to the Contract Agreement.

The Conditions of Contract

The "Conditions of Contract" are defined as meaning "*the General Conditions as amended by the Particular Conditions*".

Note that the Particular Conditions will take precedence over the General Conditions.

General Conditions

The General Conditions as discussed within this book comprise 21 clauses as follows:

CONDITIONS OF CONTRACT FOR CONSTRUCTION (THE "RED BOOK")

1 General Provisions
2 The Employer
3 The Engineer
4 The Contractor
5 Subcontracting
6 Staff and Labour
7 Plant, Materials and Workmanship

8 Commencement, Delays and Suspension
9 Tests on Completion
10 Employer's Taking Over
11 Defects after Taking Over
12 Measurement and Valuation
13 Variations and Adjustments
14 Contract Price and Payment
15 Termination by Employer
16 Suspension and Termination by Contractor
17 Care of the Works and Indemnities
18 Exceptional Events
19 Insurance
20 Employer's and Contractor's Claims
21 Disputes and Arbitration

CONDITIONS OF CONTRACT FOR PLANT AND DESIGN-BUILD (THE "YELLOW BOOK")

1 General Provisions
2 The Employer
3 The Engineer
4 The Contractor
5 Design
6 Staff and Labour
7 Plant, Materials and Workmanship
8 Commencement, Delays and Suspension
9 Tests on Completion
10 Employer's Taking Over
11 Defects after Taking Over
12 Tests after Completion
13 Variations and Adjustments
14 Contract Price and Payment
15 Termination by Employer
16 Suspension and Termination by Contractor
17 Care of the Works and Indemnities
18 Exceptional Events
19 Insurance
20 Employer's and Contractor's Claims
21 Disputes and Arbitration

CONDITIONS OF CONTRACT FOR EPC/TURNKEY PROJECT (THE "SILVER BOOK")

1 General Provisions
2 The Employer
3 The Employer's Administration
4 The Contractor
5 Design
6 Staff and Labour
7 Plant, Materials and Workmanship

8 Commencement, Delays and Suspension
9 Tests on Completion
10 Employer's Taking Over
11 Defects after Taking Over
12 Tests after Completion
13 Variations and Adjustments
14 Contract Price and Payment
15 Termination by Employer
16 Suspension and Termination by Contractor
17 Care of the Works and Indemnities
18 Exceptional Events
19 Insurance
20 Employer's and Contractor's Claims
21 Disputes and Arbitration

Particular Conditions

The FIDIC contracts include "Guidance for the Preparation of Particular Conditions".

"Particular Conditions" is now defined as comprising of two parts:

- Part A – Contract Data,
- Part B – Special Provisions.

NB: The "Appendix to Tender" document, under the 1999 FIDIC contracts, is now referred to as "Contract Data".

The Contract also refers to the five FIDIC Golden Principles:

- GP1: The duties, rights, obligations, roles and responsibilities of all the Contract Participants must be generally as implied in the General Conditions, and appropriate to the requirements of the project.
- GP2: The Particular Conditions must be drafted clearly and unambiguously.
- GP3: The Particular Conditions must not change the balance of risk/reward allocation provided for in the General Conditions.
- GP4: All time periods specified in the Contract for Contract Participants to perform their obligations must be of reasonable duration.
- GP5: Unless there is a conflict with the governing law of the Contract, all formal disputes must be referred to a Dispute Avoidance/Adjudication Board (or a Dispute Adjudication Board, if applicable) for a provisionally binding decision as a condition precedent to arbitration.

Errata added in the words "Unless there is a conflict with the governing law of the Contract, …"

The FIDIC contracts include substantial guidance on the use of Particular Conditions together with samples, including references to the equivalent Clause/Sub-Clauses within the General Conditions.

The Employer may also wish to "copy and paste" provisions within one of the FIDIC 2017 books into another, though great care must always be taken in amending contracts.

Drafting and amending contracts is a very skilled process, so it is critical that Employers, and their advisors, properly review and understand the General Conditions, and only amend through the Particular Conditions, what is required for the Employer's specific needs, or perhaps to suit legal requirements within the country where the contract is to be used.

It is also important that any changes made, for example, to definitions are properly incorporated and referenced into the various documents that form the Contract, not just the General Conditions and Particular Conditions.

All additions/amendments to the General Conditions should be included within the Particular Conditions, rather than amending the General Conditions themselves, so the General Conditions remain in the Contract, but amended within the Particular Conditions.

It is also critical that when drafting a Particular Condition, it must be clearly stated what happens to the original core clause, for example, it is deleted. If the original General Condition is not deleted, then it is more likely than an inconsistency.

It is also important to note that a Particular Condition may modify or add to the General Condition, to suit any risk allocation or other special requirements of the particular contract. However, changes should be kept to a minimum, consistent with the objective of using industry standard, impartially written contracts.

It must be remembered that if an Employer amends a contract to allocate a risk to the Contractor which may have been intended to be held by the Employer, the Contractor has a right to price it in terms of time and price, therefore the practice of Employers amending contracts to pass risk to Contractors without considering who is best able to price, control and manage those risks can in many cases prove to be unwise and uneconomical.

It is important to spend time considering whether a Particular Condition is appropriate in each case, then when that decision has been made, that the clause is drafted correctly and aligned to the drafting principles and convention of the original contract.

It is critical that whoever drafts Particular Conditions is fully aware of the FIDIC drafting philosophy.

Suggested Particular Conditions

The FIDIC contracts include a comprehensive list of suggested Particular Conditions.

Particular Conditions bring some flexibility to the contract as they amend the General Conditions.

For example, it is quite common for the Red Book to be made a lump sum contract.

Whilst it is not proposed to consider or review <u>all</u> of the suggestions that have been included within the guidance on Particular Conditions for the FIDIC 2017 Red, Yellow and Silver Books, there are some worth mentioning, which are already detailed, plus some further suggestions.

Some further points have been raised under the specific Clauses/Sub-Clauses in previous chapters of this book.

General Provisions

Whilst there is a comprehensive single alphabetical list within the FIDIC 2017 Contracts (previously it was divided into sections rather than a single alphabetical list) of defined terms within the Contracts, the Employer may wish to include additional definitions, or to include additional information and/or clarification about the definitions already included, for example, the Site is defined as "*the places where the Permanent Works are to be executed and to which Plant and Materials are to be delivered, and any other places specified in the Contract as forming part of the Site*" and could include pieces of land adjacent to the site for storage, lay down areas for pre-assembly, etc. What constitutes the "Contract" could also be further explained within the Particular Conditions.

Further details or requirements on respect of Contractor's and Employer's Documents may also feature within the Particular Conditions.

There may also be additional requirements in terms of contractual communications, in that they have to be sent in paper form in addition to, or instead of, electronic. The author has worked on contracts where the Employer stipulated (within the Particular Conditions) that EVERY electronic communication the Contractor submitted had to also be submitted in hard copy to the Engineer and the Employer within 48 hours.

There may also be a specified form of software onto which communications must be uploaded, for example, CEMAR.

FIDIC 2017 includes various time scales which the drafters feel are appropriate and reasonable for the Parties to carry out various obligations and communications, and in some cases a Party is time barred if they do not take the action, or there is a deeming provisions, in that something is deemed to happen anyway, for example, No-Objection.

If the Employer wishes to change any of these time scales within the Contract by extending or reducing them, they can do so through the Special Provisions, but it is important to maintain appropriate and reasonable time scales.

Also, the meaning and status of various communications such as "Notices", "No-objection", Acceptance, and Reviews may be amended or further clarified.

It is common for Employers to amend the priority of documents that comprise the Contract, particularly anything provided by the Employer takes precedence over anything provided by the Contractor.

Entry into a formal written Contract Agreement may be mandatory under the applicable law, before the contract exists, and before any work can commence.

There may be specific terminology or procedures to be adopted within the Contract dependent on the relevant Laws under which the Contract is operated, for example, the definition and/or standards to be applied in terms of negligence and also limitation of liability can be different under different legislations.

There may be also be specific requirements in respect of the Contractors that comprise a JV and the status of that JV e.g. Incorporated v Unincorporated requirements.

The Employer

The status of the Employer may be clarified within the Particular Conditions, particularly where an overseas Employer is partnering, say with a local sponsor.

There may also be specific requirements for the Contractor to have right of access to Site at certain times.

The use of Employer-Supplied Materials may also need to be further developed, including liability for those Materials, and at what point title passes from the Employer to the Contractor (if it does).

The Engineer (Red and Yellow Books)/Employer's Representative (Silver Book)

The status of the Engineer/Employer's Representative and the delegation to Engineer's Representatives and/or assistants and what can and cannot be delegated may be clarified or amended within the Particular Conditions.

The principle under the General Conditions is that the Engineer/Employer's Representative has the full authority to act for the Employer on all matters; however, the Employer may wish to limit that matters of a certain nature or value must be referred directly to the Employer for a decision, before actioning it with the Contractor.

Engineer's/Employer's Instructions are required to be given in writing electronically/hard copy. There may also be rules on the giving and/or receiving of oral instructions and also the use of CVIs (Confirmation of Verbal Instructions), and subsequent deeming of instructions may be defined.

The Contractor

There may be specific requirements in terms of the forms of Performance Security and conditions as to where it should be obtained from.

There are often words within the Particular Conditions to cover the Contractor's Documents, but bear on mind that the Specification indicates what has to be prepared and when it has to be submitted, and how long the Review Period is.

If the Contractor is required to supply Operation and Maintenance Manuals then these are covered by the General Conditions with reference to the Specification together with the need for spare parts and/or parts with limited life.

There may be a requirement for the Contractor to provide training in which case it may be identified by the Employer on the Specification, and the actual training details may be included by the Contractor within his tender.

There may also be a need for the Contractor to have a Quality Manager employed on the Site, again should be stated in the Specification (Key Personnel).

Regarding health and safety, for example, in the UK (which is not a great user of FIDIC contracts) there may be a requirement to state the role and responsibilities of the Principal Designer and the Principal Contractor within the Particular Conditions.

With regard to Unforeseeable Physical Conditions, the author has experienced many instances where the information available at tender stage may be qualified, for example the Employer may disclaim all liability for its accuracy, or in some cases may warrant that it is correct, in which case it makes it easier for determining the Contractor's entitlement.

Sometimes, in respect of Unforeseeable Physical Conditions arise, the Employer may amend the wording of Sub-Clause 4.12 to provide what he feels is a more "equitable" remedy, for example, 50% of Cost incurred by the Contractor rather than "the Cost".

Alternatively, the Employer may not allow the Contractor any remedy in respect of Unforeseeable Physical Conditions.

With regard to Transport of Goods, the Employer may wish the Contractor to get express permission, prior to delivery of Goods to the Site.

The Employer may provide Equipment for the Contractor's use, so the wording of the Sub-Clause may be amended to include the Contractor's car of the Employers' property.

Dependent on the relevant legislation and environmental law provisions, there may be a requirement to prepare documentation such as environmental management plans, for review by the Engineer and/or the Employer/Employer's Representative and approval by regulatory authorities, these plans being described within the Specification together with the review/approval process associated with each plan.

In respect of temporary utilities, there may be specific requirements to be provided by the Employer or Contractor. If the Employer provides temporary utilities and their use is free of charge, the Specification should state that is the case, alternatively if the Employer is intending to charge the Contractor for its provision and use, again this should be stated within the Specification.

With regard to security of the Site, if the Contractor is sharing occupation of the Site with others, including other Contractors, then it may not be appropriate for the Contractor to be held solely responsible for Site security.

If the Contractor is sharing occupation of the Site with others, it is recommended that the Contract is amended in terms by identifying and allocating responsibility for clearance and removal from the Site of any wreckage, rubbish, hazardous waste, etc., as stated within the Sub-Clause.

Sometimes the Contractor may be required to carry out certain parts of the Works by a specified date, though the Employer will not take over that part of the Works at that date.

If that is the case, the Employer may include "Milestones" by referring to them within the Specification, and amending the General Conditions within the Particular Conditions.

Milestones have not been included within General Conditions of the FIDIC contracts, only completion of "the whole of the Works" or specified "Sections" at which point the Employer takes over the whole of the Works, or that particular part of the Works.

To include the Milestone within the Particular Conditions there should be a definition of a Milestone, and also there may be a certificate or some other document issued by the Engineer or Employer's Representative confirming that fact.

The complete date for the Milestone should be linked back to the Commencement Date e.g. within 36 weeks of the Commencement Date, and also there may be delay damages for a Milestone stated in the Contract Data, so that if the Contractor fails to complete the works of the Milestone within the time for completion of the Milestone (with any extension under this Sub-Clause) the Contractor will be liable for delay damages, possibly with a limit to how much can be claimed as a maximum.

The Contractor should also reflect Milestones within his programme submissions, including anything he may require from the Employer/Employer's Representative/Engineer to be able to complete the Milestones.

The provision for Extensions of Time and the new Advance Warning should also be accommodated by reference to the Milestones.

Subcontracting

It is fairly normal for the Contractor to employ a minimum percentage of its domestic Subcontractors from local labour/businesses in carrying out the Works, with a percentage stated within the Contract Data.

However, nominated Subcontractors are chosen by the Employer/Employer's Representative/Engineer.

If the nominated Subcontractors are named within the Specification, the Contractor is aware of them and can price the risk, although they may not be appointed until after contract award.

If the Employer wishes to name Subcontractors but for them not to be classified as "nominated Subcontractors" then the Particular Conditions should deal with it.

Design

It is quite common for Employers to require Contractors at tender stage to identify the name and credentials of their design team, and whether they are internal staff or external consultants. Clearly the Contract requires them to be identified, but it could be a key part of the decision making process prior to contract award.

The status and content of the Employer's Requirements (Yellow and Silver Books) may also be set out and/or revised, including if the Employer has already employed a design team, and possibly wishes to novate that design team to the successful Contractor.

The requirement and/or format of the Contractor's documents may be developed including the requirements for as-built records after Completion.

The Review period and also the Review process may be further developed as part of the Particular Conditions.

Local restrictions

Clearly, dependent on location and relevant law there may need to be restrictions included within the Contract for restrictions on consumption or sale of alcoholic liquor or drugs.

Also, the Contract may need to incorporate religious and other festivals and observances such as Ramadan, with its associated shortened working hours and restrictions on consumption of food and drink, etc.

Programme

The Contractor is required to include certain records within its Progress Reports. These records may have to be a particular type or format. Again, the Particular Conditions should reflect this.

Clause 8 (Commencement, Delays and Suspension) should reflect on the Works and/or Sections if the Works are to be taken over in stages; also it is important that each stage is defined as a Section, and the Time for Completion of each Section is stated, in the Contract Data.

Sub-Clause provides for the programming software to be stated within the Specification, so this must be clearly defined and drawn to the attention of tenderers in the Instructions to Tender.

For lower value, less complex projects, the Employer should either consider using the Green Book or scale down the Red, Yellow or Silver Book, particularly with respect to programme requirements, which, under FIDIC 2017, are significantly more than under FIDIC 1999.

The provisions under Sub-Clause 8.5 may also be amended in respect of the definitions of the causes and entitlements for Extensions of Time.

Note also the reference to concurrency, which is a new provision within FIDIC 2017, which has been previously discussed within Chapter 7 and which may need to be further developed in terms of assessment and/or entitlement.

As there is reference to the Society of Construction Law, it is recommended that they Employer appoints programming/delay analyst specialists to advise on these aspects. In preparing the Special Provisions, therefore, it is strongly recommended that the Employer be advised by a professional with extensive experience in construction programming, analysis of delays and assessment of Extension of Time in the context of the governing law of the Contract.

In many countries delay damages can only be set at "a genuine pre-estimate of likely loss" and cannot constitute a penalty. It is vital that particularly if the delay damages are set at a high level that the Employer can justify the level, and if it is set too high it may be judged unenforceable under certain jurisdictions.

Also, if the Accepted Contract Amount is to be quoted in more than one currency, it may be preferable to define delay Damages (per day) as a percentage to be applied to each of these figures. If the Accepted Contract Amount is expressed in the Local Currency, the damages per day may either be defined as a percentage or be defined as a figure in Local Currency.

For projects where it is to the Employer's advantage to have a project completed early, it may be appropriate to incentivise the Contractor by including a provision for a bonus for early completion of the Works or a Section, again set similar to delay damages as an amount per day and applicable to each of the currencies which may the Accepted Contract Amount.

If the Works are to be completed and taken over in stages it is important that each stage is defined as a Section in the Contract Data.

The Contract Data should include a table to define for each Section:

- The relevant percentage for release of Retention Money,
- The Time for Completion,
- The applicable delay damages.

The need for specific records, and who is responsible for keeping those records in respect of measurement and valuation of the Works can be defined within the Particular Conditions. If they are by the Employer they should be stated within the Specification, if by the Contractor they should be stated within the Tender.

Provisional Sums may be required for parts of the Works which cannot be defined at tender stage. It is essential to what is intended to be covered by each Provisional Sum, and it is recommended that this be included in a Schedule prepared by the Employer, since the amount of each Provisional Sum corresponding to the defined scope will then be excluded from the other elements of the Accepted Contract Amount.

The provision for adjustments for changes in cost may be required if it would be unreasonable for the Contractor to bear the risk of escalating costs due to inflation, and may require amendment depending on what the Employer will pay for.

Payment

The procedures and also the timing for making payments under Clause 14 should be checked and if necessary amended to ensure that they are acceptable to both the Employer and any financing institution the Employer may be using to fund the project.

The Red Book is based on a bill of quantities, but the Particular Conditions may be amended to provide for a differing contract strategy, for example a lump sum, cost reimbursable contracts.

If the Employer wishes to proceed with a cost reimbursable (cost plus overheads/profit) option then what is "Cost" must be defined, and also how the overheads and profit are to be calculated and paid e.g. as a percentage or a fixed lump sum.

Lump sum contracts are preferred where the scope of work can be accurately defined.

However, as noted in the Introduction above, if significant design input by the Contractor is required, it is recommended that the Employer considers using the FIDIC 2017 Yellow or Silver Book.

For a lump sum contract, the tender documents should include a schedule of payments (see Sub-Clause 14.4 [Schedule of Payments]).

To change the Red Book to a lump sum contract:

- Delete the second paragraph of Sub-Clause 8.5 [Extension of Time for Completion]. This is in respect of a significant change in quantities.
- Delete Clause 12 (Measurement and Valuation), which is only applicable to the Red Book.
- Delete other references in other Clauses and Sub-Clauses to Clause 12, for example.

Under Sub-Clause 13.3.1 (Variation by Instruction): delete the words *"by valuing the Variation in accordance with Clause 12 [Measurement and Valuation], with supporting particulars (which shall include identification of any estimated quantities and, if the Contractor incurs or will incur Cost as a result of any necessary modification to the Time for Completion, shall show the additional payment (if any) to which the Contractor considers that the Contractor is entitled)*, and replace with "*with supporting particulars*".

- Delete sub-paragraph (a) of Sub-Clause 14.1 and replace with *"(a) the Contract Price shall be the lump sum Accepted Contract Amount and be subject to adjustments in accordance with the Contract"*.
- Delete sub-paragraph (c) of Sub-Clause 14.1 and replace with *"(c) any quantities which may be set out in a Schedule are estimated quantities and are not to be taken as the actual and correct quantities of the Works which the Contractor is required to execute; and"*.

Termination

Regarding termination provisions, the Employer should verify that the wording of the relevant Sub-Clauses is consistent with the law governing the Contract.

For example, as stated earlier in the book, in many jurisdictions it may not be permissible for the Employer to terminate the Contract for his own convenience without a material breach by the Contractor.

Exceptional Events

As has been stated, Exceptional Events were previously referred to as "Force Majeure" which is an internationally known term. As with Force Majeure, Exceptional Events should be interpreted correctly in terms of what the event is, and their direct impact upon the Contractor's progress.

Insurance

Regarding insurance provisions, the Employer should review what is included within the General Conditions, and if it wishes to change them should do so within the Particular Conditions, for example the inclusion of Decennial Liability provisions. The Employer should also take professional advice in wording those additional conditions so that it does not inadvertently carry liabilities for which the Employer is neither prepared nor covered by its own insurance.

Disputes and Arbitration

In respect of Disputes and Arbitration, there may be some other provisions that could be considered.

Regarding the role of the DAAB, as the DAAB is formed at the start of the project and stays in place throughout, they could provide a higher level dispute avoidance function and the Parties should consider using the DAAB before matters become disputes.

FIDIC strongly recommends that the DAAB should be appointed at the start of the Contract and remain in place for the duration of the Contract.

However, the Parties may prefer the DAAB to be appointed on an "ad-hoc" basis, only when a Dispute arises, and its role to be limited to resolution of the particular Dispute. It would then have no role to play in the avoidance of Disputes between the Parties, and its appointment would cease when it had given its decision on that Dispute. Should a new Dispute arise, a new ad-hoc DAAB would be appointed.

If the Parties wish to provide for an "ad-hoc DAB", rather than the recommended "standing DAAB", then the following amendments will be needed:

- Delete the definitions under Sub-Clauses 1.1.22 and 1.1.23 and replace with "Dispute Adjudication Board" or "DAB" means the person or three persons (as the case may be) so named in the Contract or appointed under Clause 21." and replace all references in the Conditions of Contract to "DAAB" with "DAB".

  Sub-Clause 21.1, plus the Appendix General Conditions of Dispute Avoidance/Adjudication Agreement with its Annex – DAAB Procedural Rules, should be amended to comply with the wording contained in the corresponding provisions of FIDIC's *Conditions of Contract for Plant and Design-Build*, First Edition 1999.
- Delete Sub-Clause 21.3 [Avoidance of Disputes];

  In the first paragraph of Sub-Clause 21.4 [Obtaining DAAB's Decision], after the words "If a Dispute arises between the Parties then, add "after a DAB has been appointed"; delete the words "(whether or not any informal discussions have been held under Sub-Clause 21.3 [Avoidance of Disputes])"; and delete sub-paragraph (a) of Sub-Clause 21.4.1 [Reference of a Dispute to the DAAB].

To facilitate the appointment of the member(s) of an "ad-hoc DAB", it is important that the Employer and the Contractor each avail themselves of the opportunity at the tender stage of the Contract to name three potential members in the Contract Data.

In respect of Amicable Settlement where the DAAB has given its decision but one or both Parties is/are dissatisfied with the decision, the provisions for Amicable Settlement are intended to encourage the Parties to settle a Dispute amicably, without the need for arbitration.

Rather than considering the 28-day period stated in this Sub-Clause as a "cooling-off period", FIDIC recommends that the Parties avail themselves of this opportunity to actively engage with each other with a view to settling their Dispute.

Such active engagement may be by, for example:

- Direct negotiation by senior executives from each of the Parties;
- Mediation (please see below);
- Expert determination (using, for example, the Expert Rules published by the International Chamber of Commerce (the "ICC", which is based at 33–43 Avenue du Président Wilson, 75116 Paris, France) https://iccwbo.org/publication/icc-expertrules-english-version/);
- Or other form of alternative dispute resolution that is not as formal, time-consuming and costly as arbitration.

In this regard, it is recommended that consideration be given by the Parties to agree to a longer time period than the period of 28 days stated in this Sub-Clause in an effort to arrive at an Amicable Settlement procedure chosen by the Parties.

Amicable Settlement procedures typically depend for their success on the consensual involvement of both Parties, on confidentiality and on both Parties' acceptance of the particular procedure. Therefore, while it is recommended that both Parties engage actively to settle the Dispute amicably, neither Party should seek to impose the procedure on the other Party.

Dispute escalation provisions

Published construction contracts in the past have always included differing forms of dispute resolution provisions, including dispute adjudication boards, amicable resolution, mediation, expert determination, arbitration and legal proceedings, and in some cases adjudication by contract and/or by statute, expert determination and mediation.

In recent years, contracting Parties, and in turn contract drafters, have also looked towards dispute avoidance to try to prevent issues becoming disputes in the first place.

Dispute avoidance can be effected in the form of more balanced contract conditions, partnering agreements, alternative procurement methods such as target contracts, but also by including specific dispute avoidance provisions within the contract.

One specific dispute avoidance provision is escalating a dispute internally through senior management up to the most senior level, rather than immediately referring to a third party such as the DAAB, or other dispute resolution measures.

The senior management representatives may form a "Dispute Avoidance Board" in itself, and consist of one or three members (odd number) as identified within the Contract Data.

A potential dispute under the contract may be referred to the Dispute Avoidance Board, two to four weeks after notification to the other Party and to the Project Manager.

The Parties are required to make all relevant information available to the Dispute Avoidance Board, who visits the Site (where applicable) and helps the Parties to settle the potential dispute without formal referral as a dispute.

A dispute cannot be referred to arbitration or legal proceedings unless and until the dispute has first been referred to the Dispute Avoidance Board as a potential dispute.

A party can, following the Dispute Avoidance Board's recommendation, notify the other Party that it is dissatisfied. This is a time-barred right, as the dissatisfied Party cannot refer the dispute to the tribunal unless it is notified within four weeks of the Dispute Avoidance Board's recommendation.

What are "tiered" or "stepped" escalation clauses?

1 A tiered or stepped escalation clause is a dispute resolution clause that records the parties' agreement that any dispute between them shall be resolved on a staged basis. Each step (or tier) is designed to handle the dispute if it has not been resolved by the previous step, and each step then "escalates" the dispute management to a level above the previous step. An escalation clause typically requires each stage of the process to be engaged before the parties can move on to the next.
2 If drafted carefully, such clauses provide parties with a commercial and cost-effective dispute resolution mechanism. If drafted poorly, however, these clauses can lead to uncertainty – which in itself can give rise to a dispute – and, at worst, can leave the parties without a mechanism for proper recourse to the courts or arbitration.
3 Typically, escalation clauses involve as an initial step some form of internal/management resolution followed by a stage (or stages) of alternative dispute resolution ("ADR") (such as mediation) and conclude – as a last resort – with formal dispute resolution, either by legal proceedings or by arbitration. They can also include a provision for resolution by expert determination, which is particularly suited to technical disputes.

Other dispute resolution methods which may be adopted

Mediation

Mediation is a consensual method whereby the Parties jointly consent to refer a matter to mediation, and they also choose the Mediator who will act.

If the Parties cannot agree on the choice of an independent and impartial mediator, or if the chosen mediator is unable or unwilling to act, then either Party may immediately apply to an appointing entity such as a professional body or official named in the Contract Data to appoint a Mediator.

Once the Mediator has been appointed, the Dispute is immediately referred to the Mediator by the Parties or by each Party.

Obligations of the Parties, in addition to the Parties' obligations as set out in the provisions that follow below, during the mediation process the Parties shall engage with the Mediator and with each other in a spirit of co-operation, in a timely manner and in good faith.

Many Employers include within their contract for disputes to be referred to Mediation.

- Written submissions are made to the appointed Mediator and other parties.
- The Mediator may, by giving notice to both parties, visit the Site to inform himself in any way he thinks fit to inform himself of the nature and facts of the dispute.

- Similarly, the Mediator may call a meeting.
- The Mediator will prepare his recommendations and render his fee account to both parties, when paid in full he will deliver his recommendation to both parties.
- The Mediator shall not be appointed arbitrator on this, or any matter connected with this contract unless agreed by both parties.

The Mediator, within seven days of their appointment, or other period as may be proposed by the Mediator and agreed by both Parties, consults with the Parties to agree a timetable for the exchange of any relevant information, the date, time and venue of the Mediation meeting, and the procedure to be adopted during the meeting and generally for the negotiations.

The agreement of the timetable must have regard to the period stated under Sub-Clause 21.5 (Amicable Settlement) of the Conditions of Contract or as may be amended by the agreement of the Parties.

The Parties' negotiations facilitated by the Mediator are conducted on a strictly private and confidential basis. Neither Party shall disclose to any third party any detail of the mediation process, including but not limited to the fact of the mediation, the identity of the mediator, any matter discussed during the mediation process, any information or document exchanged with the mediator, any negotiations during the mediation meeting and/or the outcome of the mediation.

The mediation, and/or any negotiations taking place during the mediation meeting, shall not be referred to by either Party in any concurrent or subsequent proceedings, unless such negotiations conclude with a written legally binding agreement.

If the Parties accept the Mediator's recommendations, or otherwise reach agreement on the settlement of the Dispute, such agreement must be recorded in writing and, once signed by the designated representative(s) of both Parties, is binding on the Parties.

If no agreement is reached by the Parties after negotiations have been facilitated by the Mediator, either Party may invite the mediator to provide to both Parties a non-binding opinion in writing. Such opinion shall not be used in evidence in any concurrent or subsequent proceedings.

The Parties will bear their own costs of participating in the mediation, including but not limited to the costs of preparing and submitting evidence to the mediator and attending the mediation meeting.

Each Party is responsible for paying 50% of the remuneration of the Mediator (and, if the mediator has been appointed by the appointing entity or official named in the Contract Data, the remuneration of such appointing entity or official).

However, if the Mediator finds that a Party has initiated the mediation, or has engaged in the mediation, in a frivolous or vexatious manner, then the Mediator has the power to order that Party to pay the reasonable costs of the other Party for preparing and for attending the mediation meeting.

If these costs cannot be agreed, they will be assessed by the Mediator, whose assessment shall be binding on the Parties.

Adjudication

Many countries are considering introducing adjudication as a means to resolve construction disputes, in addition to or instead of a DAAB, and perhaps is worth a mention when speaking of alternative dispute resolution methods.

Adjudication is a quick and relatively inexpensive way of resolving a dispute, whereby an impartial third-party Adjudicator decides the issues between the parties.

It has been defined as "*a summary non judicial dispute resolution procedure that leads to a decision by an independent person that is, unless otherwise agreed, binding upon the parties for the duration of the contract, but which may subsequently be reviewed by means of arbitration, litigation or by agreement*", so disputing parties still have a right to refer a dispute to arbitration, litigation or of course they can form their own agreement.

It has been used very successfully within the UK for nearly 25 years, where it is also a statutory provision, with the force of law.

The role of the Adjudicator is to enforce the contract, for example, if the parties have a dispute over the quality of electrical installations being carried out, it is the Adjudicator's task to ascertain whether the electrical installations were installed <u>in accordance with the contract</u> i.e. drawings, specifications, etc., and the applicable law, not whether it was a good or bad job.

Adjudication does not necessarily achieve final settlement of a dispute because, as stated above, either of the parties has the right to have the same dispute heard afresh in court (or where the contract specifies arbitration, in arbitration proceedings).

Once a dispute has arisen between the parties either party may seek a resolution through adjudication. The Adjudicator is normally selected within 7 days and must decide the dispute within further 28 days (subject to any agreed extension).

Once the Adjudicator has made his decision, the other party must comply with it. If he does not, a court hearing to compel compliance can usually be obtained in a matter of days.

Adjudication is thus very quick in comparison with other methods of dispute resolution such as arbitration or litigation, and it can also be used during the currency of a contract. Adjudication can be a simple procedure, so in many cases there is no need to involve lawyers or other advisers. It is usually more cost-effective than arbitration or litigation.

Starting adjudication

The steps below are simple, but it is absolutely vital to get them right.

Chronologically, they are:

- Notice of Adjudication,
- Appointment of Adjudicator,
- Referral Notice.

Notice of Adjudication

Once a party is satisfied that they have a dispute arising under a construction contract, they can initiate adjudication by submitting a written Notice of Adjudication to the other party.

The notice must contain the following details:

- Nature and brief description of the dispute and the parties involved;
- When and where the dispute arose;
- Nature of the redress being sought;
- Names and addresses of the parties to the contract.

The Notice of Adjudication is an important document, as it defines what matters the Adjudicator has to decide.

Appointment of Adjudicator

All Adjudicators must be impartial, i.e. the Adjudicator should not unfairly regard with favour or disfavour the case of one of the parties, nor should there be any appearance that the Adjudicator might do so.

Any request for an Adjudicator must be accompanied by a copy of the Notice of Adjudication, and the appointment of the Adjudicator should take place within seven days of the submission of the Notice of Adjudication to the other party.

Alternatively, it may be possible to agree with the other party the name of an individual who should act as Adjudicator. However, sometimes parties are unwilling to agree anything once they are in dispute, and you may wish to balance how long you are likely to spend trying to reach agreement with how likely it is that agreement will be reached.

The Referral Notice

The next step is for the Referring Party to send a Referral Notice to both the Adjudicator and the other party.

The Referral Notice is the document which contains all the information that the Referring Party wishes the Adjudicator to consider.

It should:

- Be consistent with the Notice of Adjudication.
- Explain the nature of the dispute and how it arose.
- Detail the facts that you rely upon.
- Provide the documentary evidence to support those facts.
- Provide sufficient details of the contractual right to the remedy which is sought.
- Not include evidence that the other side has not seen before.
- List the decisions that the Referring Party requires the Adjudicator to make.

The Adjudicator then normally has 28 days from the date of your Referral Notice to make his decision.

The Adjudicator's decision

The Adjudicator's decision may be an order for the payment of money from one party to another, or it may relate to a disputed fact or technical matter (for example, whether the workmanship is to the right standard or what the specification means).

The Adjudicator may also decide that any of the parties to the dispute are liable to make a payment under the contract and decide when that payment is due and the final date for payment. The Adjudicator also has power to award interest (either simple or compound) on outstanding payments.

The Specification

Sub-Clause 1.1.76 of the Red Book (but not the Yellow or Silver Books) defines "Specification" as meaning "*the document entitled specification included in the Contract, and any additions and modifications to the specification in accordance with the Contract. Such document specifies the Works*".

Examples of entries into the Specification include:

- Requirements for inspections and tests,
- Details of rights of access to the Site,
- Employer Supplied Materials,
- Requirements for Contractor's Documents,
- Technical Standards,
- Requirements for as-built records,
- Requirements for operation and maintenance manuals,
- Requirements for training,
- Co-operation requirements,
- Activities of other contractors,
- Health and safety requirements,
- Requirement for Quality Management (QM) system/Compliance Verification System,
- Environmental Issues,
- Temporary utilities,
- Format of monthly progress reports,
- Naming of nominated Subcontractors,
- Any facilities provided by the Employer for staff and labour,
- Key Personnel,
- Programming software requirements,
- Review periods for any submissions,
- Taking over requirements,
- Clearance of Site requirements,
- Measurement requirements,
- Design requirements,
- IPR requirements.

The Drawings

Sub-Clause 1.1.30 of the Red Book (but not the Yellow or Silver Books) defines "Drawings" as meaning "*the drawings of the Works included in the Contract, and any additional and modified drawings issued by (or on behalf of) the Employer in accordance with the Contract*".

The Schedules

The Red, Yellow and Silver Books define "Schedules" as meaning

> the document(s) entitled Schedules prepared by the Employer and completed by the Contractor, as attached to the Letter of Tender and included in the Contract. Such document(s) may include data, lists and schedules of payments and/or rates and prices, and guarantees.

Examples of entries into the Schedules include:

- Bills of Quantities,
- Schedule of Payments.

The JV Undertaking (if applicable)

This is only relevant if the Contractor constitutes an unincorporated JV.

The Red, Yellow and Silver Books define "JV Undertaking" as meaning "*the letter provided to the Employer as part of the Tender setting out the legal undertaking between the two or more persons constituting the Contractor as a JV. This letter shall be signed by all the persons who are members of the JV, shall be addressed to the Employer and shall include*:

- a each such member's undertaking to be jointly and severally liable to the Employer for the performance of the Contractor's obligations under the Contract
- b identification and authorisation of the leader of the JV; and
- c identification of the separate scope or part of the Works (if any) to be carried out by each member of the JV".

The difference between an "incorporated" and a "unincorporated" JV is that the unincorporated JV is not organised as a separate legal entity, it is purely created by Contract and exists only for the purpose of performing the obligations of the Contract.

Any other documents forming part of the Contract

Obviously there may, or many not, be further documents forming part of the Contract. Note that any such documents take the lowest precedence within the Contract.

***Conditions of Contract for Plant and Design-Build* (Yellow Book) and *Conditions of Contract for EPC/Turnkey Project* (Silver Book) Only**

The Employer's Requirements

The Employer's Requirements should include all the relevant criteria that the Contractor is obliged to meet under the Contract.

This will include quality standards, any performance criteria which the Contractor's design, and its construction, must meet, together with associated testing to prove that the criteria have been met, but need not specify any matters which would be imposed on the Works by the applicable law.

Quality should be specified in terms that are not so detailed as to reduce the Contractor's design responsibilities, but not so imprecise as to be difficult to enforce, and not subject to anyone's future opinions which tenderers may consider impossible to forecast.

The Employer sets out his or her needs in whatever degree of detail is deemed appropriate within the "Employer's Requirements". This can be anything from a performance-related statement of objectives to a detailed and prescriptive statement of what the Contractor is to provide.

The Contractor responds with the "Contractor's Proposals", which set out the way he or she proposes to design and construct the Works.

The scale of design work required from the Contractor depends on the extent to which the Employer has already commissioned such work from others.

The Contractor's design input varies from one contract to another, ranging from the development of a schematic design provided by others into a workable final design for construction, to a full design process including proposals, sketch schemes and production information based on a schedule of requirements and/or performance criteria.

There will usually be some negotiation between the Employer and the Contractor, with the aim being to settle on an agreed set of Contractor's Proposals.

Alternatively, the Employer may approach directly a design and build Contractor who employs in-house consultants.

It is this inherent flexibility that is one of the strengths of design and build as a method of procurement. This flexibility happens not just between projects, but also within projects. For example, an Employer with a prestigious office building may have some detailed requirements, comprehensively designed by an architectural team, concerning the facade and the entrance lobby to the building. Once the Employer's Requirements and the Contractor's Proposals are consistent, the Contract can be executed, and the Contractor can implement the work.

The Contractor will be responsible for undertaking the design work in the Contractor's Proposals, including the appointment of consultants if the Contractor does not have the necessary skills in-house. As stated already, the Employer may also choose to appoint consultants to monitor the various aspects of the work, essentially as a "shadow" design team, but this is not always the case.

As the Employer's Requirements will be the primary document of reference it is important that this document is as complete as possible at the point of tender.

Conditions of Contract for Plant and Design-Build **(Yellow Book) Only: The Contractor's Proposal**

The Yellow Book defines "Contractor's Proposal" as meaning "*the part of the Tender stated or implied as being the Contractor's proposals for execution of the Works as included within the Contract. Such documents may include the Contractor's preliminary design*".

Note that the Contractor's Proposal does not form part of the Silver Book.

Note also that within the *Conditions of Contract for Construction* (The "Red Book") "The Contractor's Proposal" is included within the definition, but this has been rightly deleted by the Errata, as it is only applicable to the Yellow and Silver Books.

In simple terms, the Contractor's Proposals should reflect and directly respond to the Employer's Requirements.

The Engineer/Employer's Representative must check that the Employer's Requirements clearly, concisely and correctly interpret the Employer's brief. If the Employer's Requirements are detailed, the Contractor's Proposals (Yellow Book) will be detailed, if they are vague then the Contractor's Proposals will also be vague, so the Employer may not get what he or she expected.

Note that the Silver Book does not have Contractor's Proposals, the Contractor must execute the Works in accordance with the Employer's Requirements.

Contract Data

The Contract Data is the section in the Contract which in other contracts, including FIDIC 1999, is known as the Appendix to Tender.

It contains all the bespoke elements of the Contract specific to the Contract to be entered into.

Each item on the Contract Data is linked back to a relevant Clause/Sub-Clause in the General Conditions.

Some examples include:

- Percentage (%) for profit to be added to Cost wherever the contract refers to Cost plus Profit,
- Defects Notification Period,
- The parties' contact details,
- Language of the Contract,
- Limits of Liability,
- Performance Security requirements,
- Delay damages,
- Amount of Advance Payment,
- Payment requirements,
- Insurance Amounts,
- DAAB Details.

Provisional Sums

Provisional Sums are used in contracts where there are elements of work which are not designed or cannot be sufficiently defined at the time of tender, and therefore, a sum of money is included by the Employer within the Bill of Quantities or other pricing document to cover the item.

When the item is defined or able to be properly defined, the Contractor is given the information which allows him to price it, the Provisional Sum is omitted and the price included in its stead.

The problem with Provisional Sums is that they reduce the competition amongst tenderers as they are not priced at tender stage.

The Employer should ideally decide what it wants at tender stage so it can be designed and accurately described, and the tendering contractors can properly price and programme for them.

Building Information Modelling

The FIDIC 2017 Contracts include advisory notes to users where the project uses systems for Building Information Modelling (BIM).

Building Information Modelling or "BIM" is a process for creating, and also managing information digitally on a construction project across the project life cycle from design, through construction, operation, future maintenance and eventual demolition.

The BIM is the digital description of every aspect of the built asset, and is compiled from information assembled by collaboration and updated at key stages of a project, bringing all the information about every component to one location, so anyone who requires it can easily access it for design, construction, maintenance or eventual demolition.

BIM is designed to give everyone involved with all stages of an asset access to information relevant to the whole life cycle.

Co-ordination of goals and effort for all participants and users is essential and is generally achieved by a "BIM Protocol" and a "BIM Execution Plan", both key documents to access and understand.

BIM is founded on a team approach, and successful projects utilising BIM encourage collaboration.

It should be noted that the improved quality of information in projects utilising BIM can result in a significant reduction in Variations with the associated lessening of delays and other impacts upon the construction process and associated budgets.

At the completion of the project the model may need to be brought up to as-built status. This involves not only the drawing elements but also the embedded data. Experience shows that this is a significant effort, so responsibilities for completing this task should be clear and appropriate allowances provided.

FIDIC state that for construction or building projects involving BIM, in addition to considering the general principles as introduced above, the following (non-exhaustive) list of Sub-Clauses of the General Conditions of Contract for Construction (Red Book) should be thoroughly reviewed when drafting the Particular Conditions:

1.1	Definitions
1.3	Notices and Other Communications
1.5	Priority of Documents
1.10	Employer's Use of Contractor's Documents
1.11	Contractor's Use of Employer's Documents
1.15	Limitation of Liability
2.3	Employer's Personnel and Other Contractors
2.5	Site Data and Items of Reference
3.2	Engineer's Duties and Authority
4.1	Contractor's General Obligations
4.4	Contractor's Documents
4.6	Co-operation
4.7	Setting Out
4.9	Quality Management and Compliance Verification Systems
4.20	Progress Reports
6.8	Contractor's Superintendence
6.10	Contractor's Records
8.3	Programme
9.1	Contractor's Obligations (Tests on Completion)
13.3	Variation Procedure
17.3	Intellectual and Industrial Property Rights
17.4	Indemnities by Contractor
17.5	Indemnities by Employer
17.6	Shared Indemnities
19.2.6	Other insurances required by Laws and by local practice

There is a similar list of Sub-Clauses within the Yellow and Silver Books.

Partnering arrangements

The FIDIC contracts do not specifically provide for partnering arrangements, though they can be set up by including suitable amendments within the Contracts.

The past 20–25 years have seen the growth of partnering and framework agreements with the global construction industry.

The US Construction Industry Institute defines partnering as:

> A long-term commitment between two or more organisations for the purpose of achieving specific business objectives by maximising the effectiveness of each

participant's resources... the relationship is based upon trust, dedication to common goals and an understanding of each other's individual expectations and values.

Partnering is a medium- to long-term relationship between contracting parties, whereby the Contractor, and in turn Consultants and various other parties may not be required to tender for each project, but are awarded the work, normally by negotiation.

The key with partnering is that it is intended to be a collaborative arrangement between the parties as if they were one entity.

Advantages of partnering agreements are:

- No rebidding for future projects, a long-term arrangement can extend to several projects.
- Long-term relationships are developed and based on trust.
- Contractors are appointed earlier and can then contribute to the design and procurement process.
- There is greater cost certainty, particularly where a number of projects have been carried out by the same team.
- There is continuous improvement to process, time and quality by transferring learning from one project to another.
- There are improved working relationships.
- There is continuous workflow.
- Procurement is far quicker.

The construction industry is known to be a high risk business, and many projects can suffer unexpected cost and time overruns frequently resulting in disputes between the parties.

The risks within a project are initially owned by the Employer, who may choose to adopt a "risk transfer" approach where the risks are assigned through the Contract to the Contractor who has the opportunity to price and programme for them, or a "risk embrace" approach where the Employer retains the risks.

In reality, most contracts are a combination of the two. The traditional approach to risk management is that of risk transfer, which is fine if the scope of work is clear and well defined; however, in recent years clients have become more aware that they can achieve their objectives better by adopting a more "old fashioned" risk embrace culture.

EU Procurement Directives

Whilst not wishing this book to be geographically specific, it is worth some consideration of the European Union (EU) Procurement Directives that apply to all EU members procuring within the public sector, above minimum monetary thresholds.

These thresholds are reviewed on a regular basis, and subject to EU-wide principles of non-discrimination, equal treatment and transparency. The regulations affect government departments, local authorities and health authorities, and also utilities companies operating in the energy, water and transport sectors.

The purpose of the procurement rules is to open up the public procurement market and to ensure the free movement of supplies, services and works within the EU.

The rules are enforced by the Members States' Courts and the European Court of Justice, and require that all public procurement must be based on value for money, defined as "*the optimum combination of whole-life cost and quality to meet the user's requirement*", which should be achieved through competition, unless there are compelling reasons to the contrary.

In addition to the EU Member States, the benefits of the EU public procurement rules also apply to a number of other countries outside Europe because of an international agreement negotiated by the World Trade Organization (WTO) titled the "Government Procurement Agreement" (GPA).

The EU Procurement Directives set out the legal framework for public procurement when public authorities and utilities seek to acquire supplies, services or works (e.g. civil engineering or building), procedures which must be followed before awarding a contract when its value exceeds set thresholds which are reviewed on a regular basis, unless the contract qualifies for a specific exclusion, for example, on grounds of national security.

The Directives require contracting authorities to provide details of their proposed procurement in a prescribed format, which are then published in the *Official Journal of the European Union (OJEU)*.

All companies replying to an *OJEU* advertisement have an equal opportunity to express interest in being considered for tendering. As in all tendering exercises, clients must ensure that those companies selected to tender receive exactly the same information on which to make their bid.

The notice may be issued electronically, and this service may be provided through an intermediary organisation.

Generally, contracts covered by the Regulations must be the subject of a call for competition by publishing a Contract Notice in the *OJEU*.

The *OJEU* Notice process has associated minimum timeframes for receipt of tenders or requests to participate, dependent on procurement procedure, from the date the contract notice is sent.

The Pre-Tender Planning Stage

The Contractor Pre-Tender Planning Stage generally follows the following steps:

- Considering the procurement strategy (see above),
- Compiling the shortlist of tenderers,
- Drafting criteria for contractor selection,
- Preparing information for contract tender enquiries,
- Issuing tender documents,
- Receipt of tenders,
- Comparison of tenders,
- Contract award.

Each of the above stages should be dealt with in the form of a checklist so that all matters are dealt with.

Receipt of tenders to contract award

It is critical that tenderers are given sufficient time in which to prepare and submit their tenders.

The time and place for the delivery of tenders needs to be clearly stated. It should also be made clear to all tenderers that late tender returns are invalid and should not be considered, unless all are given a further opportunity to re-tender. Preferably, late tenders should be returned to the tenderer unopened.

The tender documentation should also make clear the period over which a tender may remain open for acceptance. This would normally be three months, giving the Client time to assess the tenders and make the appropriate award.

Opening tenders

Tenders should not be opened until after the date and time for submission to avoid any suspicion of collusion. No tenderer should be allowed to alter the terms of a tender after the closing time.

The opening of tenders in public, or in front of all the tenderers is not recommended, as it can lead to misunderstandings. If it must be done, the presence (but not the content) of any qualification should also be indicated.

After opening the tenders and supporting documents they should be evaluated. Each tender is confidential, and no details should be released to any other tenderer or to anyone with a financial interest in the tendering.

Tender Analysis

By carrying out detailed checks pre-tender, when tenders are received, the evaluation criteria should be very straightforward as all the tenders should be capable of being accepted, so it will normally just include any further information gained from the bid itself:

- Price,
- Contractor's ability to complete on time (though this should ideally have been established prior to issuing tenders for pricing),
- Contractor's ability to achieve quality standards (again this should ideally have been established prior to issuing tenders for pricing),
- The Contractor's proposed team and team structure.

Note that with Design and Build tenders, the comparison of bids is far more complex as the tenders are not being judged on the basis that each has priced the same project!

There has been a trend in the past 10–15 years for tenders to be accepted largely on the basis of <u>quality</u> elements of the bid, rather than on price alone.

A tender analysis should be prepared showing details of each of the tenders submitted including:

- Price submitted,
- Any exclusions or qualifications to the price submitted,
- Any arithmetical errors,
- Any technical issues such as design requirements.

There should be clear guidance on how to deal with late, incomplete or qualified bids.

The tender assessment report should present a clear and logically reasoned case for acceptance of the recommended tender. If the inclusion of technical detail is essential, sufficient explanation – in plain language – should also be included.

The following may be included in the report:

- A tabular statement of the salient features of all the tenders received; e.g. the name of the tenderer, the tender sum, whether valid (received by due date, complying with instructions for tendering, etc.), and qualifying conditions;
- Reference to any discussions that may have taken place with any tenderers regarding the tender;
- A concise summary of the findings following the examination and analysis of each tender, reasons for considering any tender invalid, discussion of the programme and methods of execution proposed, and reasons for considering any of these unsatisfactory;
- Comments on any rates, which appear exceptionally high or low, and forecasts of the possible effects on the contract price in cases where there are large differences;
- A comparison of the recommended tender sum with the cost plan or budget;
- Recommendation of the most acceptable tender;
- Recommendations for dealing with any qualifications in the recommended tender prior to acceptance.

Only after a tender has been accepted, should unsuccessful tenderers be advised of the name and tender price of the successful tenderer.

REMEMBER – THE CHEAPEST PRICE DOES NOT ALWAYS MEAN THE CHEAPEST COST!

Notification of tender results

It is important that all competing tenderers be informed of the results of their tender as soon as possible after the tender submission date. This particularly applies to those whose tenders are not successful, or which are not being given further consideration, as it allows them to concentrate on other opportunities with other Employers.

Generally, those who represent the Employer (but who are not actually the Employer), for example Consultants, have no authority to accept tenders, as ultimately the contract is between the Employer and the Contractor, so it is normal to recommend a choice to the Employer and allow him to make the final decision and inform the tenderers accordingly only if directly instructed by the Employer.

The representative of the Employer should submit a report to the Employer, which lists the tenders received, comments on them and recommends a particular one for acceptance. The scope and detail of this report will vary according to the circumstances.

Awarding the contract

Once the tenders have been assessed, and a decision made as to which tender is successful and the contract awarded, it is essential that the contract be executed as soon as practicable, and also that unsuccessful tenderers be advised accordingly.

At this stage, it is useful to follow a checklist approach to ensure that all the pre-construction issues are covered.

These will include:

- Confirm successful (and unsuccessful tenders).
- Issue letter of intent (if applicable).
- Prepare contracts for signing.
- Agree key dates, sequence of works and programme including integration with the project programme.
- Agree timing for issue of drawings or approval of design information.
- Confirm requirements for insurances.
- Agree or obtain Employer/Employer's Representative approval of subcontracts or packages where there is a contractual requirement to do so.
- Agree the provision of samples, sample panels and mock ups to be submitted for Employer approval.
- Agree the facilities to be provided by the Contractor, for example, cabins, stores, offloading, scaffolding, skips, removing rubbish, etc.
- Agree dates for interim certificates and procedure for requesting payment.
- Agree notification requirements for instructions, variations, dayworks, etc.
- Sign contract (not necessarily the last, and certainly not the least important thing to do, but check that it has been done).

In order to satisfy the test of offer and acceptance, the acceptance of a tender must be unconditional. So, a letter of acceptance should not introduce additional conditions like "your tender is accepted subject to our obtaining the necessary finance within one month".

Introducing new conditions such as this would, under most legislations, be a counter offer, cancelling the original offer and in turn requiring acceptance by the tenderer.

While some negotiation between the parties is not unusual, eventually an unconditional acceptance of a current offer (or counter-offer) is needed.

Letter of Acceptance of tender

This is a letter from the Employer to the Contractor accepting the offer. If negotiations between Employer and Contractor take place before a tender is finally accepted, copies of any other letters giving evidence of the conditions agreed to should also be incorporated in the documents. In order to form a contract, there must be a well-defined offer to which an unconditional acceptance has been given.

Formalising the contract

In some legislations, there is not a strict requirement to sign a contract before it comes into effect, and the Contractor can commence work, but it is clearly something that needs to be done as soon as possible.

It is important to ensure that the signed documents incorporate all the documentation on which tenders and the final offer and acceptance are based i.e. they form the "entire agreement".

At the time of signing of the agreement, the Parties should initial every page of all such other documents, together with initialling any amendments, which may have been made to printed documents.

When arranging the signing of contract documents, it is critical that a check be made that the signatories have legal authority to make contracts on behalf of their organisations, and that the formalities required by the organisation's rules are followed.

Also, companies, incorporated societies or statutory bodies, typically, may require the placing of the organisation's seal and two authorised signatures.

Appendix

Errata to the FIDIC Conditions of Contract 2017

Important Note

Readers who may be reading this book alongside a copy of a FIDIC 2017 Contract should be aware that, following publication of the *Conditions of Contract for Construction* (The "Red Book"), the *Conditions of Contract for Plant and Design-Build* (The "Yellow Book") and the *Conditions of Contract for EPC/Turnkey Project* (The "Silver Book"), FIDIC published an Errata citing various significant changes/errors which had not been included within their published contracts.

Several minor typographical errors and layout irregularities had also been found but were not included in the Errata due to their insignificance with regard to the content of the contracts.

The author has taken into account the Errata (where applicable) when writing this book, and has indicated where text has changed from the published Contract, for example, in the *Conditions of Contract for Construction* (The "Red Book"):
Page 1 Sub-Clause 1.1.10: On the third and fourth lines, delete "the Contractor's Proposal"………a note has been made within Chapter 10 on Tendering.

Conditions of Contract for Construction (The "Red Book")

General Conditions

- Page 1 Sub-Clause 1.1.10: On the third and fourth lines, delete "the Contractor's Proposal".
- Page 6 Sub-Clause 1.1.77: On the second line after "Payment Certificate under", add "Sub-Clause 14.2.1 [Advance Payment Guarantee] (if applicable),".
- Page 7 Sub-Clause 1.1.81: On the first line, delete "the Contractor's Proposal,".
- Page 28 Sub-Clause 4.6: On the second-last line of the first paragraph before "Contractor's", add "of the".
- Page 29 Sub-Clause 4.7.3: In the second bullet-point of sub-paragraph (b), before "if the items of reference", add "when examining the items of reference within the period stated in sub-paragraph (a) of Sub-Clause 4.7.2," and on the second and third lines, delete "and the Contractor's Notice is given after the period stated in sub-paragraph (a) of Sub-Clause 4.7.2".
- Page 37 Sub-Clause 4.22: On the third line of the second paragraph before "4.17", add "Sub-Clause".

- Page 38 Sub-Clause 5.2.2: In sub-paragraph (a) on the first line before "Subcontractor", add "nominated".
- Page 70 Sub-Clause 14.2.1: On the fifth and sixth lines of the first paragraph, replace "based on the sample form included in the Tender documents" with "in the form annexed to the Particular Conditions".
- Page 78 Sub-Clause 14.12: On the seventh line of the first paragraph, replace "Sub-Clause 21.6 [Arbitration]" with "Clause 21 [Disputes and Arbitration]".
- Page 87 Sub-Clause 17.1: On the fourth and fifth lines of the first paragraph, replace "Date of Completion of the Works" with "issue of the Taking-Over Certificate for the Works".
- Page 88 Sub-Clause 17.3: On the first line of the second paragraph, replace "notice" with "a Notice".
- Page 94 Sub-Clause 19.2.1: On the last line of the second paragraph, delete "Clause 12 [Tests after Completion]".

Appendix – General Conditions of Dispute Avoidance/Adjudication Agreement

- Page 107 Title: Replace "General Conditions of Dispute Avoidance/Adjudication Agreement" with "General Conditions of DAAB Agreement".
- Page 107 Sub-Clause 1.2: On both the first and third lines, replace "DAA Agreement" with "DAAB Agreement".
- Page 107 Sub-Clause 1.3: On the first line, replace ""Dispute Avoidance/Adjudication Agreement" or "DAA Agreement" means"" with ""DAAB Agreement" is as defined under the Contract and is"; on the first line of sub-paragraph (c), replace "DAA Agreement" with "DAAB Agreement"; in sub-paragraph (c) (ii), replace "chairman" with "chairperson".
- Pages 107 to 115 Sub-Clause 1.7 to Clause 12: Replace all instances of "DAA Agreement" with "DAAB Agreement".
- Page 115 Sub-Clause 11.1: On the second line, delete the text: ", or in the case of a three-member DAAB the Other Members jointly,".

Annex – DAAB Procedural Rules

- Page 119 Rule 4.2: On the fourth line, replace "chairman" with "chairperson".
- Page 121 Rule 8.3: On the sixth line, replace "chairman" with "chairperson".

Guidance for the preparation of Particular Conditions

- Page 8 INTRODUCTION Insert in the beginning of GP5: "Unless there is a conflict with the governing law of the Contract, …" allowing GP5 to read "GP5: Unless there is a conflict with the governing law of the Contract, all formal disputes must be referred to a Dispute Avoidance/Adjudication Board (or a Dispute Adjudication Board, if applicable) for a provisionally binding decision as a condition precedent to arbitration".

- Pages 16 to 47 Guidance for Sub-Clauses 1.6, 1.14, 4.2, 14.2, 14.9, 14.15 and 21.1: Replace all instances of "Sample Forms" with "Forms".
- Page 35 Guidance for Sub-Clause 9.1: On the second line of the first paragraph, replace "Contractor's Proposal" with "Tender".

Form of Letter of Acceptance

- Page ii Footnote: Replace "1.1.51" with "1.1.50".

Form of Dispute Avoidance/Adjudication Agreement

- Pages iv and v: Replace all instances of "DAA Agreement" with "DAAB Agreement".
- Page iv Recital C: On the first line of sub-paragraph (c), replace "chairman" with "chairperson".

Conditions of Contract for Plant and Design-Build (The "Yellow Book")

Explanatory Charts

First page, item 2: Replace "above diagram" with "diagram below".

General Conditions

- Page 4 Sub-Clause 1.1.42: On the third line, replace "Specification" with "Employer's Requirements".
- Page 6 Sub-Clause 1.1.69: On the second line after "payment under", add "Sub-Clause 14.2.1 [Advance Payment Guarantee] (if applicable),".
- Page 6 Sub-Clause 1.1.74: On the second line, replace "Specification" with "Employer's Requirements".
- Page 7 Sub-Clause 1.1.78: On the first line, replace "any change to the Works" with "any change to the Employer's Requirements or the Works".
- Page 24 Sub-Clause 4.5.1: In sub-paragraph (a) on the first line before "Subcontractor", add "nominated".
- Page 25 Sub-Clause 4.6: On the second-last line of the first paragraph before "Contractor's", add "of the".
- Page 32 Sub-Clause 4.22: On the third line of the second paragraph before "4.17", add "Sub-Clause".
- Page 37 Sub-Clause 6.1: On the first line, replace "Specification" with "Employer's Requirements".
- Page 65 Sub-Clause 13.6: In the fourth paragraph, at the end of sub-paragraph (ii), delete "(with detailed supporting particulars)", and at the end of this fourth paragraph, add "(with detailed supporting particulars)".
- Page 67 Sub-Clause 14.2.1: On the fifth and sixth lines of the first paragraph, replace "based on the sample form included in the Tender documents" with "in the form annexed to the Particular Conditions".

- Page 75 Sub-Clause 14.12: On the seventh line of the first paragraph, replace "Sub-Clause 21.6 [Arbitration]" with "Clause 21 [Disputes and Arbitration]".
- Page 84 Sub-Clause 17.1: In the first paragraph, on the fourth and fifth lines, replace "Date of Completion of the Works" with "issue of the Taking-Over Certificate for the Works"; on the seventh line, delete "or Part"; on the last line, delete "or Part".
- Page 85 Sub-Clause 17.3: On the first line of the second paragraph, replace "notice" with "a Notice".
- Page 86 Sub-Clause 17.4: On the third line of the last paragraph, delete "or Part".
- Page 91 Sub-Clause 19.2.3: In the first paragraph, on the first line, delete ", if any," and replace "part of the Permanent Works under Sub-Clause 4.1 [Contractor's General Obligations], and/or any other design under the Contract" with "the Permanent Works".

Appendix – General Conditions of Dispute Avoidance/Adjudication Agreement

- Page 103 Title: Replace "General Conditions of Dispute Avoidance/Adjudication Agreement" with "General Conditions of DAAB Agreement".
- Page 103 Sub-Clause 1.2: On both the first and third lines, replace "DAA Agreement" with "DAAB Agreement".
- Page 103 Sub-Clause 1.3: On the first line, replace ""Dispute Avoidance/Adjudication Agreement" or "DAA Agreement" means" with ""DAAB Agreement" is as defined under the Contract and is"; on the first line of sub-paragraph (c), replace "DAA Agreement" with "DAAB Agreement"; in sub-paragraph (c) (ii), replace "chairman" with "chairperson".
- Pages 103 to 111 Sub-Clause 1.7 to Clause 12: Replace all instances of "DAA Agreement" with "DAAB Agreement".
- Page 110 Sub-Clause 11.1: On the second line, delete the text: ", or in the case of a three-member DAAB the Other Members jointly,".

Annex - DAAB Procedural Rules

- Page 114 Rule 4.2: On the fourth line, replace "chairman" with "chairperson".
- Page 116 Rule 8.3: On the sixth line, replace "chairman" with "chairperson".
- Page 118 After Rule 10, insert Rule 11: "Rule 11 Challenge Procedure 11.1 If and when the objecting Party challenges a DAAB Member, within 21 days of learning of the facts upon which the challenge is based, the provisions of this Rule shall apply. Any challenge is to be decided by the International Chamber of Commerce (ICC) and administered by the ICC International Centre for ADR. 11.2 The procedure for such challenge and information on associated charges to be paid are set out at http://fidic.org and http://iccwbo. org."

Guidance for the preparation of Particular Conditions

- Page 8 INTRODUCTION Insert in the beginning of GP5: "Unless there is a conflict with the governing law of the Contract, ..." allowing GP5 to read "GP5: Unless there is a conflict with the governing law of the Contract, all formal disputes must be referred to a Dispute Avoidance/Adjudication Board (or a Dispute Adjudication Board, if applicable) for a provisionally binding decision as a condition precedent to arbitration".

- Pages 16 to 50 Guidance for Sub-Clauses 1.6, 1.14, 4.2, 14.2, 14.9, 14.15 and 21.1: Replace all instances of "Sample Forms" with "Forms".

Form of Dispute Avoidance/Adjudication Agreement

- Pages iv and v: Replace all instances of "DAA Agreement" with "DAAB Agreement".
- Page iv Recital C: On the first line of sub-paragraph (c), replace "chairman" with "chairperson".

Conditions of Contract for EPC/Turnkey Project (The "Silver Book")

General Conditions

- Page 5 Sub-Clause 1.1.48: On the second line, replace "Specification" with "Employer's Requirements".
- Page 7 Sub-Clause 1.1.79: On the second line after "Payment Certificate under", add "Sub-Clause 14.2.1 [Advance Payment Guarantee] (if applicable),".
- Page 7 Sub-Clause 1.1.84: On the second line, replace "Specification" with "Employer's Requirements".
- Page 7 Sub-Clause 1.1.88: On the first line, replace "any change to the Works" with "any change to the Employer's Requirements or the Works".
- Page 27 Sub-Clause 4.5.1: In sub-paragraph (a) on the first line before "Subcontractor", add "nominated".
- Page 28 Sub-Clause 4.6: On the second-last line of the first paragraph before "Contractor's", add "of the".
- Page 37 Sub-Clause 4.22: On the third line of the second paragraph before "4.17", add "Sub-Clause".
- Page 42 Sub-Clause 6.1: On the first line, replace "Specification" with "Employer's Requirements".
- Page 64 Sub-Clause 12.1: In sub-paragraph (b) (ii), on the second line, delete "been".
- Page 71 Sub-Clause 13.6: In the fourth paragraph, at the end of sub-paragraph (ii), delete "(with detailed supporting particulars)", and at the end of this fourth paragraph, add "(with detailed supporting particulars)".
- Page 73 Sub-Clause 14.2.1: On the fifth and sixth lines of the first paragraph, replace "based on the sample form included in the Tender documents" with "in the form annexed to the Particular Conditions".
- Page 81 Sub-Clause 14.12: On the seventh line of the first paragraph, replace "Sub-Clause 21.6 [Arbitration]" with "Clause 21 [Disputes and Arbitration]".
- Page 90 Sub-Clause 17.1: On the fourth and fifth lines of the first paragraph, replace "Date of Completion of the Works" with "issue of the Taking-Over Certificate for the Works".
- Page 91 Sub-Clause 17.3: On the first line of the second paragraph, replace "notice" with "a Notice".

- Page 97 Sub-Clause 19.2.3: On the first line of the first paragraph, replace "the part of the Permanent Works under Sub-Clause 4.1 [Contractor's General Obligations], and/or any other design under the Contract" with "the Permanent Works".

Appendix – General Conditions of Dispute Avoidance/Adjudication Agreement

- Page 110 Title: Replace "General Conditions of Dispute Avoidance/Adjudication Agreement" with "General Conditions of DAAB Agreement".
- Page 110 Sub-Clause 1.2: On both the first and third lines, replace "DAA Agreement" with "DAAB Agreement".
- Page 110 Sub-Clause 1.3: On the first line, replace " "Dispute Avoidance/Adjudication Agreement" or "DAA Agreement" means" with " "DAAB Agreement" is as defined under the Contract and is"; on the first line of sub-paragraph (c), replace "DAA Agreement" with "DAAB Agreement"; in sub-paragraph (c) (ii), replace "chairman" with "chairperson".
- Pages 110 to 119 Sub-Clause 1.7 to Clause 12: Replace all instances of "DAA Agreement" with "DAAB Agreement"
- Page 118 Sub-Clause 11.1: On the second line, delete the text: ", or in the case of a three-member DAAB the Other Members jointly, ".

Annex – DAAB Procedural Rules

- Page 122 Rule 4.2: On the fourth line, replace "chairman" with "chairperson".
- Page 124 Rule 8.3: On the sixth line, replace "chairman" with "chairperson".

Guidance for the preparation of Particular Conditions

- Page 8 INTRODUCTION Insert in the beginning of GP5: "Unless there is a conflict with the governing law of the Contract, ..." allowing GP5 to read "GP5: Unless there is a conflict with the governing law of the Contract, all formal disputes must be referred to a Dispute Avoidance/ Adjudication Board (or a Dispute Adjudication Board, if applicable) for a provisionally binding decision as a condition precedent to arbitration".
- Pages 15 to 51 Guidance for Sub-Clauses 1.6, 1.14, 4.2, 14.2, 14.9, 14.15 and 21.1: Replace all instances of "Sample Forms" with "Forms".

Form of Letter of Acceptance

- Page ii Footnote: Replace "1.1.51" with "1.1.50".

Form of Dispute Avoidance/Adjudication Agreement

- Pages iv and v: Replace all instances of "DAA Agreement" with "DAAB Agreement".
- Page iv Recital C: On the first line of sub-paragraph (c), replace "chairman" with "chairperson".

Index

acceleration 163, 199
acceptance 84, 174, 241, 244–245, 272, 274, 282, 301–303; formal 278; letter of 51, 57, 59, 65, 69, 223–224, 272, 274–278, 303, 307, 310
Accepted Contract Amount 62, 67, 82, 93, 102, 108, 114, 162, 170, 181, 286
access arrangements 97, 208
access restrictions 150
access routes 62, 104, 235, 237
accident prevention officer 111
Additional coverage for breach of warranty 226
additional payment 45, 72, 164, 240, 242–246, 248, 287
ad-hoc DAAB 288
adjudication 43, 234, 270, 291–293
Adjudicator 46, 292–293
ADR 290, 308
advance payment 170, 175, 274, 297
Advance Payment Guarantee 170, 173, 272, 274, 305, 307, 309
advance warnings 188, 190–192
advantages of design and build 143
Agreed Final Statement 178
ambiguities 59, 160, 275, 277
Amicable Settlement 259, 264–265, 289, 291
Appendix 40, 43–46, 58–61, 63, 65, 69, 71, 170, 174, 256, 305–306, 308, 310
Appendix to Tender 80, 275–276, 280
application for interim payment 171
application of laws 270
Appoint DAAB Member 256
Appointment of Adjudicator 292–293
arbitration 75–76, 89, 177–179, 216, 231, 239, 249, 251–253, 255–256, 258–265, 279–280, 288–290, 292, 306, 308–310; commencement of 259; international 259, 264; multi-party 260; place of 259
Arbitration Acts 260
arbitrator 75, 205, 252–253, 255, 258–264; appointed 291
Arbitrator's Costs 263
Arbitrator's fees and expenses 263
Archaeological and Geological Findings Clause 237
as-built programme 194; collapsed 194
as-built records 63–64, 96, 157, 159, 285, 294
As-Planned *v* As-Built 194
assignment of benefit of subcontract 96
assistants 84, 86–87, 283
audits, internal 100–101

bar charts 183–184
Base Date 53–54, 61, 68, 82, 101–102, 104, 158–159, 162, 168, 180, 215
baseline programme 183; as-planned 194
Beneficiary 273
bids 55, 149, 151, 300–301
Bills of Quantities 41, 67, 91, 144, 146, 162, 169, 269, 287, 294
BIM (Building Information Modelling) 297–298
Bond Amount 273
breach of contract 106, 164, 195, 200, 211, 229–230, 242, 251
Building Control/Building Regulation Approval 147
Building Information Modelling *see* BIM
building permit 168

care and supply of documents 96, 114
care of the Works 213–228, 236, 238, 279–280
CEMAR 282
certificate 58–59, 67, 69, 84, 115, 123, 131, 137, 159, 174, 182; interim 303; performance 66
changes in cost 171
changes in Laws 168
Chartered Institute of Arbitrators 260
claims 41–43, 45, 60, 65–66, 72–75, 87–89, 92–93, 103, 124–125, 164, 169–170, 175, 180, 182–183, 191, 193, 201–202, 214–216, 223–234, 238–256
clearance of Site 138
climactic conditions 53, 65, 101–102, 196, 235, 237

Commencement Date 51, 54, 98–100, 131, 133, 167, 169, 172, 181, 185–186, 207, 209, 211, 213–214, 284
Commencement of Works 181, 211
communications 43, 58, 85, 94, 100, 114–115, 241, 252, 282, 298; contractual 282; effective 90; electronic 282; language for 59, 95, 111, 113, 116, 157, 167
Completion 50, 56, 66–70, 74, 106–107, 118, 127–129, 131–132, 139–141, 159–160, 163, 175–177, 181–202, 206, 214, 226–227, 232, 235–238, 273, 279–280, 284–285
Completion of outstanding work and remedying defects 133–134, 141
Compliance Verification Systems 100–101, 298
concept/design stage 83
concurrent delays 197–199
Conditions of Contract for Construction 41, 47, 56–58, 118, 142, 203, 213, 269, 274, 296, 305
Conditions of Contract for EPC/Turnkey Project 42, 46–47, 57–58, 78, 203, 213, 269, 276, 295, 305, 309
Conditions of Contract for Plant-Design and Build 41, 46, 51, 57–58, 203, 213, 269, 275, 295–296, 305, 307
Consequences of Employer's Suspension 235–238
Consequences of Force Majeure 75
Consequences of Suspension 73
Contract Agreement 41, 51, 57, 59, 170, 211, 272–278, 282
Contract Data 51, 54, 79–81, 105–106, 108, 114–116, 166–167, 170–172, 175–177, 180–181, 186, 222–223, 225–228, 257, 273, 275–277, 280, 284–286, 288–291, 296; associated 223
Contract Date 188, 218
Contract Price 41, 43, 45, 48, 50, 56, 66, 68–69, 92–93, 103, 143, 145, 165–166, 168–170, 180, 189, 239–240, 243–246, 248, 279–280
contract strategy 287
Contractor 40–56, 59–88, 90–117, 119–189, 191–193, 195–244, 246–249, 251–253, 257, 259, 268–269, 272–288, 294–299, 301–303, 305
Contractor access 72
Contractor appointment stage 147
Contractor Delay 134, 194, 198
Contractor-designed Works 58, 61, 91, 269
Contractor Pre-Tender Planning Stage 300
Contractor's access routes 104
Contractor's Claims 60, 88, 193, 229–254, 256, 279–280
Contractor's Claims and Employer's Claims 239
Contractor's design 41, 91–92, 126, 142, 144, 147, 154–156, 160, 164, 268–269, 295
Contractor's Documents 63, 80, 85, 90, 92, 94–95, 114, 116–117, 137–138, 153–154, 156–158, 160, 206–208, 212–215, 298
Contractor's entitlement to extensions of time 196
Contractor's Equipment 81, 97, 105, 108, 112, 138, 186, 208, 215, 222; initial 206
Contractor's fitness for purpose obligation 163
Contractor's liability for correction of latent defects 124
Contractor's offer 278
Contractor's Overheads 233
Contractor's parent company 272
Contractor's Personnel 64, 94, 104, 107, 110–112, 154, 157, 187, 189, 219, 228
Contractor's Programme 65, 80, 182, 185, 187
Contractor's progress and reports 107, 197, 199, 288
Contractor's Proposals 51, 57, 59, 126, 133, 149–151, 153, 165–166, 187, 276, 295–296
Contractor's quality assurance certification 101
Contractor's Records 107, 112, 114, 252–253, 298
Contractor's Representative 61, 84–85, 90, 94, 112, 115, 150, 161, 173, 241
Contractor's Representative and Key Personnel 112
Contractor's responsibility for care of the Works 46, 70
Contractor's staff and labour 208, 222
Contractor's Superintendence 111, 298
Contractor's test certificate 126
Contractor to Search 74, 137, 236, 238
contractual claims 230–231, 243
cost index 169
Cost of claim preparation 233
Cost of remedying defects 133–134
Cost plus Profit 75, 234, 239–240
COVID-19, 219
CPM 184
critical path 184, 185, 187, 193, 196–197

DAAB (Dispute Avoidance/Adjudication Board – FIDIC 2017) 137–138, 171, 177–179, 204–205, 209–210, 249, 251, 253, 255–258, 264–265, 288–289, 291, 306, 309
DAAB Agreement 256–257, 306–310
DAAB fees 257
DAAB members 257, 308
DAAB Procedural Rules 257, 288, 306, 308, 310
DAAB's decision 177, 251, 256–259, 264–265
DAAB's role 257
DAB (Dispute Adjudication Board – FIDIC 1999) 40–41, 75–76, 171, 255–265, 280, 288–289, 306, 309–310
Date of Completion 106, 134, 152, 163, 176–177, 193, 214

dayworks 167, 247, 303
DBO *see* Design, Build and Operate
Decennial Liability 288
defect detection 119–120
defect prevention 120
defective work 49, 124, 135
defects 51–52, 54, 64, 66–67, 90, 114, 118–119, 122, 124, 129–131, 133–138, 155–156, 160, 214, 226, 279–280
Defects Liability 49, 56, 66, 118–141
Defects Notification Period 243, 245; *see also* DNP
defining quality 121
delay analysis, retrospective 193
delay damages 131–132, 180–181, 193, 199–201, 205, 207, 216, 267, 271, 284, 286
delayed tests 74, 128, 140, 236, 238
Delays in receipt of Design Information 232
delegation 60–61, 84, 86, 94, 283
design 41–44, 46–47, 61, 63, 76–78, 83–84, 91–92, 95, 142–149, 151–155, 157–158, 214–215, 226–227, 267–268, 270–271, 279, 295–297
Design, Build and Operate (DBO) 218, 257
design and build 142–147, 149, 151, 267
Design and Build process 142, 144
Design-Build Contract 57, 76
design life 163
design responsibility 47, 50, 78, 154, 186, 217, 269
design responsibility and liability 142–160
details of rights of access 294
determination 41–42, 86, 88–89, 93, 98, 103, 174–175, 179, 204, 208–210, 240, 242–245, 247–250
Dispute Adjudication Agreement 40
Dispute Adjudication Avoidance Agreement 272
Dispute Adjudication Board *see* DAB
Dispute Avoidance/Adjudication Agreement 256–257, 306–310
Dispute Avoidance/Adjudication Board *see* DAAB
Dispute Avoidance Board 289–290
disputes 41, 43, 46, 75–76, 87, 89, 178–179, 191, 207, 230–231, 239–240, 251–253, 255–265, 270, 288–293
disputes and arbitration 69, 72, 89, 93, 239, 249, 255–256, 279–280, 288, 306, 308–309
DNP (Defects Notification Period) 45, 66–67, 69, 96, 99, 124, 132–135, 137–138, 140–141, 175–176, 234, 240–241, 244, 246, 248
drawings 52, 57, 59, 81–82, 92, 98, 114, 116, 120–121, 123, 154, 215, 275, 294; issue of 90, 303

Early Contractor Involvement (ECI) 144
Eichleay Formula 233
electronic transmissions 58, 85, 115, 241
Elemental Trend Analysis 183–184
Emden Formula 233
Employer 40–50, 52–56, 59–62, 64–85, 87–88, 90–99, 101–111, 113–117, 124–160, 162–181, 184–189, 191–212, 214–217, 219–254, 256–260, 266–269, 272–274, 277–288, 294–299, 302–303
Employer/Consultant 200
Employer/Employer's Representative 99, 182, 264, 284, 303
Employer Risk Event 80, 197
Employer's Claims 60, 75, 234, 239–241, 247
Employer's Convenience 40, 85, 113, 117, 207–209, 212, 216
Employer's Equipment 62, 82, 106, 114, 217
Employer's Financial Arrangements 209–210
Employer's Liabilities 43–46, 69, 71, 179–180, 217, 224
Employer's Personnel 63–64, 81, 95–97, 107, 110–111, 114, 125–126, 128–129, 132–133, 137, 155, 157, 159, 215, 217
Employer's Representative 40, 43–44, 50, 52, 75, 79, 82–85, 88–89, 143, 150, 221–222, 244–251, 283–284
Employer's Requirements 50–52, 54–55, 57–59, 63, 72, 77, 90, 127–128, 138–140, 142–153, 155–160, 186, 275–277, 295–296, 307, 309
Employer's Risks 70–71, 268
Employer's Taking Over 56, 66, 118, 129, 140
Employer's use of Contractor's Documents 80, 298
Engineer 40–42, 50–56, 60–69, 71–73, 75–79, 82–90, 92–101, 103–117, 124–140, 154–179, 181–183, 185–189, 191, 199, 201–206, 208–210, 221–222, 240–251, 258, 282–284
Engineer/Employer's Representative 79, 82–86, 91, 123–124, 182, 193, 229–230, 234, 255, 283, 296
Engineering, Procurement and Construction (EPC) 47, 50–51, 78
Engineer's Determination 89, 249, 255–256
Engineer's Duties and Authority 298
Engineer's/Employer's Instructions 283
environmental law provisions 284
EOT *see* Extension of Time
EPC (Engineering, Procurement and Construction) 47, 50–51, 78
EPC Contracts 47–50, 57, 76, 78, 277
EPC/Turnkey Project 42, 46–47, 57–58, 78, 203, 213, 269, 276, 279, 305, 309
equipment 42, 62, 67, 77–78, 105–107, 123, 139–140, 187, 224, 283; major items of 105
Errors in the Employer's Requirements 235–236
example COVID-19 219
Example Form of Advance Payment Guarantee 272, 274

Example Form of Parent Company Guarantee 272
Example Form of Payment Guarantee by Employer 272, 274
Example Form of Performance Security 272–273
Example Form of Retention Money Guarantee 272, 274
Example Form of Tender Security 272–273
Example of Schedule of Advance Warnings 192
Excepted Risks 228
Exceptional Events 54, 127, 213, 215, 218–222, 227, 231, 235–238, 271, 279–280, 287
expert witnesses 261
Extension of Defects Notification Period 134
Extension of Time (EOT) 41–42, 44, 52–55, 59–60, 65, 72–75, 80, 97–99, 103–104, 125–126, 158, 164–165, 168, 188, 193–199, 202, 220–221, 229, 234–237, 240, 246

failure to pass tests 74, 141, 236, 238
failure to pass Tests on Completion 129
failure to remedy defects 126, 134
faith, bad 271
Federation International des Ingenieurs-Conseils *see* FIDIC
FIDIC (Federation International des Ingenieurs-Conseils) 40–81, 83–88, 99–100, 110–114, 124–125, 134–139, 153–156, 163–169, 171–174, 176–183, 185–189, 196–197, 202–207, 210–211, 213, 216–219, 222–224, 239–245, 255–259, 266–305
FIDIC Conditions of Contract 305–310
FIDIC contracts 42, 76, 83, 86, 181–182, 193, 195–197, 200–201, 203, 259, 267–268, 270, 272, 280–281, 283–284
Final Payment 179, 210
Final Payment Certificate *see* FPC
Final Statement 69, 177–179; draft 69, 177–179
float 185, 187; free 185; terminal 185
fluctuations 169, 185, 211, 232
Fluctuations in cost of Labour 232
Force Majeure 50, 57, 70–72, 75, 218, 271–272, 287
Force Majeure events 75, 272
fossils 62, 73, 108
FPC (Final Payment Certificate) 67, 69, 94, 138, 175, 178–180, 208
Free-Issue Materials 62, 106

Gantt chart 183
General Conditions 51, 57–59, 256–257, 274–278, 280–281, 283–284, 288, 296, 305, 307, 309
General Conditions and Particular Conditions 281
General Conditions of DAAB Agreement 256, 306, 308, 310
General Conditions of DAB Agreement 40
General Design Obligations 52, 54, 82, 155–156
Gold Book 218, 257
Green Book 42–43, 76, 285
ground conditions 50; unforeseeable 77
Guarantor 273–274

Head Office Overheads 233
Henry Laurence Gantt 183
Hudson's Formula 233
hydrological 46, 50, 53, 82, 102

ICC (International Chamber of Commerce) 289, 308
Impacted As-Planned 194
Indemnities 114, 213–228, 279–280, 298; considering 216
Indemnities by Contractor 114, 148, 216–217, 298
injury to persons and damage to property 227
inspections 101, 119, 122–123, 125, 135, 172, 186, 263, 294; appropriate 128; quantified 120; reasonable 124; timing of 186
instructions 44–45, 53–54, 59–60, 72–73, 84–85, 87, 90, 97–98, 115, 122, 126–127, 137, 163–167, 202, 205–206, 212, 215–216, 235–238, 302–303; oral 283; reasonable 206
Insurance Amounts 297
insurances 43, 46, 71, 173, 182, 213, 222–228, 271, 279–280, 288, 298
Insuring Party 71, 222–223
Intellectual and Industrial Property Rights 114, 215, 217, 298
interim payment 69, 78, 171, 173–174, 231; certified 170
Interim Payment Certificate(s) 69, 127, 170, 173, 210; *see also* IPC
International Chamber of Commerce (ICC) 289, 308
International Court of Arbitration 259
International Quality Standard ISO 120
IPC (Interim Payment Certificate) 69, 110, 127, 170–171, 173–177, 179, 210
ISO standards 77, 100, 120
Issue of FPC 208
Issue of IPC 177

Joint Agreement 260
Joint Names Insurance 223
Joint Venture *see* JV
JV (Joint Venture) 47, 51, 113, 205, 275–277, 282, 295
JV leader 113
JV Undertaking 51, 275–277, 295

Key Personnel 112–113, 283, 294

latent defects 67, 124, 138, 145
Laws 43, 59–60, 81, 99, 110–111, 116, 138, 167–168, 219–220, 222, 228, 235–238, 260–261, 270–271, 273; adjustments for changes in 167, 169–170, 235–238; applicable 58–59, 63, 105, 110, 122, 124, 149, 155, 158, 270, 292, 295
legislations 68, 75, 175, 180, 200–201, 204, 209, 212, 217, 260–261, 303; adjustments for changes in 74, 167, 271; relevant 40, 138, 151, 197, 208, 284
Liability for breach of professional duty 226
Liability for Care of the Works 214, 217, 236, 238
Limits of Liability 297
liquidated damages/delay damages 49, 65, 182, 193, 195, 200–201

materials 41–42, 68–69, 100–101, 107–108, 121–127, 134–135, 139–140, 151–153, 170–173, 202, 207–208, 221, 223–225, 247, 263, 282
Materials and Workmanship 56, 64, 118, 278–279
MDB Harmonised Edition 75
Measurement and Payment Procedures 161–180
measurement records 161
mediation/conciliation 259, 289–291
mediator 290–291
Meteorological Records 234, 244, 252
method statement and details of resource levels 65, 103
Multilateral Development Banks *see* MDBs

Naming of nominated Subcontractors 294
NEC Contracts 188–189, 191, 218
negligence 109, 147, 216–217, 226, 228, 261, 282; gross 40, 114, 138, 180, 201, 217
Network Analysis 183
neutral events 193, 199
New York Convention 259
NOD *see* Notice of Dissatisfaction
nominated subcontractors 56, 62–63, 108–110, 186, 284–285, 294
No-objection 84–86, 92, 95–96, 100, 126, 128–130, 138–140, 154, 157–158, 174, 187–188
Notice of Adjudication 292–293
Notice of Claim 240–246
Notice of Claim time bar 241
Notice of Dissatisfaction (NOD) 46, 75, 89, 115, 177, 249–250, 255–259, 264
Notice of No-objection 86, 92, 95–96, 100, 126, 128–130, 139–140, 154, 157–158, 174, 187–188
Notice of reference to Arbitration 262
Novation of designers 148

Obtaining DAAB's Decision 209, 250, 288
OJEU (Official Journal of the European Union) 300
operation and maintenance manuals 96, 100, 127, 130, 133, 139–140, 155, 158–160, 294
Optional Termination 117, 208, 212, 221–222
Ownership of Plant and Materials 127

PAFS 175, 178–179
parent company guarantee 272–273
Partially Agreed Final Statement 175, 178–179
Particular Conditions 54, 57–61, 68, 71, 80, 84, 92–93, 164, 170, 274–278, 280–288, 306–7, 309
partnering 282, 298–299
patent and latent defects 124
Pay when paid 270
payment 41–43, 45, 53–54, 67–71, 77, 113–115, 125–127, 134–135, 137–138, 141, 161–162, 166–176, 178–180, 206–212, 229–231, 239–240, 246–247, 258, 279–280, 293; advanced 69, 175; certification of 41–42; delayed 175, 209; late 175; milestone 270; normal 222; outstanding 293; periodic 220; period of 105, 171, 175; schedule of 78, 171–172, 247, 287, 294; the Schedule of 172
Payment Certificate 70, 109, 169–170, 173–174, 208–210, 222, 246, 249, 305, 309; minimum amount of 69, 174; preparing 169; previous 63, 109, 171, 174
Payment Certificates for increases/decreases in cost 169
payment of cost plus profit 52–55, 80, 90, 97, 99, 126, 137, 140, 202, 212, 215
performance bonds 272–273
Performance Certificate 67, 69, 93–94, 111, 124, 132, 136–138, 155, 159, 177, 227
performance criteria 48, 56, 58, 78, 149, 160, 295
Performance Damages 141
Performance Guarantees 49; the Schedule of 141, 160, 163
Performance Security 41, 44, 48, 60–61, 69–70, 80, 92–94, 170, 173, 178–180, 205, 208, 272–273, 283
performance specifications 48, 77, 152; brief 50
Permanent Works 91–92, 95, 143, 153–154, 168, 172, 206–207, 214, 216, 308, 310
PERT 184
physical conditions 53–54, 62, 73, 102–103; adverse 41–42; natural 53, 102
PI (Professional Indemnity) 224, 226–227, 271
Pink Book 75
Plant and Design-Build Contract 57, 76
Plant and Materials 68–69, 71, 122–124, 127, 135, 171–173, 187, 221, 247, 282
Plant and Materials on Site 122
possession 60–62, 79–80, 82, 101–102, 116, 186, 214; giving 80; partial 131

pre-commissioning tests 128
pre-contract cost planning 91
Preparation of Bills of Quantities 91
Pre-Tender Planning Stage 300
Principal Contractor 283
Principal Designer 283
Procedure for Tests after Completion 139
Procurement Directives 299
procurement method 142, 146, 149, 267, 296
procurement strategy 41, 266–268, 300; appropriate 251, 270; correct 267
Professional Indemnity (PI) 224, 271
Program Evaluation and Review Technique 184
programme 44, 50, 65, 106–107, 112, 165–166, 182–188, 194, 199, 297–299, 302–303; as-planned 194; detailed time 186; initial 185–186, 188; master 183–184; pre-tender 183; short-term 183
programme durations 143
programme float 193
progress reports 68–69, 106, 112, 171, 183, 234, 243, 285, 298; monthly 62, 106, 294; required monthly 106
project control requirements and templates 150
Project Network Analysis 184
Project Network Techniques Precedence Diagrams 183
Prolongation of Preliminaries Costs 232
Prolonged Suspension 49, 211, 235–238
provisional operation and maintenance manuals 160
Provisional Sums 68, 109, 150, 166, 170–171, 286, 297
Public liability Insurance 227

QA (quality assurance) 100, 120
QC (quality control) 119–120
QM *see* quality management
QM documents 107
QM System 100–101
quality 58, 64, 118–121, 124, 145, 149–150, 152, 163–164, 189, 191, 266–268, 292, 295, 299–300
Quality and Defects Liability 118–141
quality assurance *see* QA
quality control *see* QC
quality elements 301
quality management (QM) 100, 118–120, 294, 298
Quality Management and Compliance Verification Systems 100, 298
Quality Manager 283

Rainbow Suite 41
Rate of progress 199
reasonable skill and care 61, 91, 124, 147–148, 270
receipt of tenders 300
Recital 307, 309–310
records 88, 90, 95–96, 99–100, 112, 114, 122, 129–130, 132–133, 158–159, 233–234, 243–244, 247–248, 252–253, 285–286; comprehensive 233, 252; contemporaneous 234, 243, 252–253; contemporary 100, 114, 244; critical documents and health and safety 62, 106
Red and Yellow Book's Unforeseeable Physical Conditions 103
Red Book 41–42, 47, 56–59, 76–77, 90–92, 142, 153–154, 161, 213, 215–216, 268–269, 274–275, 277–278, 287, 293–294, 305
Referral Notice 292–293
Referring Party 258, 293
Remedying Defects 43, 45, 126, 133–134, 136, 141
replacement Engineer 60, 87
Requirements for Contractor's Documents 294
requirements for inspections and tests 294
requirements for operation and maintenance manuals 294
Resolution of Disputes 43, 255–265, 292
Responsibility for care of the Works 213–214, 217
Retention Money 69, 175–176, 274; release of 171, 175–176, 286
Retention Money Guarantee 272, 274
revised programme 65, 186–188, 199; each 186
Right of access 79, 136, 235–238

Schedule of Advance Warnings 192
SCL Delay 197
Sectional Completion 65, 70, 153
sequence and timing of inspections and tests 186
Shared Indemnities 217, 298
Silver Book 42, 46–47, 50–59, 76–79, 82–85, 87–90, 94–98, 124–134, 136–140, 155–179, 185–189, 199, 201–210, 213, 215–217, 221–222, 239–251, 255–258, 276–279, 293–296
Site Data 52–53, 101–102, 196, 298
Site Data and items of reference 52, 82, 102, 298
Site Meeting Minutes 234, 243, 252
Society of Construction Law Delay and Disruption Protocol 197
Special Projects Office 184
Special Provisions 51, 197, 275–277, 280, 282, 286
Split Responsibilities and Insurances 153
standing DAAB 257, 288
statutory limitation period 124

Subcontractors 61, 63, 90, 94, 100, 105–106, 108–109, 147, 166–167, 170, 219, 223, 228
Sub-Contractors 75, 268
Substantial Completion 182
Surety Bond 272–273
suspension 56, 64–66, 70, 73, 127, 181, 201–203, 209, 279–280, 285
suspension and termination 203–212
suspension and termination by Contractor 56, 70, 85, 134, 203, 209–210, 236, 238, 279–280

Taking-Over Certificate 66, 68–70, 106, 108, 129–133, 160, 162, 171–172, 176, 179, 182, 213–214, 225–226, 306, 309
Taking-Over Parts 131
Temporary Works/Permanent Works 95, 97, 99, 107–108, 112, 138, 143, 153, 206, 208
Temporary Works and Contractor's Equipment 208, 222
Tender Date 104
tender documents 57–58, 91–92, 144, 154, 251, 266, 270, 275–277, 300, 306–307, 309; preparation of 91, 152
tender enquiries 300
Tender Security 272–273
termination 56, 69–70, 85, 113–114, 116–117, 203–212, 214, 216–217, 220–221, 236, 238, 271–272, 279–280; date of 94, 205–207, 212, 214, 221–222; notice of 70, 202, 210, 221
Termination by Contractor 236, 238
tests after completion 67, 139
tests in use 139
tests on completion 56, 66, 74, 118, 127–129, 131–132, 139, 159–160, 235–238, 279–280, 298; commencement of 63, 159
time at large 194–195
time bars 41–42, 72, 242–243, 245; 28-day 241; simple 242
time cost optimisation analysis 185
the Time/Price/Quality Triangle 267
TQM 119
transport of goods 104, 283
Turnkey Project 51, 78

unforeseeable 41–42, 53–54, 62, 73, 97, 101–103, 162, 215
Unforeseeable Cost 61, 97
Unforeseeable Difficulties 53–54, 103
unforeseeable events 48, 126
Unforeseeable Physical Conditions 53, 73, 101–102, 235, 237, 283
Unforeseeable shortages 65, 196, 220, 235, 237
unincorporated JV 295
unliquidated damages 201

value engineering proposals 61, 92, 154
value management 144
Variation by Instruction 54, 98, 113, 134, 156, 164–166, 216, 287
Variation by Request for Proposal 165
Variation Procedure 164, 189, 236, 238, 298
variations 45, 49, 52–53, 64–65, 67–68, 82, 85, 87, 92–94, 125, 133, 159, 161–180, 185–186, 196, 215, 235, 237, 251; evaluation of 60, 247; issue of 251; ordered 47
Variations and adjustments 56, 68, 93, 161–162, 279–280
Variations and Adjustments for changes in legislation 68
Verbal Instructions 283

warnings 189–191, 193; early/advance 45, 189, 191
weather event, extreme 103
White Book 40; updated FIDIC 40
Windows Analysis 194
World Trade Organization (WTO) 300
wreckage 107–108, 138, 284
WTO (World Trade Organization) 300

Yellow and Silver books 40, 42, 51–52, 55–56, 58, 63, 67, 118, 127–128, 138–139, 155, 163, 215–217, 226–227, 294–296
Yellow Book 41–42, 46–47, 50–59, 74, 76–77, 83, 90, 142, 155–156, 169–170, 189, 213, 268–269, 275–279, 295–296

For Product Safety Concerns and Information please contact our EU representative GPSR@taylorandfrancis.com
Taylor & Francis Verlag GmbH, Kaufingerstraße 24, 80331 München, Germany

www.ingramcontent.com/pod-product-compliance
Ingram Content Group UK Ltd.
Pitfield, Milton Keynes, MK11 3LW, UK
UKHW052224190926
475702UK00007B/1198

* 9 7 8 0 3 6 7 4 2 7 9 1 7 *